高等学校"十三五"规划教材

高分子化学

刘向东 主编

化学工业出版社

·北京·

内 容 简 介

《高分子化学》共分6章，第1章绪论，介绍高分子基础知识，第2章至第4章按逐步聚合、烯烃类单体的连锁聚合（自由基聚合、自由基共聚合、离子聚合和配位聚合）和开环聚合介绍高分子的合成反应，第5章和第6章为天然高分子和聚合物的化学反应。本书版式设计活泼生动，引入了开篇图片，章中穿插阅读资料。本书注重介绍高分子在化学工业中的实例，以培养学生的工程视角。书中配有15个讲解视频和英语阅读材料音频的二维码，有助于读者自学。为提高读者的专业英语水平，书中重要词汇以中英对照出现。

本书可作为高等院校材料类、化学类专业本科生的教材，也可供相关人员参考。

图书在版编目（CIP）数据

高分子化学/刘向东主编. —北京：化学工业出版社，2021.1（2025.1重印）
高等学校"十三五"规划教材
ISBN 978-7-122-37903-0

Ⅰ.①高⋯ Ⅱ.①刘⋯ Ⅲ.①高分子化学-高等学校-教材 Ⅳ.①O63

中国版本图书馆CIP数据核字（2020）第197493号

责任编辑：宋林青　　　　　　　　文字编辑：刘　璐　陈小滔
责任校对：赵懿桐　　　　　　　　装帧设计：关　飞

出版发行：化学工业出版社（北京市东城区青年湖南街13号　邮政编码100011）
印　　装：涿州市般润文化传播有限公司
787mm×1092mm　1/16　印张15　字数371千字　2025年1月北京第1版第3次印刷

购书咨询：010-64518888　　　　　　售后服务：010-64518899
网　　址：http://www.cip.com.cn
凡购买本书，如有缺损质量问题，本社销售中心负责调换。

定　价：45.00元　　　　　　　　　　　　　　　版权所有　违者必究

序

自 1920 年德国科学家施陶丁格首次提出高分子概念以来,高分子科学及相关产业的发展令人瞩目。在百年发展中,高分子材料的巨大进步强有力地推动了社会发展,深深影响和改变了人们的生活。目前,纤维、塑料和橡胶等高分子材料不仅成为现代社会生活中不可或缺的组成部分,也已经广泛渗透到航空航天、信息电子等国民经济的各个领域。

20 世纪中后期,高分子材料在品种数量和生产规模两个方面都曾蓬勃发展。进入 21 世纪后,高分子产业仍保持着增长趋势,科技研发也呈现出新的发展动态。整体而言,高分子材料研究正在从量大面广的通用材料研究朝着功能化、精细化、复合化、可持续发展的方向转进。例如,生物医用材料、光电转换材料对高分子的分子结构不断提出更高要求,精密电子产业中的光刻胶也不断精细化发展以适应光刻技术从微米级向纳米级的升级换代,航空航天领域大量使用的高分子基复合材料变得越来越轻、越来越强,以生物质转化产物为单体合成的高分子材料近期不断走入人们的生活。对应这种新的发展趋势,我国高分子学科进步迅速,科研论文和知识产权方面的增长速度在全球居于前列,但由于多种原因,高分子学科的教材更新较慢,甚至严重滞后于学科及产业发展。

高分子化学是高分子相关专业的重要专业基础课程,相关的教材也应与时俱进,不仅要在知识的深度和广度方面不断拓展,也应及时地把高分子学科领域的最新研究成果吸收进来。另外,教材中适当引入英语学习资料,有助于提高学生的专业外语水平。

我校刘向东老师主编的这本高分子化学教材,注重引入高分子化学领域最新成果,着力拓宽知识广度,通过精选外语阅读和习题资料,为专业外语学习提供便利。此外,本教材针对工程教育要求,相比已有教材增加了高分子合成工程技术的相关知识,提高了对工程教育的适应性。这本教材在形式和内容上创新性强,是我校在不断尝试教材更新、教学改革过程中迈出的坚实一步。我认为这是一本值得向其他兄弟院校推荐的教材,希望它能够对相关科技人员、教育工作者以及本科学生的学习和工作提供有益的帮助,并为我国高分子科学事业发展做出积极贡献。

中国工程院院士
浙江理工大学教授
2020 年 6 月 28 日

前 言

自施陶丁格（H. Staudinger）于 1920 年提出高分子概念以来，至今已有一百年历史。当前，高分子科学已经成为一个广泛涉及化学工程、材料科学等诸多学科领域的重要学科。高分子材料不仅已经成为我们日常生活中不可缺少的组成部分，也逐渐成为人工智能、移动通信、大数据网络、汽车制造、航空航天、医疗卫生、新能源、环境产业领域中的重要材料。因此，高分子化学不仅是高分子材料相关专业中非常重要的一门课程，也是化学类专业学生应该修读的一门基础课程。

市场上有多个版本的高分子化学，其中不乏非常著名和经典的教材。尽管如此，编者在多年的高分子化学教学实践中发现，高分子专业及其相关的工科类本科学生仍然需要一本针对性更强，注重工程教育理念的教材，以达到改善教学效果，提高学生学习效率的目的。此外，笔者也深刻体会到适合高分子化学双语教学的教材较少。这成为编者编写此本教材的初始动机。

自 2016 年开始编写本教材以来，笔者浏览了不同语言的多本教材，追踪阅读了很多高分子学科的学术文献，力争吸收海内外高分子化学教材的优点长处，尽可能在教材中引入已被实践验证的较新的学术理论进展。在教材编写过程中，笔者较多地引入了高分子在化学工业中的实例，导入了一定数量的可以培养学生解决复杂工程问题的思考题目，设定了一些需要学生组成团队解决的问题，增添了许多利于社会可持续发展的知识点。在双语教学方面，教材中标注了常见的英语专业词汇，增加了适合阅读和翻译的英语短文和英语习题。

本教材结构可分为三大部分，第一部分为第 1 章绪论，介绍高分子基础知识；第二部分为本教材中心内容，由第 2 章至第 4 章构成，按逐步聚合、烯烃类单体的连锁聚合和开环聚合分别介绍由小分子单体合成高聚物的聚合反应；第三部分由第 5 章和第 6 章构成，分别介绍天然高分子和包括聚合物降解老化的聚合物化学反应。

为了便于教学，编者录制了精品课程的视频和部分英语短文的朗读音频资料，读者可扫码学习。本书配套课件和习题解答也已制作完毕，使用本书作教材的教师可以向出版社免费索取：songlq75@126.com。

嘉兴学院李海东教授、杭州师范大学由吉春教授、浙江工业大学欧阳密教授、安徽工程大学刘新华教授参与初稿修订工作，并提出了宝贵意见。浙江理工大学傅雅琴教授主审了本书。

高分子化学涉及范围较广，其中的每一章节都可能是一个已经发展了数十年的学术领域，编者难以对每一章节的内容都精准把握，故疏漏之处难免，敬请读者指正。

刘向东
浙江理工大学材料科学与工程学院
2020 年 3 月 6 日

目 录

第1章 绪 论 /1

1.1 高分子的基本概念 ⋯⋯⋯⋯⋯⋯⋯⋯ 2
1.2 聚合反应与单体 ⋯⋯⋯⋯⋯⋯⋯⋯⋯ 3
    1.2.1 按单体结构的变化分类 ⋯⋯⋯⋯ 4
    1.2.2 按聚合机理分类 ⋯⋯⋯⋯⋯⋯⋯ 4
1.3 聚合物的命名 ⋯⋯⋯⋯⋯⋯⋯⋯⋯⋯ 5
    1.3.1 根据聚合物的来源命名 ⋯⋯⋯⋯ 5
    1.3.2 结构系统命名法 ⋯⋯⋯⋯⋯⋯⋯ 6
1.4 聚合物的分子量及分布 ⋯⋯⋯⋯⋯⋯ 6
    1.4.1 数均分子量 ⋯⋯⋯⋯⋯⋯⋯⋯⋯ 6
    1.4.2 重均分子量 ⋯⋯⋯⋯⋯⋯⋯⋯⋯ 7
    1.4.3 黏均分子量 ⋯⋯⋯⋯⋯⋯⋯⋯⋯ 8
    1.4.4 分子量分布 ⋯⋯⋯⋯⋯⋯⋯⋯⋯ 8
1.5 高分子的微结构和物理态 ⋯⋯⋯⋯⋯ 10
    1.5.1 结构单元的键接方式 ⋯⋯⋯⋯⋯ 10
    1.5.2 手性构型和几何构型 ⋯⋯⋯⋯⋯ 10
    1.5.3 聚合物的分子链形态 ⋯⋯⋯⋯⋯ 12
    1.5.4 聚合物的分子链构象 ⋯⋯⋯⋯⋯ 12
    1.5.5 聚合物的聚集态 ⋯⋯⋯⋯⋯⋯⋯ 14
1.6 聚合物的分类及应用 ⋯⋯⋯⋯⋯⋯⋯ 17
    1.6.1 聚合物的分类 ⋯⋯⋯⋯⋯⋯⋯⋯ 17
    1.6.2 高分子材料的应用 ⋯⋯⋯⋯⋯⋯ 18
1.7 高分子化学发展简史 ⋯⋯⋯⋯⋯⋯⋯ 23
英语读译资料 ⋯⋯⋯⋯⋯⋯⋯⋯⋯⋯⋯⋯ 26
习题 ⋯⋯⋯⋯⋯⋯⋯⋯⋯⋯⋯⋯⋯⋯⋯⋯ 30
英语习题 ⋯⋯⋯⋯⋯⋯⋯⋯⋯⋯⋯⋯⋯⋯ 31

第2章 逐步聚合 /34

2.1 线形逐步聚合 ⋯⋯⋯⋯⋯⋯⋯⋯⋯⋯ 35
    2.1.1 官能团等活性理论 ⋯⋯⋯⋯⋯⋯ 35
    2.1.2 线形缩聚反应产物的分子量以及
          分子量分布 ⋯⋯⋯⋯⋯⋯⋯⋯⋯ 36
    2.1.3 线形缩聚动力学 ⋯⋯⋯⋯⋯⋯⋯ 42
    2.1.4 逐步聚合的实施方法 ⋯⋯⋯⋯⋯ 45
2.2 体形缩聚和凝胶化作用 ⋯⋯⋯⋯⋯⋯ 46
    2.2.1 Carothers法凝胶点预测 ⋯⋯⋯⋯ 47
    2.2.2 Flory统计凝胶点计算 ⋯⋯⋯⋯⋯ 48
2.3 重要缩聚物 ⋯⋯⋯⋯⋯⋯⋯⋯⋯⋯⋯ 50
    2.3.1 聚酯 ⋯⋯⋯⋯⋯⋯⋯⋯⋯⋯⋯⋯ 50
    2.3.2 聚碳酸酯 ⋯⋯⋯⋯⋯⋯⋯⋯⋯⋯ 59
    2.3.3 聚酰胺 ⋯⋯⋯⋯⋯⋯⋯⋯⋯⋯⋯ 59
    2.3.4 聚酰亚胺 ⋯⋯⋯⋯⋯⋯⋯⋯⋯⋯ 61
    2.3.5 聚苯并咪唑类聚合物 ⋯⋯⋯⋯⋯ 63
    2.3.6 聚氨酯 ⋯⋯⋯⋯⋯⋯⋯⋯⋯⋯⋯ 64
    2.3.7 环氧树脂 ⋯⋯⋯⋯⋯⋯⋯⋯⋯⋯ 66
    2.3.8 酚醛树脂 ⋯⋯⋯⋯⋯⋯⋯⋯⋯⋯ 67
    2.3.9 氨基树脂 ⋯⋯⋯⋯⋯⋯⋯⋯⋯⋯ 70
    2.3.10 聚苯醚 ⋯⋯⋯⋯⋯⋯⋯⋯⋯⋯⋯ 71
    2.3.11 聚砜 ⋯⋯⋯⋯⋯⋯⋯⋯⋯⋯⋯⋯ 71
    2.3.12 聚芳醚酮 ⋯⋯⋯⋯⋯⋯⋯⋯⋯⋯ 72
    2.3.13 聚苯硫醚 ⋯⋯⋯⋯⋯⋯⋯⋯⋯⋯ 73
英语读译资料 ⋯⋯⋯⋯⋯⋯⋯⋯⋯⋯⋯⋯ 74
习题 ⋯⋯⋯⋯⋯⋯⋯⋯⋯⋯⋯⋯⋯⋯⋯⋯ 78
英语习题 ⋯⋯⋯⋯⋯⋯⋯⋯⋯⋯⋯⋯⋯⋯ 80

第3章 烯烃类单体的连锁聚合 /84

3.1 自由基聚合 ⋯⋯⋯⋯⋯⋯⋯⋯⋯⋯⋯ 84
    3.1.1 加成聚合和连锁聚合 ⋯⋯⋯⋯⋯ 85
    3.1.2 烯烃类单体的结构及聚合特性 ⋯ 85
    3.1.3 自由基活性及化学反应 ⋯⋯⋯⋯ 86

- 3.1.4 自由基聚合机理 ———————— 87
- 3.1.5 链引发反应和引发剂 ———————— 88
- 3.1.6 聚合速率 ———————— 94
- 3.1.7 动力学链长和聚合度 ———————— 96
- 3.1.8 阻聚作用和阻聚剂 ———————— 97
- 3.1.9 活性自由基聚合 ———————— 99
- 3.1.10 聚合方法 ———————— 101
- 3.1.11 自由基聚合重要聚合物 ———————— 108
- 英语读译资料 ———————— 112
- 习题 ———————— 116
- 英语习题 ———————— 118
- 3.2 自由基共聚合 ———————— 122
  - 3.2.1 共聚物的类型和命名 ———————— 123
  - 3.2.2 研究共聚反应的意义 ———————— 123
  - 3.2.3 二元共聚物的组成 ———————— 124
  - 3.2.4 竞聚率 ———————— 128
  - 3.2.5 单体活性和自由基活性 ———————— 129
  - 3.2.6 Q-e 概念 ———————— 131
  - 3.2.7 重要烯烃类共聚物 ———————— 132
- 英语读译资料 ———————— 133
- 习题 ———————— 136
- 英语习题 ———————— 136
- 3.3 离子聚合 ———————— 138
  - 3.3.1 阳离子聚合 ———————— 139
  - 3.3.2 阴离子聚合 ———————— 143
- 英语读译资料 ———————— 149
- 习题 ———————— 151
- 英语习题 ———————— 152
- 3.4 配位聚合 ———————— 154
  - 3.4.1 配位聚合概述 ———————— 155
  - 3.4.2 配位聚合物的立体规整性 ———————— 155
  - 3.4.3 配位聚合催化剂 ———————— 156
  - 3.4.4 配位聚合的定向机理 ———————— 159
  - 3.4.5 配位聚合的反应历程及反应动力学 ———————— 162
  - 3.4.6 重要配位聚合产物 ———————— 163
- 英语读译资料 ———————— 168
- 习题 ———————— 170
- 英语习题 ———————— 171

第 4 章 开环聚合 / 172

- 4.1 开环聚合单体的结构特征 ———————— 173
- 4.2 开环聚合机理 ———————— 174
- 4.3 开环聚合引发剂及溶剂 ———————— 175
- 4.4 开环聚合高分子材料 ———————— 175
- 4.5 易位聚合反应 ———————— 180
- 英语读译资料 ———————— 181
- 习题 ———————— 184
- 英语习题 ———————— 184

第 5 章 天然高分子 / 186

- 5.1 多糖 ———————— 187
  - 5.1.1 糖的基础知识 ———————— 187
  - 5.1.2 多糖 ———————— 188
  - 5.1.3 其他多糖 ———————— 194
- 5.2 蛋白质 ———————— 196
  - 5.2.1 氨基酸 ———————— 196
  - 5.2.2 蛋白质分级结构 ———————— 196
  - 5.2.3 重要天然蛋白质 ———————— 198
- 5.3 核酸 ———————— 199
- 5.4 木质素 ———————— 201
- 5.5 天然橡胶 ———————— 205
- 英语读译资料 ———————— 206
- 习题 ———————— 210
- 英语习题 ———————— 210

第 6 章 聚合物的化学反应 / 212

- 6.1 聚合物侧基反应 ———————— 214
  - 6.1.1 高分子侧羟基反应 ———————— 214
  - 6.1.2 高分子侧氨基反应 ———————— 215
  - 6.1.3 高分子侧酰基反应 ———————— 216
  - 6.1.4 芳香环侧基反应 ———————— 217
  - 6.1.5 环化反应 ———————— 217

| 6.1.6 氯化反应 ·········· 218
| **6.2** 分子量增大较多的聚合物反应 ········ 218
| 6.2.1 接枝聚合 ·········· 218
| 6.2.2 扩链 ·········· 220
| 6.2.3 交联 ·········· 221
| **6.3** 降解与老化 ·········· 221
| 6.3.1 热降解 ·········· 222
| 6.3.2 力化学降解 ·········· 224
| 6.3.3 水解和生化降解 ·········· 224
| 6.3.4 氧化 ·········· 224
| 6.3.5 光降解和光氧化降解 ·········· 225
| 英语读译资料 ·········· 226
| 习题 ·········· 229
| 英语习题 ·········· 229

参考文献 / 231

第 1 章 绪 论

高分子化学简介

当代大型航空客机制造中大量使用高分子材料，一些新型客机的制造材料如果按质量计算，高分子材料已经超过金属材料。高分子材料被用作制造机头、发动机罩、机翼、水平翼、尾舵、客舱内饰中的许多部件。大量高分子材料的使用，帮助客机实现了轻量化，能够负载更多燃料增大航程。

本章目录

1.1 高分子的基本概念 / 2
1.2 聚合反应与单体 / 3
1.3 聚合物的命名 / 5
1.4 聚合物的分子量及分布 / 6
1.5 高分子的微结构和物理态 / 10
1.6 聚合物的分类及应用 / 17
1.7 高分子化学发展简史 / 23

重点要点

高分子，结构单元，重复单元，聚合度，高分子平均分子量及其测定方法，高分子分子量分布，聚合反应及单体种类，聚合物分类，高分子链构型，高分子链构象，非晶态高分子，结晶态高分子，高分子聚集态及热转变，玻璃化转变温度，塑料、橡胶、纤维的力学性能，高分子科学发展简史。

1.1 高分子的基本概念

高分子是指分子量（molecular weight）很大的化合物。传统的高分子定义含有两个基本条件：一是这种化合物的分子量要足够大；二是组成高分子化合物的原子由共价键连接。

在英语中，高分子用"macromolecule"和"polymer"表示。其中"macromolecule"由"macro"和"molecule"两部分构成，其含义是巨大的分子；"polymer"由"poly"和"mer（meros）"两部分构成，它们在希腊语中分别是"很多"和"单元部分"的意思，它说明了高分子的另一个特点：高分子由数目众多的简单结构单元重复连接而成。例如，聚丙烯的化学结构如图1-1所示，聚丙烯分子就好像珍珠项链一样，由许多相同的重复单元连接成长链结构。这些重复单元有时也称作链节，高分子链（polymer chain）中重复单元的个数称为聚合度（degree of polymerization，DP）。

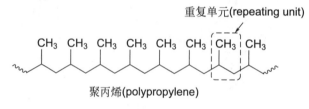

图 1-1 聚丙烯分子结构及其重复单元

高分子科学（polymer science）里并没有清晰的界限划定高分子和小分子的范围。一般而言，分子量低于1000的化合物属于小分子，分子量高于10000才开始表现出高分子的特征，分子量在1000～10000之间的过渡区域，常被称作低聚物（oligomer）。

表1-1中列出的烷烃链长和性能的关系有助于我们理解高分子的概念。甲烷是最简单的有机化合物，由一个碳原子和四个氢原子构成，在室温下呈气态。当烷烃链中的碳原子个数增加到五时，正戊烷的沸点为36℃，可以用作发泡剂和溶剂。直链烷烃的沸点和熔点随着碳原子个数的增加而增加，正二十八烷的沸点超过300℃，熔点超过36℃，在室温下为蜡状固体状态。正二十八烷以后，碳链中增加一个碳原子给烷烃带来的性能差异越来越小。碳原子个数在50～700范围的烷烃，室温下都是白色的蜡状固体，几乎无法将它们提纯分离，只能使用平均的碳原子个数描述这些烷烃混合物的分子链长度。聚乙烯（polyethylene，PE）是一种分子量巨大的烷烃混合物。制作垃圾袋的聚乙烯的分子链中平均含碳原子7000个左右。聚乙烯薄膜的拉伸强度随平均分子量的增加而缓慢增加。当聚乙烯分子链中平均碳原子数超过10万时，聚乙烯薄膜变得特别强韧，一般被称作超高分子量聚乙烯，可以用来制作防弹衣或防刺手套。如果将聚乙烯薄膜的强度和聚乙烯的平均链长关联，会得到如图1-2所示的曲线。其中链长超过一定数值，材料强度增加较平缓的区域是高分子科学的主要研究领域。

表 1-1　直链烷烃链长与常规性能

平均碳原子数	沸点/℃	名称	室温下形态	典型用途
1~4	<30		气态	热用燃料
5~10	30~180	汽油	液态	汽车燃料
11~12	180~230	煤油	液态	航空燃料、热用燃料
13~17	230~300	轻柴油	液态	柴油发动机燃料、热用燃料
18~25	305~400	重柴油	黏稠液态	热用燃料
26~50	裂解	石蜡	蜡状	蜡烛
50~1000	裂解		韧性蜡或固体	一次性餐具用涂料
1000~5000	裂解	聚乙烯	固体	塑料瓶或其他容器
>5000	裂解	聚乙烯	固体	垃圾袋、汽车零件、周转箱

高分子化合物有别于小分子化合物，具有高强度、高黏度、分子结构复杂等特性。这些特性都可归因于高分子的长链结构。由于每个高分子都是一个长链，分子间作用力要比小分子大得多，甚至超过高分子链中原子与原子之间的化学键能。在高温条件下，高分子链趋于裂解，但是整个高分子不会被气化。大分子之间的巨大分子间作用力是高分子化合物具有较高机械强度的内在因素。

随着高分子科学的发展，各种各样的高分子不断涌现，一些主链通过配位键等连接的高分子物质也有报道。因此，传统高分子定义中原子间的共价键连接方式也被扩展，现代高分子中的化学键应该是共享成键电子的价键连接。也就是说，组成高分子链的价键还包括配位键等化学键。

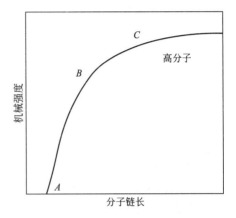

图 1-2　PE 薄膜强度与分子链长关系

1.2　聚合反应与单体

将小分子物质合成高分子的化学反应称作聚合反应（polymerization）（图 1-3）。聚合反应多种多样，并且不断发展。下面讲述的聚合反应的分类方法主要针对传统的聚合方法。

$$\text{单体}\ (\text{monomer}) \xrightarrow{\text{聚合反应 (polymerization)}} \text{高分子}\ (\text{polymer})$$

图 1-3　聚合反应

高分子化学的主要内容就是研究如何将小分子化合物合成为高分子物质，即聚合反应。合成聚合物的小分子化合物称作单体（monomer），单体能够通过聚合反应，转变成高分子的基本结构组成单元。与单词"polymer"对应，"monomer"即"单个的单元部分"之意。

单体转变成高分子链中众多的重复单元中的一个时，可称作单体单元。聚合反应主要有两种分类方法：一是按单体结构的变化分类；二是按聚合反应的机理分类。

1.2.1 按单体结构的变化分类

按单体结构的变化类型，可将聚合反应分成两类：官能团间的缩聚；小分子单体的加聚。这一分类比较简明，目前仍在沿用。

（1）缩聚

缩聚（condensation polymerization）是缩合聚合的简称，是官能团不少于两个的多官能团单体（polyfunctional monomer）通过各种有机缩合反应（condensation）形成聚合物的聚合反应。除形成缩聚物外，还有水、醇、氨或氯化氢等小分子副产物产生，缩聚物的结构单元要比单体少若干个原子。例如，二元胺和二元酸进行缩合反应得到聚酰胺［式(1-1)］就是缩聚的典型例子。当单体反应物分别是己二胺和己二酸时，反应生成的聚己二酰己二胺是大家所熟知的尼龙-66（nylon-66），广泛用作纤维（fiber）和塑料（plastic）。此外，聚酯（polyester）、聚碳酸酯（polycarbonate）、酚醛树脂（phenol-formaldehyde resin）、脲醛树脂（urea-formaldehyde resin）等都由缩聚而成。这些聚合反应将在第 2 章详细讨论。

$$H_2N-R_1-NH_2 + HOOC-R_2-COOH \longrightarrow \left[HN-R_1-NH-CO-R_2-CO \right]_n + H_2O \quad (1-1)$$

（2）加聚

小分子单体通过加成反应聚合成为高分子，没有小分子副产物生成的聚合反应称作加聚反应（addition polymerization），产物称作加聚物。大部分加聚物由含碳碳双键的单体聚合得到，这类单体统称为乙烯基单体（vinyl monomer）。氯乙烯加聚成聚氯乙烯就是加聚反应的实际例子［式(1-2)］。加聚物结构单元的原子组成与其单体相同，仅仅是电子结构有所变化。

$$CH_2=CHCl \longrightarrow \left[CH_2-CHCl \right]_n \quad (1-2)$$

1.2.2 按聚合机理分类

20 世纪中叶，Flory 根据反应机理和动力学，将聚合反应分成逐步聚合（step growth polymerization）和连锁聚合（chain growth polymerization）两大类。这两类聚合反应的转化率和聚合物分子量随时间的变化均有很大的差别。个别聚合反应可能介于两者之间。

（1）逐步聚合

多数缩聚反应都属于逐步聚合，其特征是小分子间的官能团逐步反应，每步反应的速率和活化能大致相同。两个单体分子反应，形成二聚体；二聚体与单体反应，形成三聚体；二聚体相互反应，则成四聚体。反应初期，单体很快聚合成二、三、四聚体等。短期内单体转化率很高，反应基团的转化率却很低。随后，低聚物间继续相互缩聚，分子量缓慢增加，直至基团转化率很高时，分子量才达到较高的数值。在逐步聚合反应体系中，任意不同分子量的聚合中间体之间的官能团都可以发生反应。

（2）连锁聚合

多数烯类单体的加聚反应属于连锁机理。连锁聚合需要一个具有高度反应性的活性中心，活性中心可以是自由基、阴离子或阳离子。连锁聚合通过活性中心与单体的加成反应进行。与逐步聚合反应不同，聚合物的链增长只能通过单体与反应活性中心的反应进行。活性中心的破坏意味着聚合反应的终止。连锁聚合在所有的转化率下都存在高分子量的聚合产物，随着转化率的提高，聚合物的平均分子量不断增加。

1.3 聚合物的命名

聚合物的命名法尚有许多待改进之处。有机化学中小分子的命名体系很难适用于结构复杂的聚合物。一个聚合物往往有几个名称，有的基于聚合物结构，有的基于聚合物合成原料，有时也会使用商品名。1972 年，国际纯粹与应用化学联合会（IUPAC）对线形聚合物提出了结构系统命名法，使聚合物命名的标准化前进了一大步。

1.3.1 根据聚合物的来源命名

这种命名方法主要适用于由单一单体合成的聚合物，聚合物名称常以单体名为基础。烯类聚合物以烯类单体前冠以"聚"字来命名，例如乙烯、氯乙烯的聚合物分别称为聚乙烯、聚氯乙烯。

由两种单体合成的共聚物，常摘取两单体的简名，后缀"树脂"二字来命名，例如苯酚和甲醛的缩聚物称为酚醛树脂。合成橡胶往往从共聚单体中各取一字，后缀"橡胶"二字来命名，如丁（二烯）苯（乙烯）橡胶、乙（烯）丙（烯）橡胶等。

杂链聚合物还可以进一步按其特征结构来命名，如聚酰胺、聚酯、聚碳酸酯、聚砜等。这些都代表一类聚合物，具体品种另有专名，如聚酰胺中的己二胺和己二酸的缩聚物学名为聚己二酰己二胺。

常见合成纤维俗称：

涤纶/PET 纤维（详见表 1-3）；

腈纶/聚丙烯腈纤维；

氨纶/聚氨酯纤维；

维纶/聚乙烯醇缩甲醛纤维；

丙纶/聚丙烯纤维；

锦纶/尼龙-6 纤维；

氯纶/聚氯乙烯纤维；

芳纶/全芳香聚酰胺纤维。

常见再生纤维素类纤维俗称：

黏胶纤维/黄原酸酯法再生纤维素纤维；

铜氨纤维/铜氨溶剂法再生纤维素纤维；

Lyocell 纤维/N-甲基吗啉-N-氧化物（NMMO）为溶剂，湿法纺制的再生纤维素纤维；

醋纤/醋酸纤维素纤维。

我国习惯以"纶"字作为合成纤维商品名的后缀，如聚对苯二甲酰乙二醇酯称涤纶，聚丙烯腈称腈纶，聚乙烯醇纤维称维尼纶等，其他如丙纶、氯纶则代表聚丙烯纤维、聚氯乙烯纤维。

1.3.2　结构系统命名法

为了作出更严格的科学系统命名，IUPAC 对线形聚合物提出下列命名原则和程序：先确定重复单元结构，排好其中次级单元次序，给重复单元命名，最后冠以"聚"字，就成为聚合物的名称。IUPAC 系统命名法比较严谨，但有些聚合物，尤其是缩聚物的名称过于冗长，难以掌握。

为方便起见，许多聚合物都有缩写符号，如聚甲基丙烯酸甲酯的符号为 PMMA。书刊中第一次出现不常用符号时，应注出全名。在学术性比较强的论文中，虽然并不反对使用能够反映单体结构的习惯名称，但鼓励尽量使用系统命名，并不希望用商品俗名。

1.4　聚合物的分子量及分布

高分子材料良好的力学性能主要归因于它的高分子量。在聚合物合成和加工成型过程中，分子量总是评价聚合物的重要指标。

如图 1-2 所示，高分子材料的力学强度随分子量增加而增加，A 点是初具强度的最低分子量，以千计。A 点以上的强度随分子量增加而迅速增加，到临界点 B 后，强度变化趋缓。C 点以后，强度不再显著增加。关于 B 点的聚合度，聚酰胺约 150，纤维素约 250，乙烯基聚合物则在 400 以上。常用缩聚物的聚合度约 100～200，而烯类加聚物则在 500～1000，相当于分子量 2 万～30 万，天然橡胶和纤维素均超过此值。

聚合物的分子量与有机小分子的分子量截然不同，高分子科学中讨论的聚合物是由一系列分子量（或聚合度）不等的同系物高分子组成，这些同系物高分子之间的分子量差为重复结构单元分子量的倍数，这种同种聚合物分子长短不一的特征称为聚合物的多分散性（molecular weight polydispersity）。在表征聚合物的分子量时，我们需要注意两个重要参数：平均分子量（average molecular weight）和分子量分布（distribution of molecular weight），两者是决定高分子产品性能的重要因素。

有多种方法可以测定高分子样品的平均分子量，如基于依数性、光散射性、黏度等，不同方法得到的平均分子量并不相同。

1.4.1　数均分子量

数均分子量（number-average molecular weight）是按聚合物样品中聚合物分子的数量来统计平均的分子量，通常用 $\overline{M}_n$ 来表示。例如，体系中分子量为 $M_1, M_2, M_3, \cdots, M_x$ 同系物的分子数为 $n_1, n_2, n_3, \cdots, n_x$，以各系高分子的分子数为统计单元，$\overline{M}_n$ 可由下式计算：

$$\overline{M}_\mathrm{n} = \frac{n_1 M_1 + n_2 M_2 + n_3 M_3 + \cdots + n_x M_x}{n_1 + n_2 + n_3 + \cdots + n_x}$$

$$= \frac{\sum n_i M_i}{\sum n_i} \tag{1-3}$$

式中，求和符号表示从 $i=1$ 到 $i=\infty$ 求和，n_i 是分子量为 M_i 的分子数。

测定数均分子量通常采用测量高分子溶液依数性的方法，诸如，蒸气压降低法、冰点降低法、沸点升高法、渗透压法等，但是这些方法仅适合于分子量不高于 50000 的高分子。如果分子量过大则无法准确测定。端基分析对于测量一些链端具有特殊化学结构的高分子的数均分子量有效。例如，聚酯的端羧基可以采用碱滴定法测量，碳碳双键的端基可以采用 ^{1}H NMR 进行分析。

1.4.2 重均分子量

重均分子量（weight-average molecular weight）是按照聚合物的质量进行统计平均的分子量，通常使用 $\overline{M}_\mathrm{w}$ 来表示。

$$\overline{M}_\mathrm{w} = \frac{m_1 M_1 + m_2 M_2 + m_3 M_3 + \cdots m_x M_x}{m_1 + m_2 + m_3 + \cdots m_x}$$

$$= \frac{\sum m_i M_i}{\sum m_i} \tag{1-4}$$

式中，求和符号表示从 $i=1$ 到 $i=\infty$ 求和，m_i 是分子量为 M_i 的质量。

因 $m_i = n_i M_i$，式(1-4)还可以继续演变为，

$$\overline{M}_\mathrm{w} = \frac{\sum m_i M_i}{\sum m_i} = \frac{\sum n_i M_i^2}{\sum n_i M_i} \tag{1-5}$$

$$= \sum w_i M_i \tag{1-6}$$

式中，w_i 是分子量为 M_i 的质量分数。重均分子量是采用测量高分子溶液的光散射性质的方法来测定的。光散射法测定分子量不存在分子量上限，但是当高分子平均分子量小于 10000 时，因光散射值太小而无法准确测定。

> 光散射法测定重均分子量：高分子重均分子量可由光散射法测定，分子量有效测定范围为 $5 \times 10^3 \sim 10^7$。当一束光通过某介质时，在入射光方向以外的各个方向也能观察到光强的现象称为光散射现象。光散射法研究高分子稀溶液性质时，由于高分子间距离较大，一般情况下不产生分子之间的散射光的外干涉。但是当高分子尺寸与入射光在介质里的波长处于同一个数量级时，高分子链各链段所发射的散射光波有干涉作用，这就是高分子链散射光的内干涉现象。散射光强度与散射角、溶剂、浓度、分子量等因素有关。利用外推法分析在极限情况下的（浓度和散射角无限小）高分子溶液的光散射，可以获得高分子重均分子量。因光散射法可以直接测定高分子平均分子量，与其他对比测定的方法不同，在高分子研究中占有重要地位，也是研究高分子电解质在溶液中的形态的有力工具。

1.4.3 黏均分子量

黏均分子量（viscosity-average molecular weight，$\overline{M}_v$）可通过测定高分子溶液的黏度得到。

$$\overline{M}_v = \left(\frac{\sum m_i M_i^\alpha}{\sum m_i}\right)^{1/\alpha} \tag{1-7}$$

式中，α 为常数，当 $\alpha=1$ 时，黏均分子量等于重均分子量。对于大多数聚合物来说，α 通常为 0.5～0.9，随高分子结构和溶剂的不同而不同。

α 来自于实验，属于经验常数。在高分子稀溶液中，特性黏度（intrinsic viscosity）和黏均分子量之间符合 Mark-Houwink 方程：

$$[\eta] = K\overline{M}_v^\alpha \tag{1-8}$$

式中，$[\eta]$ 代表高分子稀溶液的特性黏度；$\overline{M}_v$ 为黏均分子量。

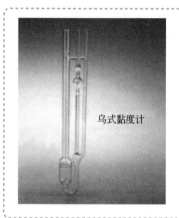

乌式黏度计

黏度法测定黏均分子量：高分子的分子链长度远大于溶剂分子，加上溶剂化作用，使其在流动时受到较大的内摩擦阻力，因此高分子溶液的黏度远远高于溶剂的黏度。高分子溶液的黏度取决于高分子的分子量、浓度、溶剂、测定温度等因素。测定黏均分子量时，通过在恒定温度下测定多个浓度的高分子稀溶液的黏度，并通过稀释外推法计算出特性黏度 $[\eta]$。特性黏度是高分子溶液无限稀释时的黏度，其值仅取决于溶剂的性质及聚合物分子的大小和形态。

表 1-2 为高分子分子量的测定方法及测定范围。

表 1-2 高分子分子量的测定方法及测定范围

	平均分子量	绝对/相对	测定范围
末端定量法	$\overline{M}_n$	绝对	$M<10^4$
物理化学法①	$\overline{M}_n$	绝对	$M<10^4$，$10^4<M<10^6$
黏度法	$\overline{M}_v$	相对	$10^3<M<10^7$
沉降分离法	$\overline{M}_w,\overline{M}_v$	绝对	$10^4<M<10^7$
光散射法	$\overline{M}_w$	绝对	$10^2<M<10^7$
凝胶渗透色谱法(GPC)	$\overline{M}_n,\overline{M}_w,\overline{M}_v$	相对	$10^2<M<10^7$
MALDI/TOF-MS	$\overline{M}_n,\overline{M}_w,\overline{M}_v$	绝对	$M<10^4$

① 指沸点上升、凝固点下降、蒸气压下降、膜渗透压等方法。其中膜渗透压法可测定 $10^4<M<10^6$ 范围的分子量，其他为 $M<10^4$。

1.4.4 分子量分布

对于所有分子量都相同的单分散聚合物而言，上述三种平均分子量相等。但是单分散聚合物在高分子科学中并不多见。

几乎所有的人工合成高分子，由于聚合反应中概率的原因，合成高聚物的分子量或聚合度以及分子链长都是不均一的，因而存在分子量分布（molecular weight distribution）问题。上述三个平均分子量数值各不相同，仔细考察它们的计算公式（1-4）～式（1-7），就会发现，三种平均分子量按照数均、黏均、重均的顺序递增。各平均分子量间的差值随分子量分布宽度的增大而增大，比较典型的分子量分布曲线如图1-4所示。

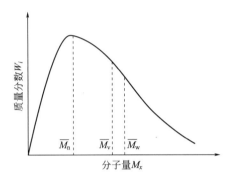

图1-4　分子量分布典型曲线

三种平均分子量中，黏均分子量和重均分子量十分接近。因此，在较多情况下，我们更关注聚合物样品的数均分子量和重均分子量。前者侧重于低分子量组分，后者侧重于高分子量组分。两者的比值$\overline{M}_w/\overline{M}_n$与分子量分布曲线的宽度关系较大，通常被作为聚合物多分散性的量度，称为分布指数（distribution index）。合成聚合物分布指数可在1.5～2.0以及20～50之间，随合成方法而定。比值愈大，则分布愈宽，分子量愈不均一。

分子量分布也是影响聚合物性能的重要因素。从高分子材料的角度来看，比较重要的是高聚物的分子量分布对加工性能和使用性能的影响。

高分子的加工性能与高聚物熔体的流变性有关，因而具有较强的分子量依赖性，且受分子量分布的影响较大。高分子材料的使用性能取决于高分子的聚集态结构和分子链的热运动，因此也具有显著的分子量和分子量分布依赖性。一般而言，高分子材料的分子量分布太窄，对加工不利；分子量分布过宽又会影响使用性能。

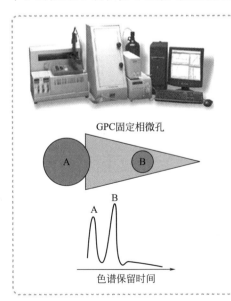

凝胶色谱法（GPC）是测定高聚物分子量分布最方便、最广泛的方法。它利用固定相中孔径不同的微孔对流动相中尺寸不同的高分子所起的体积排阻作用，使分子尺寸不同的高分子在淋洗过程中的流径不同，使不同分子量的高分子得以分离。色谱保留时间与分子量之间的关系可以通过接近单分散的高分子标样标定。目前凝胶色谱法可以测定从低聚物到分子量为几百万的高聚物的分子量分布，同时还可以得到重均分子量和数均分子量，成为当前重要的高分子表征手段。一套GPC系统包含进样、色谱柱、溶剂泵、检测、数据分析处理等设备。

1.5 高分子的微结构和物理态

1.5.1 结构单元的键接方式

与小分子不同,高分子具有多层次微结构。结构单元及其键接方式,包括结构单元本身的结构、结构单元相互键接的顺序、结构单元的空间排布等,均可以造成高分子材料性质的大幅变化。

以聚氯乙烯高分子(图1-5)为例,该线形高分子内的结构单元间可以有多种键接方式,各结构单元之间可以是头-尾键接,也可以是头-头或尾-尾键接。另一个例子是两种天然高分子——纤维素和直链淀粉(图1-6),两者的结构单元虽然完全相同,但是由于结构单元的键接方式完全不同,使得它们的性质存在非常大的差异(详见第5章)。

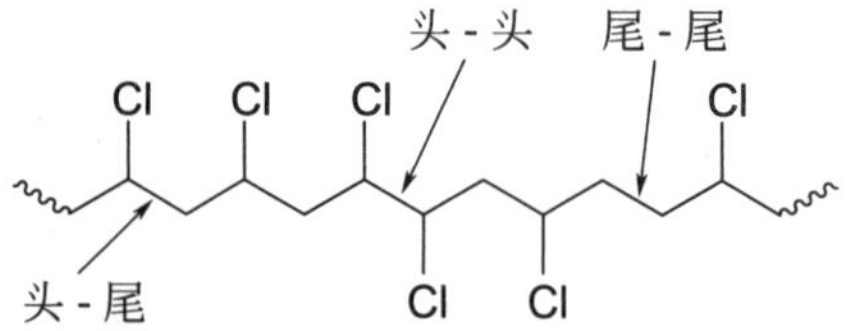

图1-5 聚氯乙烯结构单元的键接方式

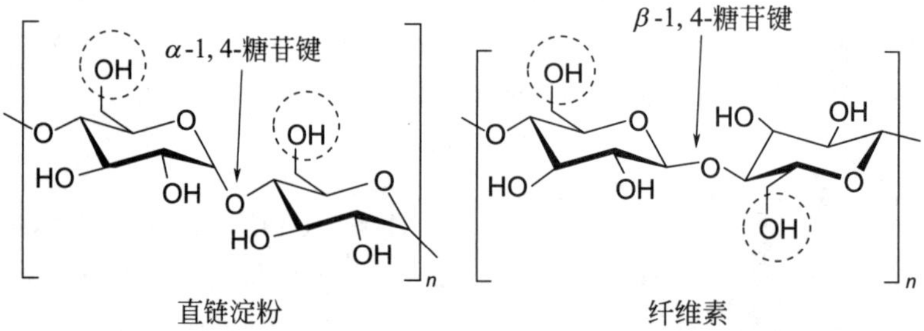

图1-6 直链淀粉和纤维素的结构单元键接方式

1.5.2 手性构型和几何构型

大分子链上结构单元中的取代基在空间上可能有不同的排布方式,主要有手性构型和几何构型两类。

(1) 手性构型

在烯烃的聚合反应中,如果双键的一个碳原子是单取代的,就会出现手性异构现象。对于连续的立体中心，其构型的规整性决定了聚合物链的有序程度，即立体规整度（tacticity）。为方便说明，将主链拉直成锯齿形，排在一平面上，如取代基 R 全部处在平面的上方，则形成全同构型（isotactic）；R 如规则相间地处于平面的两侧，则形成间同构型（syndiotactic）；如 R 无规排布在平面的两侧，则形成无规构型（atactic）。

例如，聚丙烯中的叔碳原子具有手性特征，甲基在空间的排布方式如图 1-7。上述 3 种构型的聚丙烯，性能差别很大（详见第 3 章配位聚合）。聚合物的立体构型主要由引发体系来控制。

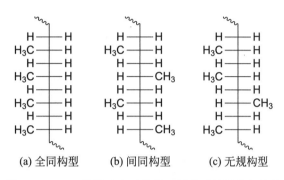

(a) 全同构型　　(b) 间同构型　　(c) 无规构型

图 1-7　Fisher 投影式表示的聚丙烯的手性异构规整性

(2) 几何构型

共轭双烯进行聚合反应，所得聚合物的分子链结构会更加复杂，除上述手性异构的可能性外，还有可能形成几何构型上的差异。例如，丁二烯进行加成聚合可能形成三种不同的结构单元（图 1-8）。其中 1,2-加成结构存在全同、间同、无规等不同手性异构，而 1,4-加成结构中因有双键存在，形成了顺式和反式两种几何异构体。异戊二烯聚合时更复杂，可能形成四种不同的结构单元（图 1-8）。不同结构单元的数量不同导致聚合物的性能发生显著的改变。顺式结构含量高的聚异戊二烯是性能优良的橡胶，而反式结构含量高的聚异戊二烯则是半结晶的塑料。

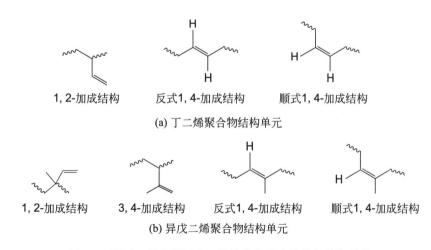

图 1-8　聚丁二烯和聚异戊二烯结构单元中的几何异构现象

1.5.3 聚合物的分子链形态

大分子中结构单元的键接方式还有可能造成高分子长链形状上的差异（图1-9）。若高分子链中的结构单元按一维线形方式连接，则会形成线形高分子（linear polymer）。这里的"线形"并不是指高分子链的形态真的像直线，而是指不含分支，但高分子链的形态可以卷曲，甚至是卷曲成线团。线形聚合物可能带有侧基，但侧基并不能称作支链。支链形高分子（branched polymer）的分支结构一定含有重复的结构单元。图1-9中支链仅仅是简单的示意图，实际上，还可能有星形、梳形、树枝形等更复杂的结构。

线形　　　　支链　　　　交联

图1-9　高分子链的形态

缩聚反应中的两个单体分子分别含有两个反应性官能团时，往往会形成线形高分子。例如，二元醇和二元酸缩合成聚酯。烯烃类单体经连锁聚合大多会形成线形高分子，环状单体经开环聚合也会形成线形高分子。一些烯烃类单体在进行聚合时会发生链转移反应，导致支链的出现。有时，为了改变线形高分子的性质，会有目的地在主链上接枝不同结构的聚合物分子链，形成接枝共聚物。

线形或支链形高分子可溶于适当溶剂中，加热熔融塑化，冷却则固化成型，这类聚合物称作热塑性聚合物，聚乙烯、聚氯乙烯、聚苯乙烯、涤纶、尼龙等都属于热塑性高分子。支链形聚合物不容易结晶，高度支化的高分子甚至难溶解，只能溶胀。

交联高分子（cross-linked polymer）可以看作许多线形大分子由化学键连接而成的体形结构。交联程度浅的网状结构，受热时尚可软化，但不熔融；适当溶剂可使其溶胀，但不溶解。交联程度深的体形结构，受热时不再软化，也不易被溶剂溶胀，而是刚性固体。酚醛树脂、脲醛树脂、醇酸树脂等可制备热固性树脂。天然橡胶、丁苯橡胶等原来都是线形高聚物，加入适当交联剂（如硫或有机硫）使之交联，交联程度不深时具有良好的弹性。

1.5.4 聚合物的分子链构象

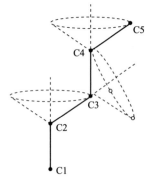

图1-10　高分子链中单键旋转示意图

与小分子有机化合物相似，聚合物分子链中的单键也可以发生旋转。分子中的原子或原子团因围绕单键旋转而产生的不同空间排列顺序称为构象（conformation）。聚合物分子形状（macromolecular shape）即聚合物分子的宏观构象是由分子中的所有单键的内旋转所引起的构象变化决定的。图1-10为高分子链中单键旋转示意图。

如同在有机化学中讲述的一样，对分子中的单键而言，C3—C2键可以C2—C1键为轴按固定角度旋转，即C3可以在图中所示虚线圆周上的任意位置。但由于与C1上取代基之间的排斥作用,当它们形成重叠构象时,能量最高,最不稳定,形成交叉构象时,能量最低,构象最稳定。

若分子中再增加C4—C3键,那么C4的可能位置就会更多。考虑到C1和C4两个碳原子间的排斥作用,如图1-11所示,两个碳原子距离最远的反式交叉构象能量最低,两种旁式交叉构象的能量次之。高分子的分子链中含有大量单键,因此,可形成的稳定构象数也很多。

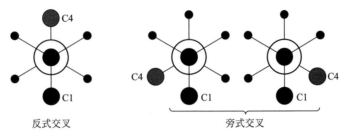

图1-11 丁烷的三种稳定构象

无论有机小分子还是高分子,能量最低的构象与能量较高的构象之间都存在能量差,称为构象能。由一种构象转变为另一种构象时必须克服旋转势垒。一般而言,旋转势垒低,高分子链的柔顺性好。

具体而言,高分子链的分子结构对柔顺性的影响主要有如下几个方面。

(1) 主链结构

当主链中含碳氧键、碳氮键、硅氧键时,由于氧、氮原子的键合原子数比碳原子少,其内旋转的空间阻碍小,旋转势垒低,柔顺性一般好于以碳碳键为主链的高分子。硅氧共价键的键长大于碳氧键,旋转势垒更低,因而具有更好的柔顺性。当主链中含有非共轭双键时,虽然双键本身不能旋转,但是与双键相邻的单键上原子间的距离比一般共价键长,使旋转的空间阻碍变小,内旋转更容易进行,柔顺性更好。当主链由共轭双键组成时,双键本身不能旋转,分子链柔顺性变差。

经常进入世界五百强的海内外高分子化工企业

美国:陶氏(DOW),埃克森(Exxon),3M,杜邦(DuPont),亨斯曼(Huntsman);

德国:巴斯夫(BASF),拜耳(Bayer),朗盛(LANXESS),汉高(Henkel),赢创(Evonik);

法国:道达尔(Total);

英资:ICI;

日本:三井化学(Mitsui Chemicals),帝人(TeiJin),东丽(Toray),旭化成(AsahiKASEI);

韩国:LG;

荷兰:阿克苏诺贝尔(Akzo Nobel);

瑞士:汽巴精化(Ciba Specialty Chemicals);

瑞典:北欧化工(Borealis);

沙特:SABIC;

比利时:索尔维集团(Solvay S. A.);

中国:中国石油化工集团公司(Sinopec Group),台湾远东新世纪股份有限公司(Far Eastern New Century Corporation),台湾台塑集团(Formosa Plastic Group)。

(2) 侧基结构

对于极性侧基，极性越大，数目越多，分子链内以及分子链间的相互作用越强，分子链内旋转越困难，柔顺性越差。非极性侧基的体积越大，内旋转空间阻碍作用越大，分子链柔顺性越差。

(3) 氢键

当高分子分子链内或链间能形成氢键时，氢键的作用力较强，能比极性作用更显著地影响分子链的构象，会显著增加分子链的刚性。

1.5.5 聚合物的聚集态

高分子链的柔顺性仅仅是影响高分子材料性能的一个因素。聚合物的聚集态（morphology）对高分子材料性能的影响也很大。例如，聚乙烯的分子链是很柔顺的，但是如果结构规整的聚乙烯分子链发生结晶（crystallization），分子链的运动就会受到晶格的限制，从而失去分子链的柔顺性。结晶度（degree of crystallinity）较高的聚乙烯是相对刚性的材料。

聚合物的聚集态涉及高分子固态结构的多种形式，如大分子取向、相分离、结晶和其他相转变等。聚合物的聚集态结构能从不同层次上影响高分子材料的强度、弹性等机械性能，以及耐热性、耐溶剂性、老化等多种性能。聚合物的聚集态可以粗分成非晶态（无定形态，amorphous morphology）和晶态（crystalline morphology）两类。少数聚合物结构规整，结晶度高，能得到比较完善的结晶形态。例如，聚乙烯甚至可在特定的条件下获得聚合物单晶。大多数聚合物的聚集态为半结晶态（semi-crystalline），规整的晶区相互之间通过无取向、无规构象的分子链构成的非晶区连接（图1-12）。低结晶度聚合物中，少量的不完善的结晶微区分散在非晶态的基体中。有些分子链结构规整性差的聚合物则是完全非晶态的，如无规聚苯乙烯、无规聚甲基丙烯酸甲酯等。

(1) 无定形态

聚合物的无定形聚集态是指聚合物中分子链的堆砌不具有长程有序性，完全是无序的。无定形态是一个比晶态更普遍的高分子聚集态。目前有关聚合物的无定形态主要有两种理论模型。无规线团模型认为高分子无定形区域中，高分子链之间相互贯穿、纠缠，当其分子量足够大时，相互穿透的分子链就会形成稳定的纠缠结构，这种纠缠结构使聚合物分子链的运动受限，就像一团杂乱堆放的毛线，毛线越长，越容易产生纠缠，越难将其彼此分开。由于这种纠缠作用，聚合物分子链运动时不是整个分子链的刚性运动，而是以含若干个链单元的链段为运动单元的蠕动。还有一种模型称为两相球粒模型，它认为无定形态聚合物并不是完全无序的，而是存在着局部有序区域。其中的有序部分是由聚合物分子链折叠而成的"球粒"或"链结"，其尺寸约为数纳米，排列比较规整，但比晶态的有序性要低得多。有序球粒之间由无规高分子链连接（图1-12）。

如图1-13所示，无定形态高分子在低温条件下，高分子链和链段都不能运动，只有分子链中的原子或基团在其平衡位置进行振动，高分子科学中称为玻璃态（glassy state）。随着温度的升高，高分子链中以数个重复单元为单位的高分子链段开始运动，但是整个高分子

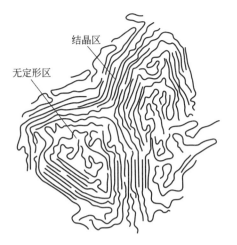
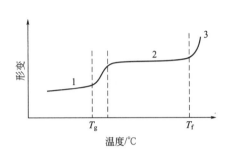

图 1-12 高分子聚集态的两相球粒模型　　图 1-13 无定形高分子的形变-温度曲线图
　　　　　　　　　　　　　　　　　　　　　　1—玻璃态；2—高弹态；3—黏流态

链却不能移动，高分子材料表现出高弹性质，称为高弹态（rubbery state）。温度再升高，整个高分子链开始运动而表现出黏流性质，高分子体系进入黏流态（viscous state）。其中玻璃态向高弹态的转变温度为玻璃化转变温度（glass transition temperature），表示为 T_g，它的另一个含义是链段能够运动或主链中价键能扭转的温度。高弹态向黏流态的转变温度称为黏流温度（viscous flow temperature），表示为 T_f，它是高分子链开始运动的温度。

（2）结晶态

聚合物可形成单晶（single crystal）和球晶（spherulites）等多种形态。聚合物单晶是具有规则几何外形的薄片状晶体，如聚乙烯的单晶为菱形。聚合物的单晶只能从极稀溶液（0.01%～0.1%，质量分数）中缓慢结晶而成。常压下形成的单晶片的厚度不超过 50nm，远远小于高分子链长（通常达数百纳米），因此聚合物分子链在单晶中是折叠排列。质量分数超过 0.1% 时将发展为多层晶片，质量分数大于 1%，则只能得到球晶。

球晶是聚合物最常见的结晶形态，大小一般为数十到数百微米，一般从聚合物浓溶液中析出或由熔体冷却时形成。一般认为球晶生成初期是以折叠链晶片为主（图 1-14），由于熔体迅速冷却或其他条件所限，这些小晶片来不及规整地堆砌成单晶，为了减小表面能而以某些晶核（多层片晶）为中心，逐渐向外扩张生长，经历捆束状阶段后同时向四周扭曲生长，形成支化且具有较大尺寸的片晶，成为球晶中的片状原纤。在片状原纤之间，存在着片间原纤（又称系带分子）。球晶中的无定形部分包括折叠链结构中的缺陷、系带分子和缠结部分等。在结晶程度较低时，球晶分散于连续的无定形区中，随着结晶度的提高，球晶在生长过程中会与相邻球晶相互碰撞，从而互相挤压成为不规则的多面体。

聚合物的结晶能力与高分子链的规整性、柔性、分子间力等有关，同时还受拉力、温度等操作条件的影响。例如，线形聚乙烯分子结构简单规整，易紧密排列成结晶，结晶度高达 90%；带支链的聚乙烯结晶度就低得多（55%～65%）。聚四氟乙烯结构与聚乙烯相似，结构对称而不表现出极性，氟原子也较小，容易紧密堆砌，结晶度高。涤纶树脂分子结构简单且规整，但分子链中苯环赋予分子链一定的刚性，且无强极性基团，结晶就比较困难，需在适当的

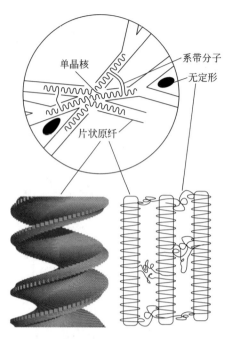

图 1-14 高分子球晶及其内部结构

温度下经过拉伸才达到一定的结晶度。聚氯乙烯、聚苯乙烯、聚甲基丙烯酸甲酯等带有体积较大的侧基，分子难以紧密堆砌，容易形成非晶态。天然橡胶和有机硅橡胶分子中含有双键或醚键，分子链柔顺，在室温下处于无定形的高弹状态。如温度适当，经拉伸，则可规则排列而暂时结晶；但拉力一旦去除，规则排列不能维持，立刻恢复到原来的完全无序状态。和无定形一样，结晶性聚合物在低温时也呈玻璃态。加热至玻璃化温度以上，高分子链段开始运动，继续受热则出现另一热转变温度——熔点（T_m）。T_g 和 T_m 均是表征聚合物聚集态的重要参数。

无定形高分子和结晶性高分子的受热变化行为有所不同。在玻璃化温度以上，无定形聚合物先从硬的橡胶转变成软的、可拉伸的弹性体，再转变成胶状的，最后成为液体。每一转变都是渐变的，并无突变。而结晶聚合物的行为却有所不同，在玻璃化温度以上，熔点以下，保持着橡胶的高弹态或柔韧状态，温度升高到熔点以上，则直接液化。晶态聚合物的结晶存在缺陷，并且一般具有较宽的分子量分布，因此熔点是一个熔融范围，并不是具体的某一温度（图 1-15）。

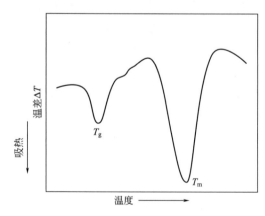

图 1-15 典型的高分子 DSC 曲线

T_g 和 T_m 可用来评价聚合物的耐热性。玻璃化温度是无定形聚合物的使用上限温度，熔点则是晶态聚合物的使用上限温度。实际使用时，一般要求无定形塑料的 T_g 比室温高 50~70℃；对于晶态塑料，T_g 可以低于室温，但是 T_m 一定要高于室温。橡胶处于高弹态，T_g 为其使用下限温度，一般要求 T_g 比室温低大约 70℃。大部分合成纤维是结晶性聚合物，如尼龙、涤纶、维尼纶、丙纶等，其 T_m 往往比室温高 150℃以上，便于烫熨。在大分子中引入芳杂环、引入极性基团和交联是提高玻璃化温度和耐热性的三大重要措施。

还有一类结构特殊的液晶高分子，这类晶态高分子受热熔融或被溶剂溶解后变为液体，但是其中的晶态分子仍然保留着有序排列，呈各向异性，兼具晶体和液体的双重性质，称为液晶态。

液晶高分子除 T_g 和 T_m 外，还有清亮点温度 T_i。液晶高分子升温到一定温度时，先转变成能流动的浑浊液晶相，随温度继续升高到另一临界温度，液晶相消失，成为透明液体，这一转变温度即为清亮点温度 T_i。清亮点温度可用于评价液晶的稳定性。

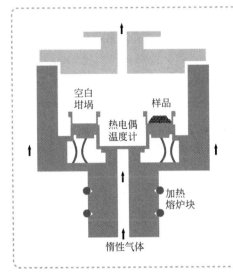

差示扫描量热法（Differential Scanning Calorimetry，DSC），也称差示热分析，是在温度程序控制下，测量试样与参比物之间的温度差随温度变化的技术。试样在升温或降温过程中，若发生物理变化和化学变化，就有吸热（或放热）效应发生，试样和参比物之间就出现温度差，测量这种温度差随温度变化的曲线，是研究物质在加热（或冷却）过程中发生各种物理或化学变化的重要手段。冷溶、蒸发、升华、解吸、脱水多为吸热效应；吸附、氧化、结晶等多为放热效应。

1.6 聚合物的分类及应用

1.6.1 聚合物的分类

从原料来源角度，高分子可分为天然高分子、合成高分子、改性高分子。按用途可粗分成塑料、橡胶、纤维、涂料、胶黏剂等。按热行为可分为热塑性聚合物和热固性聚合物。从高分子主链结构考虑，可分为碳链聚合物、杂链聚合物和元素有机聚合物，在此基础上，再进一步细分，如聚烯烃、聚酯、聚酰胺等。

（1）碳链聚合物

高分子主链完全由碳原子组成，绝大部分烯类和二烯类的加成聚合物属于这一类，如聚乙烯、聚氯乙烯、聚丁二烯、聚异戊二烯等。

（2）杂链聚合物

高分子主链中除了碳原子外，还有氧、氮、硫等杂原子，如聚醚、聚酯、聚酰胺等缩聚物和杂环开环聚合物，天然高分子多属于这一类。这类聚合物都有特征基团，如醚键、酯键、酰胺键等。

(3) 元素有机聚合物

元素有机聚合物也称为半有机高分子，大分子主链中没有碳原子，主要由硅、硼、铝和氧、氮、硫、磷等原子组成，但侧基多半是有机基团，如甲基、乙基、乙烯基、苯基等。聚硅氧烷（有机硅橡胶）是典型的例子。如果主链和侧基均无碳原子，则称为无机高分子，如硅酸盐类。

H. Staudinger（1881～1965）：德国化学家，1912年于苏黎世工业大学任化学教授。1920年发表"论聚合反应"的论文，提出高分子的概念。1932年出版《高分子有机化合物》，被尊称为高分子鼻祖，1953年获诺贝尔化学奖。

1.6.2 高分子材料的应用

合成塑料、合成纤维、合成橡胶是三大高分子材料，也是高分子化学工业中规模巨大的三个基础产业领域。此外还有涂料、胶黏剂、功能高分子三个类别。涂料和胶黏剂是合成树脂的特殊应用形式，与塑料、纤维、橡胶三个基础产业相比，其规模相对较小，但种类繁多。材料按用途可分为结构材料和功能材料两大类。高分子结构材料用于制造受力构件，应该具有良好的力学性能。功能高分子材料具有特殊的电、磁、光、热、声学、力学、化学、生物医学性能，能够实现不同能量形态的相互转化，目前广泛应用于各类高科技领域。除了

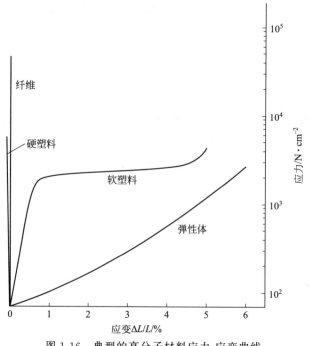

图 1-16　典型的高分子材料应力-应变曲线

本身固有的特殊功能以外，功能高分子材料往往对机械强度也有一定的要求。

聚合物的力学性能可以用应力-应变（stress-strain）性能表征，应力-应变性能指聚合物材料从施加拉伸应力使其伸长（应变）直至断裂过程的行为，其结果通常以应力-应变曲线表示。几种应力-应变曲线如图 1-16 所示，其中如下四个参数是表征高分子材料力学性能的重要手段。

① 弹性模量（modulus）　代表物质的刚性，对变形的阻力，以起始应力除以相对伸长率来表示，即应力-应变曲线的起始斜率。

② 拉伸强度（tensile strength）　使试样破坏的应力。

③ 断裂伸长率（ultimate elongation）　试样断裂时的伸长程度。

④ 弹性伸长率（elastic elongation）　试样的可逆伸长程度。

Carothers（1896～1937）：美国有机化学家。1924 年获伊利诺伊大学博士学位后，先后在该大学和哈佛大学担任有机化学的教学和研究工作。1928 年应聘在美国杜邦公司进行有机化学研究。首先合成了氯丁二烯及其聚合物，为氯丁橡胶的开发奠定了基础。1935 年以己二酸与己二胺为原料制得聚合物，1939 年实现工业化后定名为尼龙，是最早实现工业化的合成纤维品种，奠定了合成纤维工业的基础。他的学生中有著名高分子学者 Flory。

分子量、热转变温度（T_g、T_m）、微结构、结晶度往往是聚合物合成阶段需要表征的参数，而力学性能则是聚合物成型制品的质量指标，与上述参数密切相关（表 1-3）。一般而言，极性、结晶度、玻璃化温度愈高，则机械强度愈大，伸长率愈小。

表 1-3　主要高分子结构及性质

聚合物	符号	重复单元结构	T_g/℃	T_m/℃
聚乙烯	PE	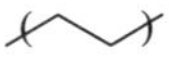		
高密度聚乙烯	HDPE		−120	130
低密度聚乙烯	LDPE		−120	110
聚丙烯	PP	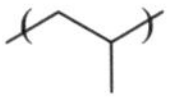	−10	全同 176
聚异丁烯	PIB	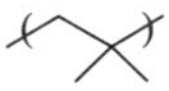	−73	44
聚苯乙烯	PS	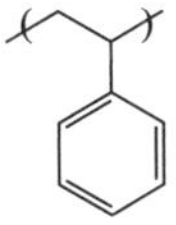	95	全同 240
聚氯乙烯	PVC	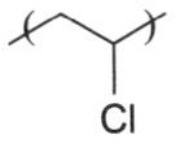	81	

续表

聚合物	符号	重复单元结构	T_g/℃	T_m/℃
聚偏氯乙烯	PVDC	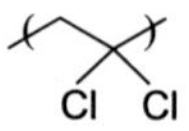 —CH₂—CCl₂—	−17	198
聚四氟乙烯	PTFE	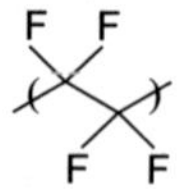 —CF₂—CF₂—		327
聚丙烯酸	PAA	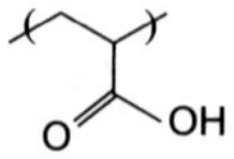 —CH₂—CH(COOH)—	106	
聚丙烯酰胺	PAM	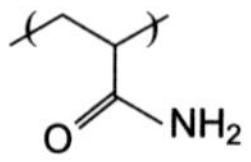 —CH₂—CH(CONH₂)—	6	
聚丙烯酸甲酯	PMA	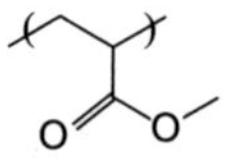 —CH₂—CH(COOCH₃)—	10	
聚甲基丙烯酸甲酯	PMMA	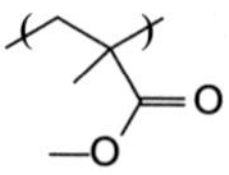 —CH₂—C(CH₃)(COOCH₃)—	105	
聚丙烯腈	PAN	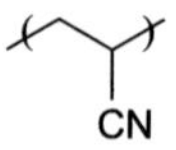 —CH₂—CH(CN)—	97	317
聚醋酸乙烯酯	PVAc	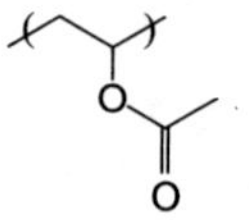 —CH₂—CH(OCOCH₃)—	28	
聚乙烯醇	PVA	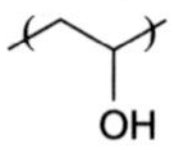 —CH₂—CH(OH)—	85	258
聚丁二烯	PB	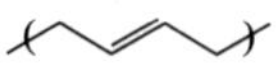 —CH₂—CH=CH—CH₂—	−108	2
聚异戊二烯	PIP	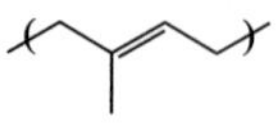 —CH₂—C(CH₃)=CH—CH₂—	−73	
聚甲醛	POM	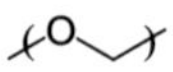 —O—CH₂—	−82	175

20 高分子化学

续表

聚合物	符号	重复单元结构	T_g/℃	T_m/℃
聚环氧乙烷		 ―(O―CH₂―CH₂)―	−67	66
聚苯醚		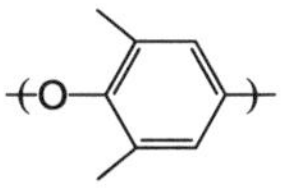 (2,6-二甲基苯醚重复单元)	220	480
涤纶树脂	PET	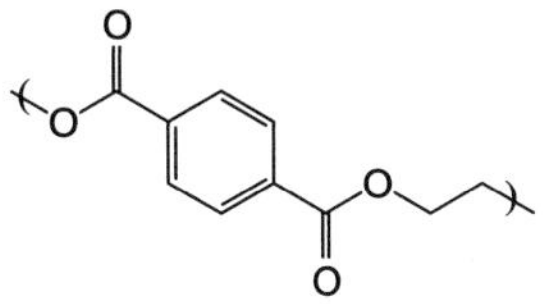 (对苯二甲酸乙二醇酯重复单元)	69	267
聚碳酸酯	PC	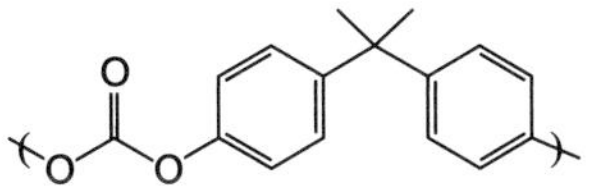 (双酚A碳酸酯重复单元)	150	
尼龙-66	PA66	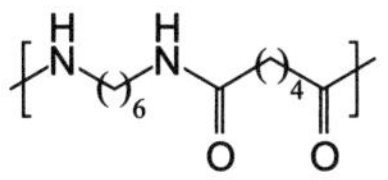 ―[NH―(CH₂)₆―NH―CO―(CH₂)₄―CO]―	50	250
尼龙-6	PA6	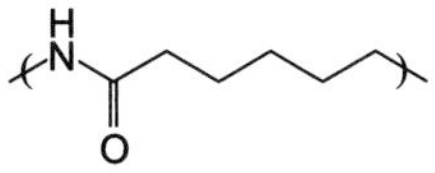 ―[NH―(CH₂)₅―CO]―	50	230
聚氨酯	PU	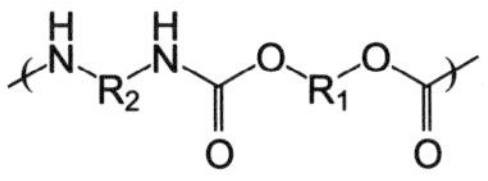 ―[NH―R₂―NH―CO―O―R₁―O―CO]―		
聚脲		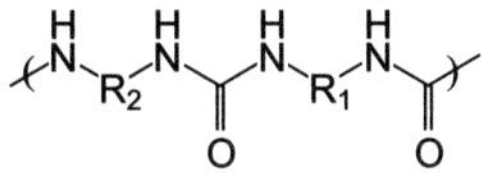 ―[NH―R₂―NH―CO―NH―R₁―NH―CO]―		
双酚A聚砜		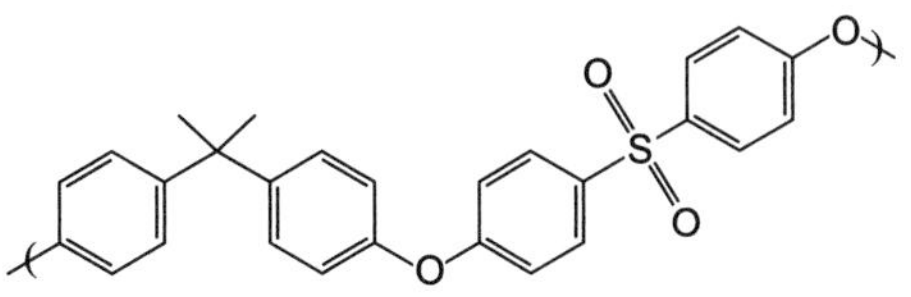 (双酚A聚砜重复单元)	150	180

第1章 绪 论

续表

聚合物	符号	重复单元结构	T_g/℃	T_m/℃
酚醛树脂		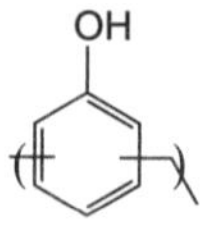 (苯酚-OH 结构)		
脲醛树脂		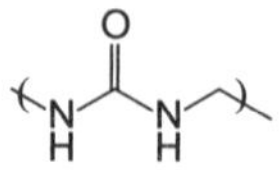 (脲基 —NH—CO—NH— 结构)		
硅橡胶		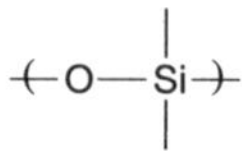 $\mathrm{+O{-}Si\!+\!}$	−123	

橡胶、纤维、塑料的结构和性能有很大的差别，可从应力-应变曲线上看出。

(1) 橡胶

橡胶具有高弹性，很小的作用力就能产生很大的形变（500%～1000%），外力除去后，能立刻恢复原状。橡胶往往是非极性无定形聚合物，分子链柔性大，玻璃化温度低（−55～−120℃）。室温下分子链处于卷曲状态，拉伸时伸长，除去应力后回缩。少量交联可以防止大分子滑移。拉伸起始弹性模量小（<70N/cm²），拉伸后诱导结晶，弹性模量和强度增加。伸长率达到 400% 时，强度可增至 1500N/cm²；伸长率 500% 时达到 2000N/cm²。

(2) 纤维

纤维与橡胶相反，不易变形，伸长率小（0～50%），弹性模量（>35000N/cm²）和拉伸强度（>35000N/cm²）都很高。许多纤维用聚合物的重复单元中带有极性基团，能够增加高分子的分子间作用力和结晶性。纺丝过程中的拉伸力一般可以提高纤维结晶度。纤维聚合物的 T_m 一般在 200℃ 以上，以确保热水洗涤不易变形但可以烫熨整理。聚合物 T_m 如果高于 300℃，会使熔融纺丝变得困难。能溶于适当有机溶剂的聚合物可以使用溶液纺丝。纤维用聚合物的 T_g 应适中，T_g 过高不利于拉伸纺丝，而 T_g 过低则易使织物变形。尼龙-66 是典型的合成纤维，其中的酰胺基团有利于在分子间形成氢键，纺丝过程的拉伸处理又能提高纤维结晶度。尼龙-66 的 T_m 约为 265℃，T_g 为 50℃ 左右，拉伸强度（70000N/cm²）和弹性模量（5000N/cm²）都很高，而伸长率却较低（<20%）。

(3) 塑料

塑料的力学性能介于橡胶和纤维之间，范围很广，从接近橡胶的软塑料到接近纤维的硬塑料都有。聚乙烯是典型的软塑料，弹性模量为 20000N/cm²，拉伸强度为 2500N/cm²，伸长率约为 500%。

软塑料（flexible plastic）具有中等结晶度，T_m 和 T_g 范围较宽，拉伸强度（1500～7000N/cm²）、弹性模量（15000～35000N/cm²）、伸长率（20%～800%）均为中等程度。

硬塑料（rigid plastic）的特点是刚性大，难变形，拉伸强度（3000～8500N/cm²）和弹性模量（70000～350000N/cm²）较高，而断裂伸长率却很低（0.5%～3%）。多数硬塑料的高分子链为刚性链，属非晶态。例如，聚苯乙烯（T_g，95℃）和聚甲基丙烯酸甲酯（T_g，105℃）因侧基较大而刚性增加。

Flory（1910～1985）：Flory 在高分子物理化学方面的贡献，几乎遍及各个领域。他是实验家又是理论家，是高分子科学理论的主要开拓者和奠基人之一。著有《高分子化学原理》和《长链分子的统计力学》等，曾获 1974 年度诺贝尔化学奖。

齐格勒（Ziegler）和纳塔（Natta）：1953 年，齐格勒使用有机铝试剂，成功将乙烯变成丁烯，开创了长链烯烃类高聚物生产的新纪元。这类高聚物广泛用于塑料、纤维、橡胶工业等领域。纳塔应用齐格勒催化剂，试验了丙烯的聚合反应，得到了有规则分子结构、性能优良的聚丙烯。由于齐格勒和纳塔发明了乙烯、丙烯聚合的催化剂，奠定了定向聚合的理论基础，大大降低了生产成本，优化了工艺流程，为现代化学工业做出了杰出贡献。

1.7　高分子化学发展简史

高分子和人类密切相关，所有的生命体均由各种各样的天然高分子构成，食物中的蛋白

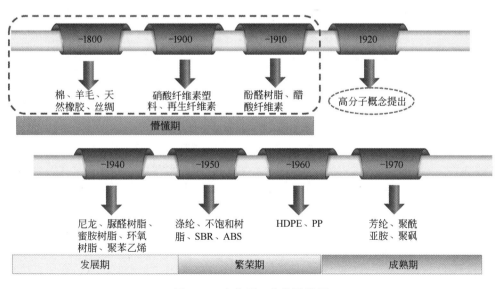

图 1-17　高分子工业发展进程

质和淀粉也是高分子。远在几千年以前，人类就使用棉、麻、丝、毛等天然高分子作织物材料，使用木材作建筑材料。人们在古代就掌握了纤维造纸、皮革鞣制以及抽丝纺纱等天然高分子的加工过程。直至二十世纪二三十年代，才有数种合成高分子材料出现。目前，高分子材料的体积产量已经远超钢铁和金属。在材料科学及应用领域，高分子材料已与金属材料、无机材料等并驾齐驱，成为国民经济中不可或缺的重要组成部分。高分子的工业化发展进程见图 1-17。高分子发展年表和高分子领域诺贝尔奖情况见表 1-4 和表 1-5。

表 1-4 高分子发展年表

年份	名称	年份	名称
1839	橡胶硫化	1939	尼龙-6
1846	硝化纤维素	1940	丁基橡胶
1851	硬橡胶	1941	低密度聚乙烯 LDPE
1860	模塑虫胶	1941	PET
1868	硝酸纤维素塑料	1942	不饱和聚酯
1889	硝化纤维素光学胶片	1943	聚四氟乙烯
1890	铜氨再生纤维素	1943	硅橡胶
1892	黏胶纤维	1945	丁苯橡胶
1907	酚醛树脂	1948	ABS 树脂
1908	醋酸纤维素	1950	聚丙烯腈纤维
1920	脲醛树脂	1953	抗冲击聚苯乙烯 HIPS
1924	醋酸纤维素纤维	1956	聚苯醚
1926	醇酸树脂	1957	高密度聚乙烯 HDPE
1927	聚氯乙烯	1957	聚丙烯
1929	脲醛树脂	1957	聚碳酸酯
1931	聚甲基丙烯酸甲酯	1961	芳纶
1931	氯丁橡胶	1962	聚酰亚胺
1934	环氧树脂	1965	聚砜
1935	乙基纤维素	1970	PBT
1936	聚醋酸乙烯酯	1975	聚乙炔
1937	聚苯乙烯	1982	聚醚酰亚胺
1937	丁腈橡胶	1991	碳纳米管
1938	尼龙-66	1995	活性可控自由基聚合
1939	密胺树脂		

表 1-5 高分子科学领域诺贝尔奖

学者	时间	研究领域
Hermann Staudinger	1953	聚合物概念
Karl Ziegler and Giulio Natta	1963	聚合物结构立体规整性
Paul Flory	1974	高分子分子链热力学
Bruce Merrifield	1984	多肽固相合成
Pierre de Gennes	1991	液晶、高分子界面结构及调控
A. J. Heeger, Alan Mac Diarmid, and H. Shirakawa	2000	导电高分子

19世纪中叶，天然高分子的化学改性开始发展，如天然橡胶的硫化（1839年），硝酸纤维素塑料的出现（1868年），黏胶纤维的生产（1893~1898年）等。20世纪初期，出现了第一种合成树脂和塑料——酚醛塑料，1909年实现工业化生产。第一次世界大战期间，丁钠橡胶问世。20世纪20年代，醇酸树脂、醋酸纤维、脲醛树脂也相继投入生产。

19世纪，人们还没有正确理解高分子的结构，直到20世纪初期，才初步确定天然橡胶由异戊二烯构成，纤维素和淀粉由葡萄糖残体构成，但还不知道它们由共价键结合，怀疑它们是小分子聚集的胶体。1890~1919年，Emil Fischer因为对蛋白质的研究，开始涉及聚合物的结构，为之后高分子概念的建立奠定了良好基础。1920年，Staudinger提出了聚苯乙烯、橡胶、聚甲醛等都是共价结合的大分子的假说，Staudinger后来于1932年发表的专著《高分子有机化合物》被视为高分子科学的起点。

20世纪30~40年代是高分子化学和工业开始兴起的时代，高分子化学和工业生产两者相互促进，不久就进入到快速发展阶段。30年代初期，Carothers着手研究聚酯和聚酰胺的缩聚反应，并于1935年成功研制出尼龙-66，1938年实现了工业化生产。20世纪30年代，一批经自由基聚合得到的烯类聚合物实现了工业化，如聚氯乙烯（1927~1937年）、聚醋酸乙烯酯（1936年）、聚甲基丙烯酸甲酯（1927~1931年）、聚苯乙烯（1934~1937年）、高压聚乙烯（1939年）等。自由基聚合的成功突破了经典有机化学的范围，与缩聚一起奠定了早期高分子化学学科发展的基础。

在缩聚和自由基聚合等基本原理指导下，20世纪40年代，高分子工业以更快的速度发展，相继开发了丁苯橡胶、丁腈橡胶、氟树脂、ABS树脂等，属于阳离子聚合的丁基橡胶也在这一时期生产，同时发展了乳液聚合和共聚合的基本理论，逐步改变了完全依靠条件摸索的技艺时代。当时，陆续实现工业化的缩聚物有不饱和聚酯树脂、有机硅、聚氨酯、环氧树脂等。由于原料问题，1940年开发成功的涤纶树脂到1950年才实现工业化。聚丙烯腈纤维也在解决了溶剂问题以后，于1948~1950年正式投产。

高分子溶液理论和分子量测定推动了高分子化学的发展。20世纪40年代初提出的Flory-Huggins理论揭示了高分子溶液与理想溶液存在巨大偏差的实质，此后，Flory开始研究排斥体积效应，50年代提出的Flory-Krigbaum稀溶液理论是该领域的代表性成果。Flory在高分子领域等多方面做出了很大贡献，于1974年获得了诺贝尔奖。

白川英树：日本著名化学家，获2000年诺贝尔化学奖（三名得主之一）。1966年获得东京工业大学化工专业博士，之后在东京工业大学资源科学研究所任助教。1976年应艾伦·黑格教授之邀赴美，在宾夕法尼亚大学担任博士研究员。1979年回筑波大学任副教授，1982年起任教授。主要研究成果是掺杂聚乙炔膜的电导性能的提高。

这个时期，物理和物理化学中的许多表征技术，如核磁、红外、X衍射、光散射等，对高分子结构的剖析和确定起了重要作用。

20世纪50~60年代，出现了许多新的聚合方法和聚合物品种，高分子工业发展得更快，规模也更大。1953~1954年，Ziegler、Natta等发明了有机金属引发体系，在较温和的条件下合成了高密度聚乙烯和等规聚丙烯，开拓了高分子合成的新领域，并因此获得诺贝尔奖。几乎同时，Szwarc对阴离子聚合和活性高分子的研制做出了贡献。这些研究为20世纪60年代以后聚烯烃、顺丁橡胶、异戊橡胶、乙丙橡胶以及SBS（苯乙烯-丁二烯-苯乙烯）嵌段共聚物（热塑性弹性体）的大规模发展提供了理论基础。

继20世纪50年代末期聚甲醛、聚碳酸酯出现以后，60年代还开发了聚砜、聚苯醚、聚酰亚胺等工程塑料。许多耐高温和高强度的合成材料也层出不穷，这为缩聚反应开辟了新的方向。可以说，60年代是聚烯烃、合成橡胶、工程塑料以及离子聚合、配位聚合、溶液聚合大发展的时期，与以前开发的聚合物品种、聚合方法一起，形成了合成高分子全面繁荣的局面。

20世纪70~90年代，高分子化学学科更趋成熟，进入了新的时期。新聚合方法，新型聚合物，新的结构、性能和用途不断涌现。除了原有聚合物以更大规模、更加高效地工业生产以外，大家更重视新合成技术的应用和高性能、特种功能聚合物的研制开发。新的合成方法有茂金属催化聚合、活性自由基聚合、基团转移聚合、丙烯酸类-二烯烃易位聚合、以CO_2为介质的超临界聚合以及大分子取代法制聚磷氮烯等。高性能涉及超强、耐高温、耐烧蚀、耐油、低温柔性等，相关的聚合物有芳杂环聚合物、液晶高分子、梯形聚合物等。此外，一些新型结构聚合物，如星形和树枝状聚合物、新型接枝和嵌段共聚物、无机-有机杂化聚合物等也相继开发。

目前，功能高分子更注重光电功能和生物功能的研究和开发。光电功能高分子（如杂化聚合物-陶瓷材料）在半导体器件、光电池、传感器、质子电导膜中起重要作用。在生物医药领域，除医用高分子材料外，还涉及药物控制释放、酶的固载、胶束、胶囊、微球、水凝胶、生物相容界面等，都是生物医用材料领域中新的研究内容。

高分子科学推动了化工、材料等相关行业的发展，也丰富了化学、化工、材料等相关学科。在高分子学科的形成过程中，离不开其他学科的基础和相关行业的推动。高分子化学还会与生物学科相互渗透。目前几乎50%以上的化学化工工作者以及材料、轻纺乃至机械等行业的众多工程师都在从事聚合物的研究开发工作。

高分子化学已经不再是有机化学、物理化学等某一传统化学学科的分支，而是整个化学学科和物理、生物乃至药物等许多学科的交叉和综合，今后还会进一步丰富和完善。

英语读译资料

(1) Molecular weight of polymers

Polymers are macromolecules built up by the linking together of large numbers of much smaller molecules. The small molecules that combine with each other to form polymer molecules are termed monomers, and the reactions by which they combine are termed polymerizations. There may be hundreds, thousands, tens of thousands, or more monomer molecules linked together in a polymer molecule. When one speaks of polymers, one is concerned with materials whose molecular weight may reach into the hundreds of thousands or millions.

The molecular weight of a polymer is of prime importance in the polymer's synthesis and application. Chemists usually use the term molecular weight to describe the size of a molecule. The more accurate term is molar mass. The interesting and useful mechanical properties that are uniquely associated with polymeric materials are a consequence of their high molecular weight. Most important mechanical properties depend on and vary considerably with molecular weight. There is a minimum polymer molecular weight, A, usually a thousand or so, to produce any significant mechanical strength at all. Above A, strength increases rapidly with molecular weight until a critical point (B) is reached. Mechanical strength increases more slowly above B and eventually reaches a limiting value (C). The critical point B generally corresponds to the minimum molecular weight for a polymer to begin to exhibit sufficient strength to be useful. Most practical applications of polymers require higher molecular weights to obtain higher strengths. The minimum useful molecular weight (B), usually in the range 5000-10000, differs for different polymers.

(2) Goodyear and rubber

Charles Goodyear grew up in poverty. He was a Connecticut Yankee born in 1800. He began work in his father's farm implement business. Later he moved to Philadelphia, where he opened a retail hardware store that soon went bankrupt. Goodyear then turned to being an inventor. As a child he had noticed the magic material that formed a rubber bottle he had found. He visited the Roxbury India Rubber Company to try to interest them in his efforts to improve the properties of rubber. They assured him that there was no need to do so.

He started his experiments with a malodorous gum from South America in debtor's prison. In a small cottage on the grounds of the prison, he blended the gum, the raw rubber called hevea rubber, with anything he could find-ink, soup, castor oil, etc. While rubber-based products were available, they were either sticky or became sticky in the summer's heat. He found that treatment of the raw rubber with nitric acid allowed the material to resist heat and not to adhere to itself. This success attracted backers who helped form a rubber company. After some effort he obtained a contract to supply the U.S. post office with 150 rubber mailbags. He made the bags and stored them in a hot room while he and his family were away. When they returned they found the bags in a corner of the room, joined together as a single mass. The nitric acid treatment was sufficient to prevent surface stickiness, but the internal rubber remained tacky and susceptible to heat.

While doing experiments in 1839 at a Massachusetts rubber factory, Goodyear accidentally dropped a lump of rubber mixed with sulfur on the hot stove. The rubber did not melt, but rather charred. He had discovered vulcanization, the secret that was to make rubber a commercial success. While he had discovered vulcanization, it would take several years of ongoing experimentation before the process was really commercially useful. During this time he and his family were near penniless. While he patented the process, the process was too easily copied and pirated so that he was not able to fully profit from his invention and years of hard work. Even so, he was able to develop a number of items.

Goodyear and his brother, Nelson, transformed NR (hevea rubber) from a heat-"softenable" thermoplastic to a less heat-sensitive product through the creation of crosslinks between the individual polyisoprene chain-like molecules using sulfur as the cross-linking agent. Thermoplastics are two-dimensional molecules that may be softened by heat. Thermosets are materials that are three-dimensional networks that cannot be reshaped by heating. Rather than melting, thermosets degrade. As the amount of sulfur was increased, the rubber became harder, resulting in a hard rubber-like (ebonite) material.

The spring of 1851 saw the construction of a remarkable building on the lawns of London's Hyde Park. The building was designed by a maker of greenhouses so it was not surprising that it had a "greenhouse-look". This Crystal Palace was to house almost 14000 exhibitors from all over the world. It was an opportunity for them to show their wares. Goodyear, then 50 years old, used this opportunity to show off his over two decades' worth of rubber-related products. He decorated his Vulcanite Court with rubber walls, roof, furniture, buttons, toys, carpet, combs, etc. Above it hung a giant 6 ft. rubber raft and assorted balloons. The European public was introduced to the world of new man-made materials.

A little more than a decade later Goodyear died. Within a year of his death, the American Civil War broke out. The Union military used about $27 million worth of rubber products by 1865, helping launch the American rubber industry.

(3) Carothers and nylon-66

Wallace Hume Carothers is the father of synthetic polymer science. History is often measured by the change in the flow of grains of sand in the hourglass of existence. Carothers is a granite boulder in this hourglass. Carothers was born, raised, and educated in the U. S. Midwest. In 1920, he left Tarkio College with his BS degree and entered the University of Illinois where he received his MA in 1921. He then taught at the University of South Dakota where he published his first paper. He returned to receive his PhD under Roger Adams in 1924. In 1926 he became an instructor in organic chemistry at Harvard.

In 1927, DuPont began a program of fundamental research "without any regard or reference to commercial objectives." This was a radical departure since the bottom line was previously products marketed and not papers published. Charles Stine, director of DuPont's chemical department, was interested in pursuing fundamental research in the areas of colloid chemistry, catalysis, organic synthesis, and polymer formation, and convinced the Board to hire the best chemists in each field to lead this research. Stine visited with many in the academic community including the then president of Harvard, one of my distant uncles, J. B. Conant, an outstanding chemist himself, who told him about Carothers. Carothers was persuaded to join the DuPont group with a generous research budget and an approximate doubling of his academic salary to $6000. This was the birth of the Experimental Station at Wilmington, Delaware.

Up to this point, it was considered that universities were where discoveries were made and industry was where they were put to some practical use. This separation between basic and applied work was quite prominent at this juncture and continues in many areas even today in some fields of work though the difference has decreased. But in polymers, most of the basic research was done in industry, having as its inception the decision by DuPont to bridge this "unnatural" gap between fundamental knowledge and application. In truth, they can be considered as the two hands of an individual, and in order to do manual work both hands are important.

Staudinger believed that large molecules were based on the jointing, through covalent bonding, of large numbers of atoms. Essentially he and fellow scientists like Karl Freudenberg, Herman Mark, Michael Polanyi, and Kurt Myer looked at already existing natural polymers. Carothers, however, looked at the construction of these giant molecules from small molecules forming synthetic polymers. His intention was to prepare molecules of known structure through the use of known organic chemistry and to "investigate how the properties of these substances depended on constitution." Early work included the study of polyester formation through reaction of diacids with diols forming polyesters. But he could not achieve molecular weights greater than about 4000 below the size at which many of the interesting so-called polymeric properties appear.

DuPont was looking for a synthetic rubber (SR). Carothers assigned Arnold Collins to carry out this research. Collin's initial task was to produce pure divinylacetylene. While performing the distillation of an acetylene reaction, in 1930, he obtained a small amount of an unknown liquid, which he set aside in a test tube. After several days the liquid turned to a solid. The solid bounced and eventually was shown to be a SR polychloroprene, whose properties were similar to those of vulcanized rubber but was superior in its resistance to ozone, ordinary oxidation, and most organic liquids. It was sold under its generic name "neoprene" and the trade name "Duprene."

In 1930, Carothers and Julian Hill designed a process to remove water that was formed during the esterification reaction. They simply froze the water as it was removed, using another recent invention called a molecular still (basically a heating plate coupled to vacuum), allowing the formation of longer chains. In April, Hill synthesized a polyester using this approach: he touched a glass stirring rod to the hot mass and then pulled the rod away, effectively forming strong fibers; the pulling helped reorient the mobile polyester chains. The polyester had a molecular weight of about 12000. Additional strength was achieved by again pulling the cooled fibers. Further reorienting occurred. This process of "drawing" or pulling to produce stronger fibers is now known as "cold drawing" and is widely used in the formation of fibers today. The process of "cold drawing" was discovered by Carothers' group.

While interesting, the fibers were not considered to be of commercial use. Carothers and his group then moved to look at the reaction of diacids with diamines instead of diols. Again, fibers were formed but these initial materials were deemed not to be particularly interesting.

In 1934, Paul Flory was hired to work with Car others to help gain a mathematical understanding of the polymerization process and relationships. Thus, there was an early association between theory and practice or structure-property relationships.

The polyamide fiber project was begun again. One promising candidate was formed from the reaction of adipic acid with hexamethylenediamine, called fiber 66 because each carbon-containing unit had six carbons. It formed a strong, elastic, largely insoluble fiber with a relatively high melt temperature. DuPont chose this material for production. These polyamide s were given the name "nylons." Thus was born nylon-66, the first synthetic material whose properties equaled or exceeded the natural analog, silk. (Although this may not be so in reality, it was believed at the time to be true.)

As women's hemlines rose in the 1930s, silk stockings were in great demand but were very expensive. Nylon changed this, as it could be woven into sheer hosiery. The initial presentation of nylon hose to the public was by Stine at a forum of women's club members in New York City on October 24, 1938. Nearly 800000 pairs were sold on May 151940 alonethe first day they were on the market. By 1941, nylon hosiery held 30% of the market but by December 1941 nylon was diverted to make parachutes etc.

From these studies Carothers established several concepts. First, polymers could be formed by employing already known organic react ions but with reactants that had more than one reactive group per molecule. Second, the forces that bring together the individual polymer units are the same as those that hold together the starting materials, namely primary covalent bonds. Much of the polymer chemistry names and ideas that permeate polymer science today were standardized through his efforts.

习 题

1. 写出聚氯乙烯、聚苯乙烯、涤纶、尼龙-66、聚丁二烯和天然橡胶的结构式（重复单元）。选择其常用分子量，计算聚合度。

2. 写出下列聚合物的单体分子式和常用的聚合反应式：聚丙烯腈，天然橡胶，丁苯橡胶，聚甲醛，聚苯醚，聚四氟乙烯，聚二甲基硅氧烷。

3. 举例说明和区别线形和体形结构，热塑性和热固性聚合物，非晶态和结晶聚合物。

4. 举例说明橡胶、纤维、塑料的结构性能特征和主要差别。

5. 举例说明玻璃化温度和聚合物的熔点如何被大分子微结构、平均分子量等因素影响？

6. 说明和区分如下名词概念：

(1) 高分子，高分子链；(2) 天然高分子，合成高分子；(3) 单体，结构单元，重复单元，链节；(4) 聚合度，平均分子量，分子量分布；(5) 加聚，缩聚。

7. 解释说明为什么对于单取代乙烯类单体（$CH_2=CHY$）而言，高分子链中头尾相连

8. 求下列混合物的数均分子量、重均分子量和分子量分布指数。
(1) 组分 A：质量＝10g，分子量＝30000；
(2) 组分 B：质量＝5g，分子量＝70000；
(3) 组分 C：质量＝1g，分子量＝100000。

9. 加氢反应可以使聚丁二烯分子链中的1,4-加成结构转变为线形高分子链结构，转变1,2-加成结构为乙基分支结构。某种聚丁二烯样品的分子量为168000，红外谱图表明其中含47.2%顺式、44.9%反式、7.9%的乙烯基结构。计算该样品的加氢反应产物的高分子链中，每两个乙基分支之间的平均碳原子个数。

10. 如下一些聚合物的 IUPAC（I）命名或习惯名称（T），写出它们的分子式和中文名称。
(1) polymethylene(I)；(2) polyformaldehyde(T)；(3) poly(phenylene oxide)(T)；
(4) poly[(2-propyl-1,3-dioxane-4,6-diyl)methylene](I)；(5) poly(1-acetoxyethylene)(I)；
(6) poly(methy acrylate)(T)。

11. 在 GPC 方法之前，离心沉降分离法曾经是测定聚合物平均分子量及分布的基本手段。如下数据来源于由癸二酸和1,6-己二醇合成的聚酯，计算其数均分子量和重均分子量，并计算其分子量分布指数。

次数	1	2	3	4	5	6	7	8	9
质量/g	1.15	0.73	0.415	0.35	0.51	0.34	1.78	0.10	0.94
$M_r \times 10^{-4}$	1.25	2.05	2.40	3.20	3.90	4.50	6.35	4.10	9.40

12. MALDI-TOF-MS（Matrix-Assisted Laser Desorption/Ionization Time of Flight Mass Spectrometry，基质辅助激光解吸电离飞行时间质谱）是近年来发展起来的一种新技术。测试样品溶于一个固体底物中形成晶体，用激光脉冲使其离子化，离子被加速后通过飞行管时分离，所有的离子均可通过飞行时间被分离检测。该技术在分子量可测范围，分辨率和所测分子量的准确性等方面都优于同类技术。请查阅相关文件，总结并讨论该技术在高分子分子量及其分布测试中的优缺点。

13. 假设你从协作橡胶厂得到1kg单分散无规聚丁二烯样品，分子量为54000，请计算这个样品中两个高分子的分子链结构完全相同的概率。如果聚丁二烯样品为高度规整（配位聚合产物），其等规度高达99.5%，那么样品中存在两个分子链结构完全相同的高分子的概率又是多少？

=== 英语习题 ===

1. Explain the fact that conversion of the amide groups —CONH— in nylon to methylol groups —CON(CH$_2$OH)— by reaction with formaldehyde, followed by methylation to ether groups —CON(CH$_2$OCH$_3$)— results in the transformation of the fiber to a rubbery product with low modulus and high elasticity.

2. Give the overall chemical reactions involved in the polymerization of these monomers, the resulting repeat unit structure, and an acceptable name for the polymer.

(a) [structure: methacrylic acid — CH2=C(CH3)C(=O)OH]

(b) H2N–(CH2)4–NH2 + Cl–C(=O)–(CH2)4–C(=O)–Cl

(c) HO–(CH2)5–C(=O)–OH

(d) O=C=N–C6H4–N=C=O + HO–(CH2)4–OH

(e) H2N–CH(CH3)–C(=O)–OH

3. ^{1}H NMR is used to attempt to quantify the molecular weight of a poly(ethylene oxide) molecule with methyoxy end groups at each terminus. If the integration of the methyl protons relative to the methylene protons gave a ratio of 1 : 20, what can you say about the molecular weight?

4. What would be M_w and M_n for a sample obtained by mixing 10g of polystyrene ($M_w=100000$, $M_n=70000$) with 20g of another polystyrene ($M_w=60000$, $M_n=20000$)?

5. A terpolymer is prepared from vinyl monomers A, B, and C; the molecular weight of the repeat units are 104, 184, and 128, respectively. A particular polymerization procedure yields a product with the empirical formula $A_{3.55} B_{2.20} C_{1.00}$. The authors of this research state that the terpolymer has "an average unit weight of 134". Verify the value. The average weight per repeat unit could be used to evaluate the overall degree of polymerization (X_n) of this terpolymer. For example, if the molecular weight of the polymer was 43000, calculate the corresponding X_n.

6. 1,2-glycol bonds are cleaved by reaction with periodate; hence poly(vinyl alcohol) chains are broken at the site of head-to-head links in the polymer. The fraction of head-to-head linkages in poly(vinyl alcohol) may be determined by measuring the molecular weight before (subscript b) and after (subscript a) cleavage with periodate according to the following formula: Fraction $= 44(1/M_a - 1/M_b)$. Derive this expression and calculate the value for the fraction in the case of $M_b=10^5$ and $M_a=10^3$.

7. The ^{1}H NMR spectrum in Figure 1-18 corresponds to a sample of polyisoprene containing a sec-butyl initiating group and a hydroxyl terminating end group. The relative peak integrations are (a) 26.9, (b) 5.22, (c) 2.00, and (d) 5.95. What is M_n for this polymer? What is the relative percentage of 1,4- and 3,4-addition?

8. One block copolymer of butadiene (B) and isoprene (I) is described as having a 2 : 1 molar ratio of B to I with the following microstructure:

B 45% cis-1,4; 45% trans-1,4; 10% vinyl.
I over 92% cis-1,4.

Draw the structure of this polymer.

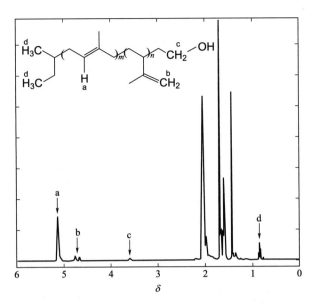

Figure 1-18　^{1}H NMR spectrum of the polyisoprene sample discussed in problem 7

第 2 章 逐步聚合

逐步聚合（一）　　逐步聚合（二）　　逐步聚合（三）　　逐步聚合（四）

　　羽毛球运动击球时，球拍弹击羽毛球头的时间为 4~6ms，球拍框、球拍中杆、球拍线在短时间内要承受较大的冲击力，球拍线要承受 15kg 左右的拉力，球拍要承受 100kg 的扭力和将近 1cm 的形变。一把中高端羽毛球拍质量为 80~100g，由高强纤维增强高分子树脂复合而成；球拍线直径大约为 0.7mm，由多层材料复合而成，主要成分为聚酯和聚酰胺。

本章目录

2.1　线形逐步聚合　/ 35
2.2　体形缩聚和凝胶化作用　/ 46
2.3　重要缩聚物　/ 50

重点要点

　　等活性理论，反应程度，平均聚合度，聚合度，过量比，缩聚反应平衡常数，熔融聚合、溶液聚合，界面缩聚，固相缩聚，官能度，凝胶点，聚酯，涤纶，不饱和聚酯，醇酸树脂，生物可降解聚酯，聚碳酸酯，聚酰胺，聚酰亚胺，聚氨酯，环氧树脂，酚醛树脂，氨基树脂

2.1 线形逐步聚合

大部分缩聚反应遵循逐步聚合机理,但是逐步聚合和缩聚反应两个名词的含义不同。如每一步聚合都是缩合反应,也就是说,每一步反应都有小分子缩合副产物析出时,就称为缩合聚合(缩聚反应,condensation polymerization)。例如,聚酯、聚酰胺的合成反应总是伴随着水等小分子的生成,属于典型的缩聚,也是典型的逐步聚合反应。也有一些逐步聚合反应,每一步都是加成反应,无小分子析出,称为逐步加成聚合反应(step growth addition polymerization)。例如,环氧化合物与多元胺生成固化树脂,异氰酸酯与多元醇生成聚氨酯的反应都属于逐步加成聚合反应,这些反应的聚合机理也同样属于逐步聚合反应。

逐步聚合是高分子合成的重要手段,在高分子化学领域有着极为重要的地位。在生物体内,绝大多数天然高分子如淀粉、纤维素、蛋白质、DNA、RNA等,都是在酶的参与下通过缩聚而成的。聚酯(polyester)、聚酰胺(polyamide)、聚氨酯(polyurethane)、环氧树脂(epoxy resin)等杂链聚合物已经大规模工业化生产,并被广泛应用,它们也是通过逐步聚合合成的。一些耐高温聚合物以含芳香环结构的高分子主链为特征,如聚酰亚胺、梯形聚合物等,也是由逐步聚合的方法制备的。除上述典型的结构高分子材料以外,利用逐步聚合还可以合成许多功能高分子,如液晶高分子、光电高分子等。

本章将剖析逐步聚合机理的共同规律,并介绍目前高分子工业中重要的逐步聚合物。

2.1.1 官能团等活性理论

在逐步聚合过程初期,聚合物分子量的增加相对比较缓慢。以二元醇和二元酸合成聚酯的缩聚反应为例:首先生成二聚体(dimer),然后是二聚体与二元醇或二元酸单体反应生成三聚体(trimer)。二聚体之间也可以进一步反应,生成四聚体(tetramer)。四聚体和三聚体继续与体系内的四聚体、三聚体、二聚体或单体反应,生成五聚体(pentamer)、六聚体(hexamer),乃至分子量更大的聚合产物。在逐步聚合反应过程中,聚合物分子量随反应时间持续增大。逐步聚合的一个显著特点是,在反应初期,二元醇和二元酸的单体就已经消失,即单体转化率接近100%,而聚合产物的平均分子量却依然较低。在反应后期,低聚体之间相互反应,使得聚合产物的平均分子量迅速增加。在任何情况下,反应混合物都由各种聚合度的二元醇、二元酸组成,任何一个含有羟基的分子与任何一个含有羧基的分子都能反应,这是逐步聚合的普遍特征。

在有机化学反应中,一元酸和一元醇的酯化反应在特定温度下只有一个速率常数。在逐步聚合反应中,体系内分子量不同的低聚物种类较多,每一种低聚体与另外一种低聚体间的化学反应速率常数都不相同,也就是说聚合反应体系内存在数目众多的反应,非常复杂,无法用动力学处理。因此,在进行逐步反应的动力学分析时,需要利用假设条件进行简化处理。这个假设认为聚合体系内的所有官能团的活性相同,通常称为官能团等活性(equal reactivity of functional groups)。对于双官能团单体(例如二元醇中的两个羟基)而言,等活性的含义是，无论另一个官能团是否已经反应，也无论低聚物的聚合度是多少，所有的同类官能团的反应活性相同。官能团等活性是逐步聚合动力学分析的重要基础。一些实验数据和理论分析也支持这个假设。

如表 2-1 所示，一元酸系列和乙醇的酯化研究表明，$n=1\sim3$ 时，速率常数迅速降低，但 $n>3$ 后，酯化速率常数几乎不变。因为诱导效应只能沿碳链传递 1~2 个原子，对羧基的活化作用也只限于 $n=1\sim2$。$n=3\sim17$，活化作用微弱，速率常数趋向定值。二元酸系列与乙醇的酯化情况也相似，并与一元酸的酯化速率常数相近。可见，在一定聚合度范围内，官能团活性与聚合物分子量大小无关。

表 2-1 有机羧酸和乙醇的酯化反应速率常数（25℃）

单位：$10^4 \text{L} \cdot \text{mol}^{-1} \cdot \text{s}^{-1}$

n	$\text{H(CH}_2)_n\text{COOH}$	n	$\text{H(CH}_2)_n\text{COOH}$
1	22.1	8	7.5
2	15.3	9	7.4
3	7.5	11	7.6
4	7.5	15	7.7
5	7.4	17	7.7

聚合体系的黏度随分子量的增加而增加，一般认为分子链的移动减弱，从而使官能团活性降低。但实际上端基的活性并不取决于整个大分子重心的平移，而与端基链段的活动有关。大分子链构象改变，链段的活动以及羧基与端基相遇的速率要比重心平移速率高得多。在聚合度不高、体系黏度不大的情况下，并不影响链段的运动，两链段一旦靠近，适当的黏度反而不利于分开，有利于持续碰撞，这给"等活性"提供了条件。但到聚合后期，黏度过大后，链段活动也受到阻碍，甚至包埋，端基活性才降低。

2.1.2 线形缩聚反应产物的分子量以及分子量分布

分子量以及分子量分布是影响聚合物性能的重要因素。控制缩聚产物的分子量及其分子量分布是线形缩聚反应中的核心问题。

(1) 缩聚物的平均聚合度

以聚对苯二甲酸乙二醇酯［poly(ethylene terephthalate)］为例，其具有两种结构单元，分别与乙二醇和对苯二甲酸这两种单体结构相呼应；其重复单元和结构单元如图 2-1 所示。我们用 X_n 表示结构单元的聚合度，用 DP 表示重复单元的聚合度，对于通常的线形高分子而言，二者具有如下关系：

$$X_n = 2 \times DP \tag{2-1}$$

(2) 影响平均聚合度的因素

① 反应程度

在缩聚反应过程中，活性官能团之间相互作用，其数目不断减少，聚合产物的分子量逐渐增加。在有机小分子物质的化学反应中经常使用的反应物的转化率在缩聚反应过程的初期就接近 100%，而在缩聚反应过程的中后期则变化不大。转化率和缩聚产物分子量之间的关

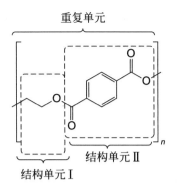

图 2-1 逐步聚合物的重复单元及结构单元

系并不明显。在缩聚反应中使用官能团的转化率来描述缩聚反应的进行程度。高分子化学中将已经完成缩聚反应的官能团数与起始官能团数目的比值称为反应程度，用 p 表示。

在线形缩聚反应中的两种官能团数相等的假设条件下，起始官能团总数为 $2N_0$，当缩聚反应进行到一定程度，体系内剩余的官能团数为 $2N$ 时，已经完成缩聚反应的官能团数为 $2(N_0-N)$，则反应程度 p 可表示如下：

$$p = \frac{\text{已反应官能团数}}{\text{起始时官能团总数}} = \frac{N_0 - N}{N_0} \tag{2-2}$$

平均聚合度为：

$$X_n = \frac{\text{缩聚初始时的单体分子数}}{\text{已经生成的缩聚物分子数}} = \frac{N_0}{N}$$

上述两个公式结合，可以得到一个重要公式：

$$X_n = \frac{1}{1-p} \tag{2-3}$$

由式(2-3)可以看出，当反应程度不高时，尽管聚合度随反应程度的增加而增大，但变化的幅度并不大。在缩聚反应后期，反应程度的较小增加都会导致缩聚产物平均聚合度的急剧增加。

反应程度的测定方法有三种：测定析出的小分子数量，如水分子、HCl 等；测定反应体系内残存的反应官能团的数量，如羧基等；测定体系的黏度变化。

假定缩聚体系内两种反应官能团数量完全相等，理论上缩聚反应可以不断进行下去，直到缩聚体系内的所有反应官能团完全消耗，整个体系成为一个大分子。但是，实际的缩聚反应是不可能进行彻底的。

② 过量官能团数量的影响

由于实际单体称量的精度限制和单体的纯度限制，两种反应官能团的数量完全相等是不可能实现的。如果一种官能团过量，势必影响高分子缩聚产物的生成。

假设 N_a 为反应官能团 a 的起始总数，N_b 为反应官能团 b 的起始总数，如果 $N_a<N_b$，设 $N_a/N_b=r$，称 r 为基团数量比，缩聚反应将进行到 a 官能团完全消耗，体系内所有缩聚产物的分子链两端将全部为 b 官能团，使缩聚反应无法继续进行，平均聚合度不再增大，也就是反应程度 $p=1$。这时缩聚体系内的分子总数为：

$$\frac{N_a - N_b}{2}$$

缩聚反应起始时的单体分子总数为：

$$\frac{N_a + N_b}{2}$$

则缩聚产物的平均聚合度为：

$$X_n = \frac{\text{缩聚反应起始时单体分子总数}}{\text{体系内缩聚产物分子总数}}$$

$$= \frac{N_a + N_b}{N_b - N_a} = \frac{1+r}{1-r} \tag{2-4}$$

显然，反应进行彻底，反应程度 $p=1$ 是一种无法达到的理想状态，计算聚合度还需要导入反应程度变量。

设定聚合反应在时间 t 后的反应程度为 p，a 官能团的残余量为 N_{at}，那么已经完成缩聚反应的 a 官能团的数量为：

$$N_a - N_{at} = N_a \times \left(\frac{N_a - N_{at}}{N_a}\right) = N_a p$$

这些已经完成缩聚反应的 a 官能团要消耗同样数量的 b 官能团，则体系内残存的官能团总数为：

$$N_a + N_b - 2N_a p$$

缩聚产物的高分子数为：

$$\frac{N_a + N_b - 2N_a p}{2}$$

因反应初始时的单体总数为：

$$\frac{N_a + N_b}{2}$$

则聚合度为：

$$X_n = \frac{N_a + N_b}{N_a + N_b - 2N_a p} = \frac{1+r}{1+r-2rp} \tag{2-5}$$

工业上利用过量分率（q）的概念更加方便，即：

$$q = \frac{N_b - N_a}{N_a} = \frac{1-r}{r} \tag{2-6}$$

则聚合度可以转换为：

$$X_n = 1 + \frac{2}{q} \approx \frac{2}{q} \tag{2-7}$$

通过式（2-7）可以看出，即使过量分率仅为 1%，聚合度 X_n 也只有 200 左右，达不到高分子材料的分子量要求。工业生产时，不仅要求单体的纯度较高，而且称量的准确性也要严格要求。值得注意的是，缩聚反应对于单体配料比的要求不仅表现在投料阶段，也必须在整个反应过程中准确维持。例如，挥发性不同的两种单体，在高温减压条件下缩聚时，单体配料比有可能在缩聚过程中发生变化，使得缩聚产物的分子量难以提高。

为了使缩聚反应的官能团数接近相等，除了考虑单体纯度和精确计量之外，还可以考虑使用两种官能团在同一分子上的单体，如羟基酸、氨基酸等。在制备聚酰胺的缩聚过程中，也有先把二元胺和二元酸中和成盐后使用的具体实例。

③ 可逆反应平衡常数的影响

在两种官能团数相等的缩聚体系中进行线形缩聚时，缩聚物的聚合度与反应程度的关系如式(2-3)所示，即聚合度随反应程度的增大而增大。但是对于可逆反应而言，缩聚反应的反应程度是由可逆反应的平衡常数决定的，而且绝大多数的缩合反应都是可逆的。下面，我们设定封闭体系和非封闭体系两种模型来讨论反应平衡常数对缩聚产物分子量的影响。

a. 封闭体系 如下的聚酯化是可逆反应，在封闭体系中，正逆反应将构成平衡，反应程度将受到限制。

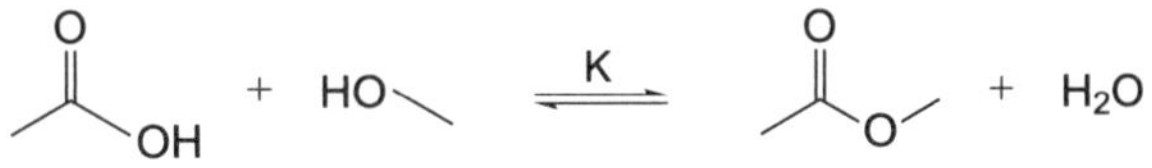

$$\text{RCOOH} + \text{HOR'} \xrightleftharpoons{K} \text{RCOOR'} + H_2O$$

起始　　c_0　　　　c_0　　　　0　　　0
平衡　　c_0-c　　　c_0-c　　　c　　　c

因 $c = p \times c_0$，则：

$$K = \frac{p^2}{(1-p)^2}$$

解得：

$$p = \frac{\sqrt{K}}{\sqrt{K}+1}$$

$$X_n = \frac{1}{1-p} = \sqrt{K}+1 \tag{2-8}$$

缩聚反应可逆的程度可由平衡常数来衡量，根据平衡常数大小，可将线形缩聚粗分为三类。

- 平衡常数 $K<100$，如聚酯化反应，$K\approx 4$，副产物水的浓度对分子量有较大影响，封闭体系内难以得到高分子量缩聚物。
- 平衡常数 $100<K<1000$，如聚酰胺化反应，$K\approx 400$，副产物水的浓度对分子量有一定影响。聚合初期，可在水介质中进行，聚合反应后期，需减压脱水，才能进一步提高分子量。
- 平衡常数 $K>1000$，可以看作不可逆，如合成聚砜一类的逐步聚合。

表 2-2 列出了几种聚合物的平衡常数 K，平衡时的反应程度 p_c 和聚合度 X_n。

表 2-2 聚合物平衡常数 K，平衡时反应程度 p_c，聚合度 X_n 的比较

聚合物	K	p_c	X_n
酚醛树脂	3000	0.9821	53
尼龙-66	400	0.9524	21
涤纶	4	0.667	3

对于平衡常数较大的合成酚醛树脂和尼龙的缩聚反应，在不除小分子水的封闭体系中，缩聚产物的聚合度也能基本满足；而对于平衡常数仅仅为 4 的聚酯的缩合反应，如果不及时除掉小分子水，则无法得到高分子产物。

b. 非封闭体系 在缩聚体系中，如果使用减压加热等手段不断除去缩聚副产物小分子，那么就可以使平衡反应向右进行，能够获得相对更高的分子量。

设定上述缩聚反应中的水浓度保持在较低水平，为 n_w：

$$K = \frac{p \times n_w}{(1-p)^2}$$

$$X_n = \frac{1}{1-p} = \sqrt{\frac{K}{p \times n_w}}$$

当 p 接近 100% 时，则

$$X_n = \sqrt{\frac{K}{n_w}} \tag{2-9}$$

式(2-9)称为缩聚平衡方程，说明缩聚体系中的小分子除得越干净，缩聚物的分子量越大。但是，在缩聚反应后期，缩聚体系往往因分子量增大而变得特别黏稠，其中的小分子副产物并不能轻易除尽。在工业化生产中常常采用真空减压、高温、强化搅拌更新蒸发表面、缩小缩聚液层厚度、通入惰性气体等方法脱除小分子。

缩聚反应以放热反应居多，升高缩聚温度会使平衡常数减小，不利于获得高分子量缩聚物，但大多缩合反应的热效应不大，通常 $\Delta H = -33 \sim -42 \text{kJ/mol}$，升高温度的影响并不大。而升高温度使体系黏度下降，有利于小分子的蒸出，所以缩聚反应平衡常数不是很大的情况下，在较高温度下有利于分子量的提高。

④ 副反应的影响

a. 线形缩聚的成环反应 线形缩聚时，缩聚单体及其中间产物具有成环倾向。一般，五、六元环化合物的结构比较稳定。例如二羟基酸，经双分子缩合后，易形成六元环乙交酯。单体浓度对成环缩聚有影响。成环是单分子反应，缩聚则是双分子反应，因此，低浓度有利于成环，高浓度则有利于线形缩聚。

b. 消去反应 二元羧酸受热会脱羧，引起原料官能团数目比的变化，从而影响到产物的分子量。羧酸酯比较稳定，用来代替羧酸，可以避免这一缺点。二元胺有可能进行分子内或分子间的脱氨反应，进一步反应还可能导致支链或交联。

c. 化学降解 聚酯化和聚酰胺化是可逆反应，逆反应水解是化学降解之一。合成缩聚物的单体往往就是缩聚物的降解剂，例如醇可使聚酯类醇解，又如胺类可使聚酰胺进行胺解。化学降解会使聚合物分子量降低，聚合时应设法避免。但应用化学降解的原理可使废聚合物降解成单体或低聚物，回收利用。例如，废涤纶聚酯与过量乙二醇共热，可以醇解成对苯二甲酸乙二醇酯齐聚物；废酚醛树脂与过量苯酚共热，可以酚解成低分子酚醇。这些低聚物都可以重新用作缩聚的原料。另一方面，从环境保护方面考虑，还可以合成易降解的聚合物。

d. 链交换反应 同种线形缩聚物受热时，通过链交换反应，将使分子量分布变窄。两种不同缩聚物（如聚酯与聚酰胺）共热，也可进行链交换反应，形成（聚酯-聚酰胺）嵌段共聚物。

(3) 线形缩聚反应中的分子量分布

聚合产物是分子量不等的大分子的混合物，分子量存在着一定的分布。Flory 根据官能团等活性理论，应用统计方法，推导出线形缩聚物的聚合度分布函数式。

我们使用最简单的例子来讨论 Flory 聚合度分布函数式。设 aAb 型单体进行自缩聚反应，A 为烷基，a 和 b 为反应官能团。羟基羧酸的缩聚反应是实例之一。

假设聚合物分子含有 x 个结构单元 A(aA_xb)，在 t 时的反应程度为 p，那么在统计方法上，p 也是每一个 a 官能团参与缩聚反应的概率，而（$1-p$）是每一个 a 官能团不反应的概率。上述 x 聚体中含有（$x-1$）个已经缩聚的 a 官能团，其生成概率为 p^{x-1}；另外含有一个未反应的 a 官能团，生成概率为（$1-p$）。因此，x 聚体的生成概率就是：

$$p^{x-1}(1-p)$$

从另一角度考虑，x 聚体的生成概率还应等于缩聚体系中 x 聚体的数量分数（N_x/N），其中 N_x 为 x 聚体的分子数，N 为大分子总数。体系内的大分子总数 N 还可以表示为：

$$N = \sum_{x=1}^{\infty} N_x = N_0(1-p)$$

因此，x 聚体的数量分布函数为：

$$N_x = N \times p^{x-1}(1-p) = N_0 \times p^{x-1}(1-p)^2 \tag{2-10}$$

如果忽略端基的质量，则 x 聚体的质量分数或质量分布函数为：

$$m_x/m = \frac{xN_x}{N_0} = xp^{x-1}(1-p)^2$$

$$m_x/m = \frac{xN_x}{N_0} = xp^{x-1}(1-p)^2 \tag{2-11}$$

式（2-10）和式（2-11）代表线形缩聚反应程度为 p 时的数量分布函数和质量分布函数，往往称作最概然分布函数（most probable distribution），或 Flory、Flory-Schulz 分布函数。

参照数均分子量的定义，数均聚合度可以写成：

$$X_n = \frac{\sum xN_x}{\sum N_x} = \frac{\sum xN_x}{N} = \sum_{x=1}^{\infty} x \frac{N_x}{N}$$

将式（2-10）代入，得：

$$X_n = \sum_{x=1}^{\infty} xp^{x-1}(1-p) \tag{2-12}$$

同理，可以导出重均聚合度为：

$$X_w = \sum x \frac{m_x}{m} = \sum_{x=1}^{\infty} x^2 p^{x-1}(1-p)^2 \tag{2-13}$$

另有：

$$\sum_{x=1}^{\infty} xp^{x-1} = \frac{d}{dp}\left[\sum_0^{\infty} p^x\right] = \frac{d}{dp}\left[\frac{1}{1-p}\right] = \frac{1}{(1-p)^2}$$

$$\sum_{x=1}^{\infty} x^2 p^{x-1} = \frac{d}{dp}\left[\frac{p}{(1-p)^2}\right] = \frac{1+p}{(1-p)^3}$$

代入式（2-12）和式（2-13），得：

$$X_n = \frac{1}{1-p}$$

$$X_w = \frac{1+p}{1-p}$$

则分子量分布宽度为：

$$\frac{X_w}{X_n} = 1 + p \approx 2$$

Flory 的最概然分布在数学上比较完善,但是在早期却很难得以验证。当代 GPC 技术的发展,Flory 分布已经在大量的缩聚反应中得以验证,包括聚酰胺化反应、聚酯化反应等,许多逐步聚合物的分子量分布指数的实测值都接近 2,说明了 Flory 统计理论分布的可靠性。

2.1.3 线形缩聚动力学

以二元酸和二元醇的聚酯化为例,不可逆和可逆条件下的线形缩聚动力学有所不同,下面进行分别论述。

(1) 不可逆线形缩聚

有机小分子的酯化反应和生成聚酯的缩聚反应均是可逆平衡反应。反应过程中如能不断排除副产物水,平衡反应可持续向右进行,反应可以按不可逆反应进行处理。

缩聚酯化的反应机理可以表示为:

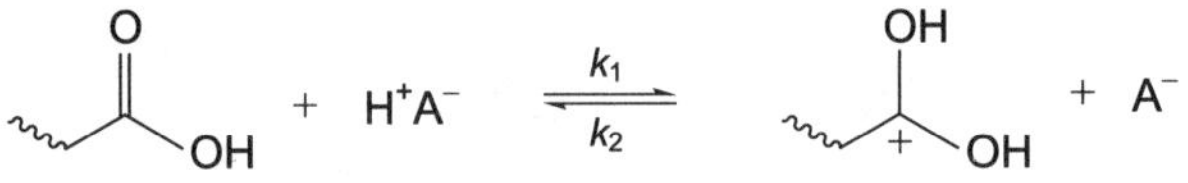

（图示反应式 (1)、(2)、(3)）

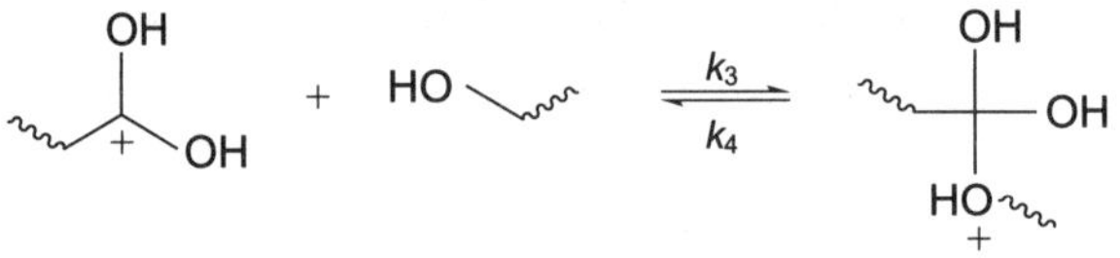

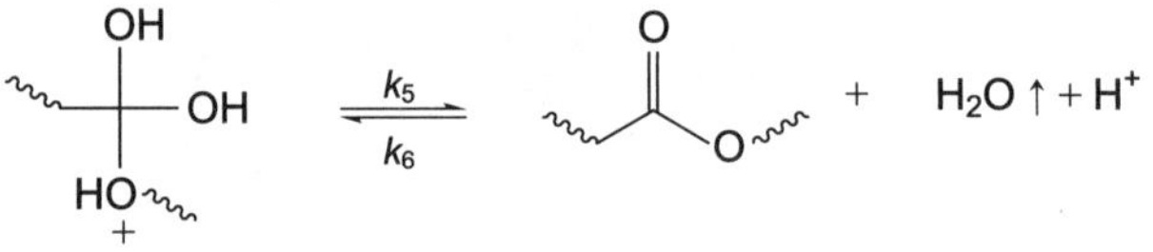

羧酸首先被质子化 (1),因碳氧双键的极化有利于亲核加成,质子化后的羧酸与醇反应生成中间体 (2),该中间体再脱水生成酯 (3)。

在不断排除反应体系中的水的条件下,反应 (3) 中的可逆反应可以忽略,即 $k_6=0$。另外因反应式 (2) 中右侧的反应中间体较不稳定,k_3 远远小于 k_4、k_5、k_2、k_1,聚酯化速率或羧基消失速率由反应 (2) 控制。则反应速率表示为:

$$r_p = -\frac{d[COOH]}{dt} = k_3[C^+(OH)_2][OH] \tag{2-14}$$

式中 $[C^+(OH)_2]$ 质子化物种的浓度难以测定,可以引入反应 (1) 中的平衡常数 K_1 的关系式加以消去。

$$K_1 = \frac{k_1}{k_2} = \frac{[C^+(OH)_2][A^-]}{[COOH][HA]} \tag{2-15}$$

将式 (2-15) 代入式 (2-14),得:

$$r_p = -\frac{d[COOH]}{dt} = \frac{k_1 k_3 [COOH][OH][HA]}{k_2 [A^-]} \tag{2-16}$$

考虑到有机弱酸 HA 的解离平衡 $HA \rightleftharpoons H^+ + A^-$,当解离平衡常数为 K_{HA} 时,得:

$$K_{HA} = \frac{[H^+][A^-]}{[HA]} \tag{2-17}$$

将式(2-17)代入式(2-16),得到酸催化的酯化速率方程:

$$r_p = -\frac{d[COOH]}{dt} = \frac{k_1 k_3 [COOH][OH][H^+]}{k_2 K_{HA}} \quad (2-18)$$

当使用强酸作为酯化缩聚的催化剂时,在缩聚过程中,外加酸或氢离子浓度几乎不变,而且远远大于低分子羧酸自催化的影响,因此,可以忽略自催化的速率。将式(2-18)中的$[H^+]$和k_1、k_2、k_3、K_{HA}合并为k',设定缩聚原料中羧基数和羟基数相等,即$[COOH]=[OH]=c$,则式(2-18)简化为:

$$-\frac{dc}{dt} = k'c^2$$

积分得:

$$\frac{1}{c} - \frac{1}{c_0} = k't$$

引入反应程度p,并将式中的官能团数N_0、N以羧基浓度c_0、c来代替,则得:

$$c = c_0(1-p) \quad (2-19)$$

将式(2-3)和式(2-19)代入,得:

$$\frac{1}{1-p} = X_n = k'c_0 t + 1$$

以上公式表明X_n与t成线形关系。

缩聚体系内无酸催化剂的条件下,聚酯缩聚反应仍能缓慢进行,缩聚反应主要依靠原料羧酸本身的酸性来催化。上述酸催化体系中的[HA]即为[COOH],将它们代入式(2-16),得:

$$r_p = -\frac{d[COOH]}{dt} = \frac{k_1 k_3 [COOH][OH][HA]}{k_2 [A^-]} = k''[COOH]^2[OH] \quad (2-20)$$

该式表明,无酸催化的缩聚体系过程对羧酸为二级,对醇为一级。如果投料时官能团比为1,即$[COOH]=[OH]=c$,则式(2-20)简化为:

$$r_p = -\frac{dc}{dt} = k''c^3$$

或

$$-\frac{dc}{c^3} = k''dt$$

$$2k''t = \frac{1}{c^2} - \frac{1}{c_0^2}$$

将式(2-19)代入,得:

$$X_n^2 = \frac{1}{(1-p)^2} = 2kc_0^2 t + 1 \quad (2-21)$$

由上述分析及实验结果可见,无论是酸催化还是自催化的聚酯缩聚体系,理论推导出的反应动力学方程都只能在反应程度较高的一个狭窄范围内适用。其根本原因在于反应过程中体系极性的变化导致了酯化反应速率常数的变化。反应初期体系中尚未反应的羧基和羟基浓度很高,极性很强,反应速率常数较大;随着反应物羧基和羟基浓度的下降,缩聚体系的极性降低,反应速率常数也随之减小。当反应程度达到0.8~0.9后,体系的极性和反应速率常数趋于稳定,理论推导出的动力学方程才适用。

（2）线形平衡缩聚反应的影响因素

在缩聚反应体系中，影响缩聚产物的分子量及其分布、聚合反应速率的因素有许多，以下进行分别论述。

① 反应温度

温度对可逆平衡化学反应的影响取决于该反应的热效应。对于吸热反应，温度与平衡常数具有正相关；对于放热反应则负相关。因为大多数缩聚反应都是放热反应，其焓值在 $-33\sim-42kJ/mol$，它们的平衡反应常数随温度升高而略有降低。

前面曾经论述过，平衡缩聚反应中缩聚产物的分子量与该反应平衡常数的平方根成正比。由此可见，升温会使缩聚产物的聚合度降低。不过，由于大多缩聚反应体系的黏度随温度的升高而减小，使缩聚副产物小分子容易排除，有利于缩聚反应和聚合度的提高。如果温度过高，会发生缩聚高分子链裂解（水解、酸解、醇解）、环化、官能团分解等各种副反应，所以必须全面考虑反应温度的影响，通过实验确定最佳反应温度。

② 压力

由于大多数缩聚反应都会伴随沸点较低的小分子副产物的生成，它们在缩聚体系内的浓度直接影响缩聚物分子量的大小。所以大多数缩聚反应中后期都需要有减压去除小分子副产物的过程。但是减压操作也有可能在反应初期导致低沸点单体的逃逸，不利于缩聚单体间等量配比的维持，从而不利于高分子缩聚产物的获得。在这种情况下，缩聚反应初期往往需要加大容器内压力，在反应的中后期才逐渐减压，最后达到较高的真空度。

③ 催化剂

线形平衡缩聚反应体系内经常使用催化剂来提高聚合反应速率。催化剂的存在并不会导致反应平衡常数的改变。聚酯的缩聚反应常常使用酸催化剂。除酸性的强弱外，选择催化剂时还需要考虑催化剂和缩聚体系的相容性和分散性。有机强酸如对甲苯磺酸是聚酯缩聚反应经常使用的一种酸催化剂。

④ 单体浓度

缩聚产物的分子量与单体浓度有关，其随着单体浓度的升高而增大。除一些必须加入有机溶剂以降低体系黏度（如涂料和黏结剂的合成）的缩聚反应外，通常不使用溶剂，以保证最高的单体浓度和聚合度。

⑤ 搅拌

对于缩聚反应而言，搅拌具有混合反应物、强化传热和控温、加速小分子副产物的排除等作用。但是，在缩聚后期，缩聚物已经具有较高分子量的情况下，强烈搅拌产生的高强度的剪切力可能导致线形高分子链的断裂，引起分子量的降低。

⑥ 惰性气体

许多缩聚反应需要在高温条件下进行，例如涤纶的缩聚温度需要控制在280℃以上。一些官能团在这种高温下容易发生氧化、分解等副反应，从而导致高分子链端反应能力丧失。此外，空气中氧气在高温条件下容易和缩聚物发生氧化反应，使体系颜色变深，影响到聚合物的使用性能。为了解决这些高温副反应的问题，缩聚反应过程中，尤其是反应中后期需要提高反应温度的时候，经常采用通入氮气等惰性气体的办法来隔绝空气中的氧气。需要注意的是，如果原料单体的沸点较低，则应该避免在反应初期通入惰性气体，以免造成低沸点单体的损失和影响缩聚体系内官能团配比的准确性。

2.1.4 逐步聚合的实施方法

逐步聚合目前可以采用熔融、溶液、界面、固相等四种聚合方法进行，其中熔融缩聚和溶液缩聚是最主要的聚合方法。能够利用界面缩聚的单体非常有限，故界面缩聚是一种比较少见的缩聚方法。固相缩聚常作为辅助聚合方法，与前三者联合使用。

(1) 熔融聚合

熔融聚合的缩聚体系中仅有聚合单体和少量催化剂，产物纯净，分离简单，缩聚物的分子量较高，而且缩聚反应器的生产效率也较高。所谓"熔融"是指缩聚温度需要设定在单体和聚合物熔点以上，使缩聚反应体系始终处于熔融状态，以保证足够的反应速率。由于大多缩聚反应放热不多，故多需外加热。对于平衡缩聚体系，还需减压脱除小分子副产物。大部分缩聚反应初期，缩聚物的分子量和体系黏度不高，物料的混合和低分子物的脱除并不困难。在缩聚反应后期（反应程度大于97%~98%），需要强化设备的传热和传质效率才能够有效排除小分子。熔融聚合法用得很广，如合成涤纶、酯交换法聚碳酸酯、聚酰胺等。在纤维生产设备中，还将熔融缩聚和纺丝装置连在一起，形成连续生产，上部是连续聚合反应器，下部是纺丝和牵引装置。

(2) 溶液聚合

溶液聚合的缩聚体系中有缩聚单体、催化剂和适量的溶剂（包括水），一般需要缩聚单体的反应活性较高。由于溶剂的加入，缩聚体系的黏度相对熔融缩聚体系低，有利于提高聚合设备的传热和传质效率，缩聚温度也可以控制在较低范围，副反应也较少，但是缩聚物的分子量较熔融缩聚法低。如为平衡缩聚，则可通过蒸馏或加碱成盐除去副产物。溶液缩聚的缺点是需要回收溶剂，缩聚产物中残余溶剂小分子的排除也往往比较困难。采用溶液缩聚的具体实例包括，聚砜聚合法和尼龙-66的聚合（前期为水溶液缩聚，后期转为熔融缩聚）等。此外，溶液缩聚法适合涂料和胶黏剂的合成，因为涂料和胶黏剂产品均以聚合物溶液状态使用，采用溶液聚合反而省去了溶解缩聚物的步骤。

(3) 界面缩聚

界面缩聚的两种单体分别溶于水和与水不相溶的有机溶剂中，缩聚单体在互不相溶、分别溶解的油水界面处接触，并快速进行聚合。能够进行界面缩聚的单体的反应活性需非常高，能够在接触瞬间快速完成缩聚反应，例如将癸二酰氯的四氯乙烷溶液放在烧杯底层，再小心地倒入己二胺的水溶液作为上层，无需搅拌，二元胺和二酰氯在室温下就能很快聚合，速率常数高达 $10^4\sim10^5\,\text{L}/(\text{mol}\cdot\text{s})$。缩聚物聚酰胺-610就在界面处迅速形成，可用玻璃棒拉出纤维。水相中需加碱，以中和副产物氯化氢，以免氯化氢与胺结合成盐，使反应减慢。碱量过多，又易使二酰氯水解成羧酸或单酰氯，降低聚合速率和分子量。界面缩聚的过程特征属于扩散控制，工业实施时，应有足够的搅拌强度，保证单体及时传递。界面缩聚的优点是缩聚温度较低、副反应少、不必严格等官能团数比、反应快、分子量较熔融聚合产物高等；缺点包括适用单体范围较窄，单体原料酰氯成本较贵，消耗的有机溶剂量较多，且回收麻烦，生产成本较高。目前实现界面缩聚工业化的实例仅有光气法合成聚碳酸酯一种。

(4) 固相缩聚

在缩聚产物的玻璃化温度以上、熔点以下的固态所进行的缩聚，称作固相缩聚。例如系船缆绳、降落伞绳、乘用车座椅安全带用的涤纶纤维需要缩聚物的分子量在3万以上，常规熔融缩聚法很难达到。将涤纶树脂在惰性气流保护下加热到熔点（265℃）以下，如220℃，继续固相缩聚，这一温度远高于玻璃化温度（69℃），链段仍能自由运动，且不妨碍继续缩聚，在高度减压或惰性气流的条件下，排除副产物乙二醇，继续提高分子量，使其符合工程塑料和帘子线强度的要求。聚酰胺-6也可以进行固相缩聚，进一步提高分子量。固相聚合是上述三种方法的补充。

2.2 体形缩聚和凝胶化作用

此节之前，我们主要介绍了线形缩聚反应理论。能够聚合成为线形聚合物的单体要具有可进行缩合反应的官能团。对单体分子而言，每个单体分子必须具有两个反应官能团，例如前面已经讨论过的涤纶聚酯的聚合反应。可以设想，如果在涤纶聚酯的缩聚反应中，具有两个羟基的乙二醇被丙三醇取代，那么产物就会成为一种在三维空间内交联的网络聚合物，即体形缩聚物。理论上，可以把这种聚合物看作是一个巨大的分子。体形聚合物具有不溶解、不熔融的特点，也比线形聚合物具有更高的机械强度。正是由于体形聚合物具有优良的力学性能和热稳定性，它成了高分子工业中一种非常重要的聚合物，在许多领域均有广泛应用。

官能度是指一个单体分子中能够参加反应的官能团的数目。单体的官能度一般容易判断，个别单体，反应条件不同，官能度也不同，如：

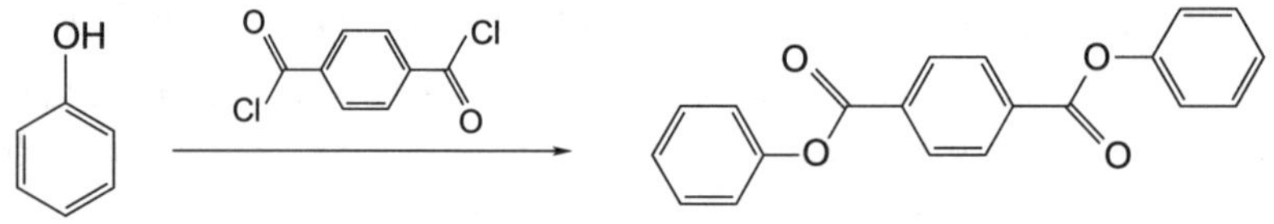

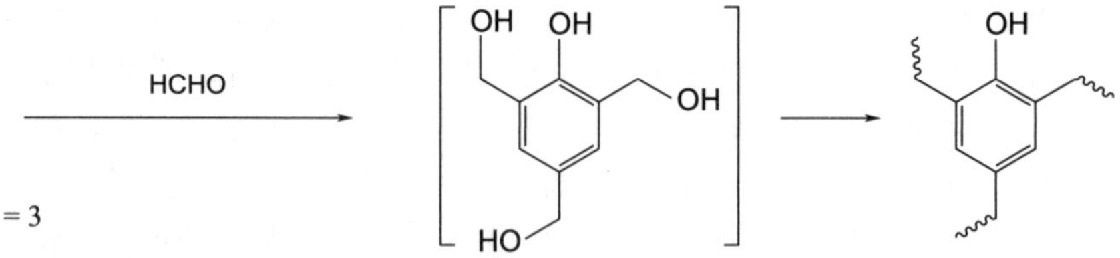

当苯酚与酰氯进行酯化反应时，仅有酚基参与，官能度为1；而在苯酚与甲醛的聚合体系中，苯酚的邻、对位芳香氢原子均有可能参与，官能度为3。

缩聚反应中，不同的官能度体系所获得的缩聚产物分子链结构有所不同。

官能度为1的单体和任何官能度的单体反应（1 vs n 体系），只能获得小分子化合物，属于缩合反应，并不属于缩聚体系。

官能度为2的单体和另一官能度为2的单体反应（2 vs 2体系），可以获得线形高分子，属于线形缩聚体系。

官能度为2的单体和另一官能度大于2的单体反应（2 vs 3、4、n 体系），可以获得体形高分子，属于体形缩聚体系。具体实例如邻苯二酸酐与甘油、邻苯二酸酐与季戊四醇等。

多官能团单体聚合到某一程度，黏度突增失去流动性，突然转变成不溶、不熔、具有交联网络结构的弹性凝胶，称为凝胶化现象。这时的反应程度称作凝胶点。凝胶点的定义为液体开始转化为固体，出现凝胶瞬间的临界反应程度。研究发现，发生凝胶化时，体系中存在凝胶和溶胶两个部分。凝胶不溶于任何溶剂中，相当于许多线形大分子交联成一个整体，分子量可以看作无穷大；溶胶则被包裹在凝胶的网络结构中，由于其分子量尚小，仍然可以溶解。在凝胶点以后，缩聚交联反应仍在进行，溶胶量不断减少，凝胶量相应增加。由于交联网络的限制，连接在网络上的官能团的运动性降低，凝胶点后的反应速率明显降低，只有在更加苛刻的条件下，如升温或延长固化反应时间等，才能达到较高的反应程度。

凝胶化过程具有突然性，无论是在实验室还是在大规模工业化的生产现场，预测凝胶点都是至关重要的。例如，体形缩聚物的生产过程常常分成预聚物（树脂）制备和成形交联固化两个阶段，这两个阶段对凝胶点的预测和控制都很重要。预聚时，如反应程度超过凝胶点，将在聚合釜内固化而造成反应报废。成型时则须控制适当的固化时间或速度。制备热固性泡沫塑料时，要求发泡速度与固化速度相协调。制造层压板时，控制适宜的固化时间，才能保证材料强度。因此凝胶点是体形缩聚中的首要控制指标。

2.2.1 Carothers 法凝胶点预测

体形缩聚体系凝胶点的理论预测需要计算参与缩聚反应的单体的平均官能度。所谓平均官能度，是指在两种及以上单体参与的缩聚反应中，反应体系内实际能够参加缩聚反应的各种官能团的数量与反应物分子数量的比值，也就是每个缩聚单体分子平均带有的可参与缩聚反应的官能团数。下面，我们分三种情况讨论平均官能度的计算方法。

(1) 等官能团数

缩聚单体 A 和 B 所持官能团数相等的情况下，设 N_a、N_b 为它们的分子数，f_a、f_b 为各自的官能度，则平均官能度为：

$$\overline{f} = \frac{f_a N_a + f_b N_b}{N_a + N_b} \tag{2-22}$$

例如，2mol 甘油（$f=3$）和 3mol 邻苯二甲酸酐（$f=2$）体系共有 5mol 单体和 12mol 官能团，故：

$$\overline{f} = \frac{f_a N_a + f_b N_b}{N_a + N_b} = \frac{2 \times 3 + 3 \times 2}{2 + 3} = 2.4$$

(2) 官能团数不相等的两组分体系

当体系中的两种官能团数不等时，体系中的有效官能团总数，抑或能够参加缩聚反应的官能团数，应该为官能团数量较少一方的两倍。因为这样才能够满足缩聚反应中两种官能团始终同步，等量消耗的原则。设 N_a、N_b 为它们的分子数，f_a、f_b 为各自的官能度，且 $f_a N_a > f_b N_b$，则平均官能度为：

$$\overline{f} = \frac{2 f_b N_b}{N_a + N_b} \tag{2-23}$$

以 1mol 甘油和 5mol 邻苯二甲酸酐体系为例：

$$\bar{f} = \frac{2f_b N_b}{N_a + N_b} = \frac{2 \times 3 \times 1}{1 + 5} = 1.0$$

(3) 官能团数不相等的多组分体系

官能团数不相等的多组分体系的平均官能度可作类似计算，计算时只考虑参与反应的官能团数，不计算未参与反应的过量官能团。以 A、B、C 三组分体系为例，三者分子数分别为 N_a、N_b、N_c，官能度分别为 f_a、f_b、f_c，若 $f_a N_a > f_b N_b + f_c N_c$，则平均官能度：

$$\bar{f} = \frac{2(f_b N_b + f_c N_c)}{N_a + N_b + N_c} \tag{2-24}$$

以 2mol 甘油和 2mol 邻苯二甲酸酐及 1mol 油酸的三元体系为例：

$$\bar{f} = \frac{2(f_b N_b + f_c N_c)}{N_a + N_b + N_c} = \frac{2 \times (2 \times 2 + 1 \times 1)}{2 + 2 + 1} = 2.0$$

Carothers 提出的体形缩聚凝胶点预测理论的核心是：在体形缩聚体系达到凝胶化的那一刻，聚合物的分子量急剧增大，甚至交联为一体。此时的聚合度定义为无穷大。

设 N_0 为体系内反应开始时的单体分子总数，则官能团总数为 $\bar{f} N_0$，在反应程度为 p 时，体系内的分子总数为 N，因每一步缩聚反应需要消耗两个官能团，则已经参加缩聚反应的官能团数为 $2(N_0 - N)$，则反应程度：

$$p = \frac{2(N_0 - N)}{N_0 \bar{f}} = \frac{2}{\bar{f}} - \frac{2N}{N_0 \bar{f}} = \frac{2}{\bar{f}} - \frac{2}{\bar{f} \bar{X}_n}$$

当 $\bar{X}_n \to \infty$，凝胶点：

$$p_c = \frac{2}{\bar{f}} \tag{2-25}$$

此式称为 Carothers 方程，表示凝胶点和缩聚单体平均官能度之间的关系。

当平均官能度 $\bar{f} > 2$ 时，可以得到合理凝胶点；$\bar{f} < 2$ 时，凝胶点大于 1，预计形不成凝胶，无固化危险。

例如，等化学计量比的甘油和邻苯二酸酐进行缩聚，平均官能度为 2.4，按 Carothers 方程计算凝胶点 $p_c = 0.833$。此外，按上式计算可得：

$$\bar{X}_n = 50 \text{ 时}, \ p = 0.816; \ \bar{X}_n = 500 \text{ 时}, \ p = 0.831$$

可见接近凝胶点时，只要反应程度稍有增加，聚合度就增加许多，可见凝胶化的突然性。

Carothers 方程的缺点是假设出现凝胶点时的聚合度为无限大，偏离了实际情况，导致凝胶点数值偏高。实际上，凝胶点时的聚合度并非无穷大，有时仅有数十大小。例如，实验结果表明，上述等化学计量比的甘油和邻苯二酸酐的缩聚体系出现凝胶时的聚合度约为 24，相应凝胶点为 0.80。尽管如此，由于在凝胶点附近聚合度变化迅速，按 Carothers 方程求出的凝胶点还是有意义的。但值得注意的是，Carothers 方程得到的是凝胶点的上限，实际值要比 Carothers 理论值小。

2.2.2　Flory 统计凝胶点计算

Flory 利用统计数学的方法研究体形缩聚凝胶化理论，建立了单体官能度与凝胶点间的关系。在体形缩聚中，官能度大于 2 的多官能团单体可以形成体形缩聚物中的支化点 (branch unit)，这种多官能团单体经缩聚进入到交联大分子链后形成支化单元，一个支化点连接另一个支化点的概率称作支化系数 (branching coefficient)，以 α 表示。支化系数也可以定义为一个支化点上的特定单元经线形高分子链连接到另一个支化点的概率。

$$A-A + B-B + A_3 \longrightarrow A-A[B-B-A-A]_i B-B-A$$

假设双官能团单体 A—A、B—B 和如上三官能团单体 A_3 进行反应。如果官能团 A 和 B 反应的反应程度为 p_A 和 p_B，已经反应的 B 有两种可能：与双官能团单体的 A 或与三官能团单体的 A 反应。设 ρ 为多官能团单体上 A 占体系中全部 A 的分数，则 B 与多官能团单体反应的概率为 ρp_B，与双官能团反应的概率为 $(1-\rho)p_B$。这样，一个支化点上的 A 通过线形分子链（聚合度 $X_n = i$）连接到另一个支化点的概率为：

$$p_A[p_B(1-\rho)p_A]^i p_B \rho$$

根据概率加法定理，一个支化点上的 A 通过线形分子链连接到另一个支化点的概率为：

$$\alpha = \sum_{i=0}^{\infty} p_A[p_B(1-\rho)p_A]^i p_B \rho = \rho p_A p_B \sum_{i=0}^{\infty}[p_A p_B(1-\rho)]^i$$

经数学换算，有：

$$\alpha = \frac{\rho p_A p_B}{1 - p_A p_B(1-\rho)}$$

此式即为 Flory 凝胶化理论基础支化系数 α 的计算公式。

设 A 和 B 的起始数目比为 γ，即，$N_A/N_B = \gamma$，则 $p_B = \gamma p_A$：

$$\alpha = \frac{\rho p_A p_B}{1 - p_A p_B(1-\rho)} = \frac{\gamma p_A^2 \rho}{1 - \gamma p_A^2(1-\rho)}$$

若 $\rho = 1$，表明缩聚体系内所有含 A 单体为三官能团单体，无双官能团 A 单体。在 $\gamma < 1$ 时：

$$\alpha = \gamma p_A^2 = \frac{p_B^2}{\gamma}$$

若 $\gamma = 1$，即 $p_A = p_B = p$，$\rho < 1$ 时：

$$\alpha = \frac{\rho p^2}{1 - p^2(1-\rho)}$$

若 $\gamma = 1$，且 $\rho = 1$ 时，则：

$$\alpha = p^2$$

Flory 凝胶化理论认为，如果支化单元的官能度为 f，某一链的一端连接一个支化单元的概率为 α，那么该支化单元可以衍生出 $(f-1)$ 个支链，而每个支链又可以 α 的概率再连接一个支化单元，则一个已经连在分子链上的支化单元再与另一支化单元相连的概率为 $(f-1)\alpha$，这 $(f-1)\alpha$ 个链中的每一个支化单元又分别可以衍生出 $(f-1)\alpha$ 个链，合计为 $(f-1)^2\alpha^2$ 个链。以此类推。

如果 $1 > (f-1)\alpha > (f-1)^2\alpha^2$，那么链的数目会越来越少，不可能产生三维网状凝胶；反之，若 $(f-1)^2\alpha^2 > (f-1)\alpha > 1$，那么链的数目会越来越多，一定会出现凝胶。因此，产生凝胶的临界点时，$(f-1)\alpha = 1$，临界支化系数为：

$$\alpha_c = \frac{1}{f-1}$$

将 α_c 代入上述各公式，就可以计算出各种情况下的凝胶点。

例如，若 $\gamma=1$，即 $p_A=p_B=p$ 时：

$$p_c = \frac{1}{\sqrt{[1+\rho(f-2)]}}$$

若 $\rho=1$，则：

$$p_c = \frac{1}{\sqrt{[\gamma+\gamma(f-2)]}}$$

若 $\rho=1$，$\gamma=1$，则：

$$p_c = \frac{1}{\sqrt{f-1}}$$

按 Flory 凝胶理论计算得到的凝胶点常常比实测值小，所以按 Flory 凝胶理论计算得到的理论值可视为实际凝胶点的下限。

2.3 重要缩聚物

2.3.1 聚酯

聚酯（polyesters）是指在高分子主链上具有酯基结构的杂链聚合物。下面四种聚酯是已经工业化产品的典型代表，它们的应用也各不相同。

- 线形芳族聚酯，如涤纶聚酯，用作合成纤维和工程塑料；
- 不饱和聚酯，如使用马来酸酐与乙二醇共聚，在高分子链中含有双键，可进一步自由基聚合，常使用苯乙烯混合稀释后，用于制作不饱和树脂；
- 醇酸树脂，属于线形或支链形无规预聚物，残留基团可进一步交联固化，常用作涂料；
- 线形饱和脂肪族聚酯，如聚酯二醇，用作聚氨酯的预聚物。

高分子聚酯的合成原理与低分子酯类相似，主要有下列四种合成路线：

(a) 使用有机酸和醇直接酯化 (esterification of diacid and diol)

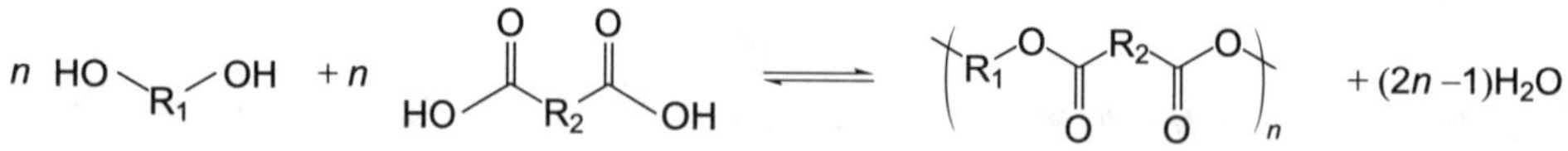

$$n\ HO-R_1-OH + n\ HOOC-R_2-COOH \rightleftharpoons \left[O-R_1-O-CO-R_2-CO\right]_n + (2n-1)H_2O$$

(b) 使用酯与醇进行酯交换 (ester interchange with alcohol or transesterification)

$$n\ HO-R_1-OH + n\ R_3OOC-R_2-COOR_3 \rightleftharpoons \left[O-R_1-O-CO-R_2-CO\right]_n + (2n-1)R_3OH$$

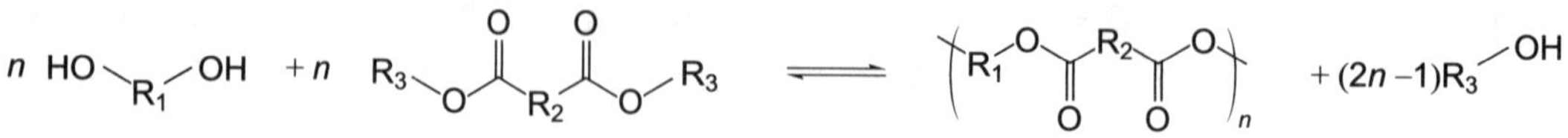

(c) 使用酰氯与醇反应 (esterification of acid chlorides)

$$n\ HO-R_1-OH + n\ ClOC-R_2-COCl \longrightarrow \left[O-R_1-O-CO-R_2-CO\right]_n + (2n-1)HCl$$

(d) 羟基羧酸的酯化反应 (esterification of hydroxycarboxylic acid)

$$n \; HO-R-COOH \rightleftharpoons +(O-R-CO)_n + (n-1)H_2O$$

(e) 通过内酯开环聚合 (lactone polymerization)

$$\text{(lactone)} \longrightarrow +(O-R-CO)_n$$

以上的 (a) 和 (b) 是可逆平衡反应, 反应速率较慢, 需加酸作催化剂, 也需在减压条件下不断排除低分子副产物, 使平衡向聚酯方向移动。(c) 大多为不可逆的快速反应, 但因原料成本较高, 在工业生产中应用较少。(d) 由于原料有限, 在工业生产中的实际应用较少。(e) 属于连锁聚合, 将在第 4 章中详细叙述。

(1) 涤纶聚酯

涤纶聚酯是聚对苯二甲酸乙二醇酯 (PET) 的商品名, 主链中的苯环提高了聚酯的刚性、强度和熔点 ($T_m = 265 ℃$), 亚乙基则赋予其柔性和加工性能, 两方面性能的综合, 使 PET 纤维获得成功, 并可用作工程塑料。涤纶聚酯还有其他一些英语俗称, 如 Mylar、Dacron 和 Terylene 等, 是当今产量最大, 应用最广的聚酯产品。PET 聚合流程见图 2-2。

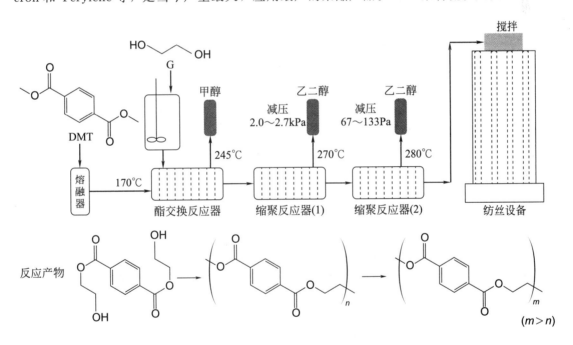

图 2-2 PET 聚合流程

PET 可由对苯二甲酸 (terephthalic acid, TA) 与乙二醇 (ethylene glycol) 缩聚而成, 遵循线形缩聚的普遍规律。在 20 世纪 60 年代, 由于 TA 熔点很高 (300℃), 加热后容易升华, 在溶剂中溶解度很小, 难以用精馏、结晶等方法来提纯, 所以很难生产纯度较高的 TA 产品。在 TA 纯度难以保障时, 两种缩聚单体的等基团数比很难达到, 生产不出具有较高分子量的 PET 产品。后期发现对苯二甲酸甲酯 (dimethyl terephthalate, DMT) 容易提纯, 与乙二醇缩聚反应的副产物是甲醇, 较水分子更易从系统中排除, 所以采用酯交换法合成

PET 的工艺率先得到发展。再后来，TA 的提纯工艺问题也得以解决，直接酯化法合成 PET 的技术也得到广泛应用。目前，酯交换法和直接酯化法两种合成技术呈并存状态。

① 酯交换法

酯交换法主要由甲酯化、酯交换、终缩聚三步组成。

a. 甲酯化　对苯二甲酸与稍过量甲醇反应，先酯化成对苯二甲酸二甲酯，蒸出水分、多余甲醇以及苯甲酸甲酯等低沸物，再经精馏，可得纯对苯二甲酸二甲酯（DMT）。

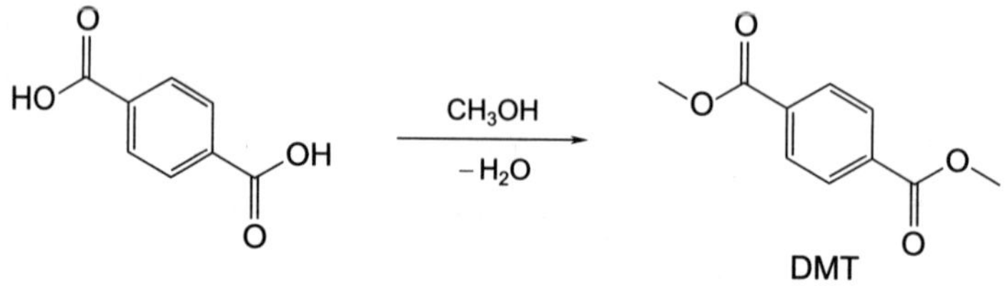

b. 酯交换　在 170~250℃下，以醋酸镉和三氧化二锑作催化剂，使对苯二甲酸二甲酯与乙二醇（物质的量之比约 1:2.4）进行酯交换反应，形成对苯二甲酸乙二醇酯低聚物。借甲醇的馏出，使反应向右移动，保证酯交换充分。

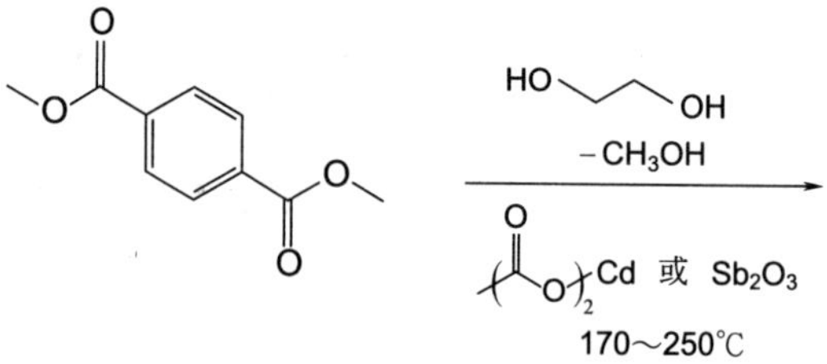

c. 终缩聚　在高于涤纶熔点下，如 280℃，以三氧化二锑为催化剂，使对苯二甲酸乙二醇酯自缩聚或酯交换，借减压和高温，不断馏出副产物乙二醇，逐步提高聚合度。

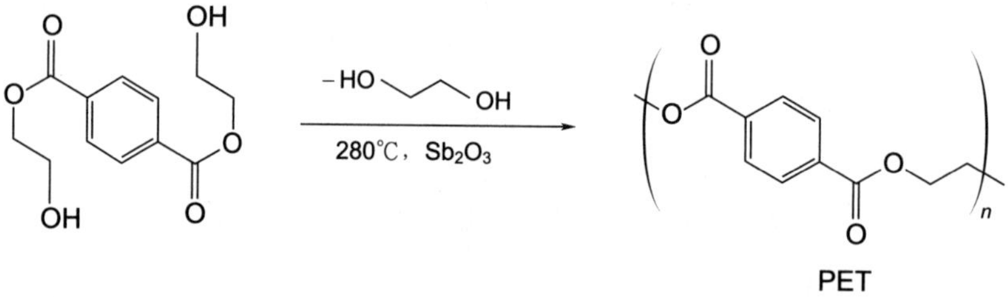

甲酯化和酯交换阶段，并不考虑等基团数比。缩聚阶段，根据乙二醇的馏出量，自然地调节两基团数的比，逐步逼近化学计量比，略使乙二醇过量，封锁分子两端，达到预定聚合度。

在 PET 聚合后期，随着缩聚反应程度的提高，缩聚体系的黏度变得很大，因此，在聚合工艺中，有意将缩聚分成两段，分别在两个反应器内进行，以便分别设定各个反应器的工艺参数，提高生产效率。一般而言，前段预缩聚的条件为 270℃，2.0~3.3kPa；后段终缩聚的工艺条件为 280℃，67~130Pa。缩聚结束后，熔体在高温下与消光颜料等填料混合，经拉伸纺丝，即成纺织工业中的重要的涤纶纤维商品。

② 直接酯化法

使用高纯 TA 原料与过量乙二醇在 200℃预先直接酯化成对苯二甲酸二羟乙酯，而后升温，按上述 DMT 工艺继续缩聚成高聚合度的 PET 产品。比较 DMT 工艺而言，TA 工艺在经济上更为合理。

无论是直接酯化法还是间接酯交换法，在这两种 PET 合成工艺中，温度的控制都非常重要。温度过高，往往导致副反应的发生，例如，PET 合成中容易出现的副反应有双醇脱水反应（dehydration of diol）和聚酯 β-裂解反应（β-scission of polyester）：

$$2\,HOCH_2CH_2OH \xrightarrow{-H_2O} HOCH_2CH_2OCH_2CH_2OH \tag{1}$$

$$\text{(聚酯} \beta \text{-裂解反应)} \tag{2}$$

上述副反应不仅会破坏缩聚体系中严格的原料配比，而且会直接影响产品质量。例如，乙二醇的脱水产物，缩二乙二醇，参与缩聚反应被引入到 PET 产品主链后，会导致 PET 产品 T_m 明显下降。PET 作为饮料或食品容器使用时，如果 PET 中含有少量乙醛杂质，会产生毒性。同时，乙醛杂质的存在也是 PET 产品变色发黄的原因。

PET 与线形饱和脂肪族聚酯相比，由于主链中增加了苯环结构，分子链的刚性增强，其强度和熔点（$T_m \approx 265\,℃$，$T_g \approx 80\,℃$）均相应提高，在 150～175℃ 使用仍然保持优良的机械强度、韧性和抗疲劳特性。PET 薄膜的拉伸强度可以和铝膜相比，接近钢材强度的一半，抗冲击强度是其他常用热塑性高分子材料的 3～5 倍。PET 的产量非常高，PET 纤维约占 PET 总产量的一半。PET 纤维的抗皱性、耐磨性和免烫性（定型性）都很好，与棉纤维或再生纤维素纤维混纺的产品有良好的手感和吸湿性。除日常生活中的服装、窗帘和室内装饰用途外，PET 在轮胎帘子线、工业过滤布等工业领域也有广泛应用。PET 的另一大用途是瓶用塑料，用量超过 PET 生产总量的三分之一。各种饮料、酒类以及一些食品的包装容器均使用 PET 原料。此外，PET 也用于各种摄影胶片、磁带、绝缘膜和工程塑料。

近年来，高分子工业中又出现了多种 PET 以外的改性聚酯品种。如聚对苯二甲酸丁二醇酯（polybutylene terephthalate，PBT），熔点较 PET 略低（$T_m \approx 230\,℃$，$T_g \approx 20 \sim 30\,℃$），熔融状态下的流动性和熔纺性能较 PET 有较大改善，结晶速率变得更快，加工性能更好，可以在 140℃ 下长期工作，具有较好的韧性和耐疲劳性，在纺织、精密仪器部件、电子电器等领域得到广泛的应用。聚对苯二甲酸丙二醇酯（polytrimethylene terephthalate，PTT）是已经实现商品化的聚酯家族中的新成员，它的熔点介于 PBT 和 PET 之间（$T_m \approx 240\,℃$，$T_g \approx 45 \sim 70\,℃$），是当前国际市场上的热门高分子新材料之一。PTT 纤维综合了尼龙的柔软性、腈纶的蓬松性、涤纶的抗污性等优点，再加上本身固有的高弹性，常温染色等优点，成为一种具有吸引力的纤维用高分子材料。目前 PTT 广泛用于衣料、产业、装饰和工程塑料等各个领域，尤其在地毯领域得到了迅猛发展。

(2) 不饱和聚酯

不饱和聚酯（unsaturated polyester）是一种由多元羧酸和多元醇缩聚而成、分子结构中含较多不饱和双键的预聚物（prepolymer），分子量通常为 1000～3000，溶于烯烃类单体稀释剂中，通过加成聚合交联固化，形成三维交联的热固性树脂。

不饱和聚酯树脂从合成制备到固化成型的工艺步骤一般可以分为三个阶段。第一阶段，多元羧酸和多元醇缩聚反应生成含有不饱和结构的聚酯预聚物。第二阶段，将上述聚酯预聚物溶解于可聚单体中稀释成为一种黏稠液体，即市售的树脂产品。第三阶段，在使用厂家添加引发剂和促进剂，并按具体产品要求添加增强材料等填料，按一定工艺条件使树脂发生交联固化，成型为最终产品。

常用的不饱和聚酯树脂通用配方由丙二醇（propanediol），马来酸酐（maleic anhydride），邻苯二甲酸酐（phthalic anhydride）在 150~200℃下缩聚合成不饱和聚酯预聚物，使用苯乙烯（styrene）作为稀释剂单体（表 2-3）。苯乙烯既是溶剂，又是供进一步交联固化的共聚单体。其中丙二醇过量的目的是为了弥补反应过程中的损失，并封锁两端。有时使用对甲苯磺酸作催化剂，以缩短反应时间；加甲苯或二甲苯作溶剂，帮助脱水；通氮或二氧化碳以防高温下氧化。为了防止不饱和聚酯树脂在储运期间发生自聚，一般加入对苯二酚（氢醌）作为阻聚剂。固化时，一般在使用现场添加少量过氧化环己酮/环烷酸钴，引发体系在室温条件下短时间内固化成型。

表 2-3 通用不饱和聚酯树脂配方

组分	物质的量之比	质量比	质量分数/%
1,2-丙二醇	2.2	171	
马来酸酐	1.0	100	
苯二酸酐	1.0	151	
失水量	1.0	−0.184	
聚酯产量		403	65.5
苯乙烯	2.0	212	34.5

不饱和聚酯树脂的合成及固化反应如下：

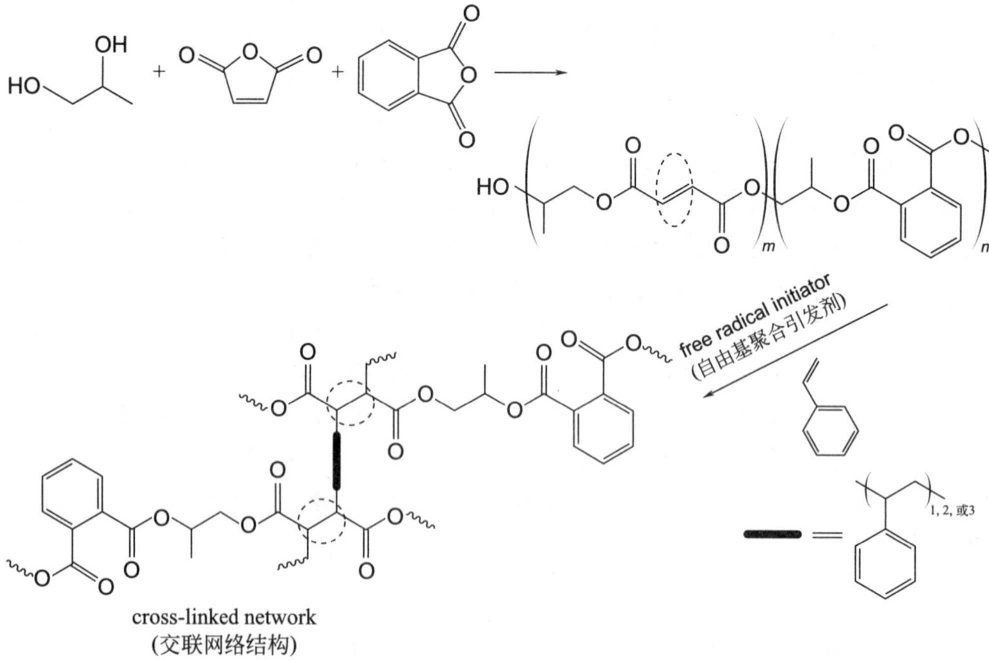

除了以上通用配方内的化合物外，还有多种二元酸、二元醇和稀释单体可用，通过改变单体种类和配比，可制得上千类不饱和聚酯树脂产品。

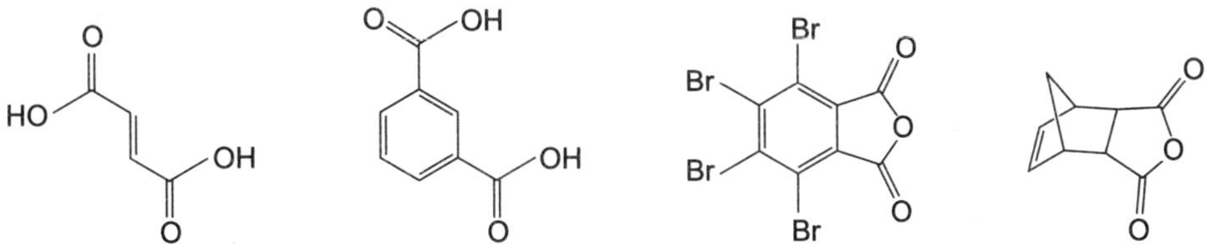

二元羧酸类化合物

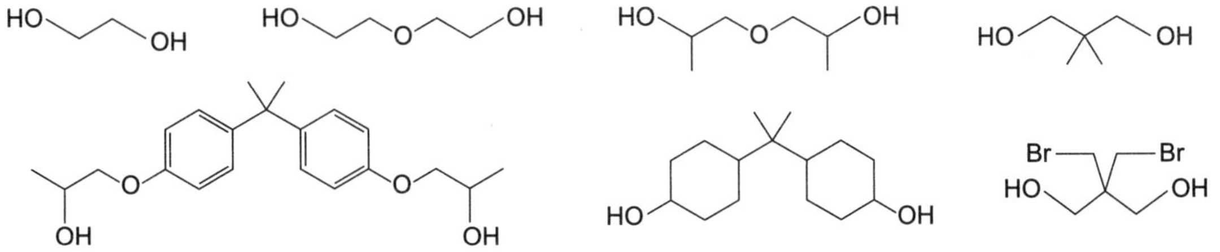

二元醇类化合物

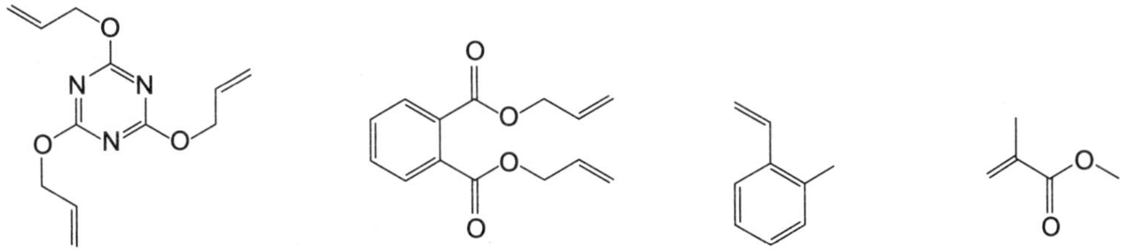

稀释单体化合物

不饱和聚酯树脂的主要用途为玻璃钢，其加工工艺简单，价格较便宜，在固化过程中无水分和其他小分子副产物生成，可以在较低压力和温度下成型。液态的不饱和聚酯树脂产品可以使用模具浇注、喷涂、模压成型、手糊成型等加工工艺，用于制作浴盆、建筑外墙、特种地板、人造大理石、化学试剂储罐、汽车车体维修、船体等。

(3) 醇酸树脂

醇酸树脂也是一种不饱和树脂，但是不饱和双键不在聚合物主链上，而是位于侧链。醇酸树脂的基本原料有三类，分别是多元醇、多元酸和单元不饱和脂肪酸。和上述不饱和树脂相似，多元酸、多元醇和不饱和脂肪酸首先缩聚生成聚酯预聚物，含有不饱和脂肪酸链的醇酸树脂预聚物与空气接触后，通过自动氧化引起的自由基反应进一步交联，转化为固体。

常用于醇酸树脂制备的二元酸类单体有邻苯二甲酸酐、间苯二甲酸、柠檬酸、己二酸、癸二酸等，多元醇包括甘油、一缩二乙二醇、季戊四醇、山梨醇等。

可用于醇酸树脂的不饱和脂肪酸很多，例如，油酸、蓖麻酸、亚油酸、亚麻酸、桐油酸等来源于天然产物的有机羧酸均可以使用。为了调整醇酸树脂的交联密度，赋予固化产物柔韧性，有时不饱和脂肪酸和月桂酸、硬脂酸、软脂酸等饱和脂肪酸混合使用。不饱和脂肪酸和饱和脂肪酸常常被分别称为干性油和不干油。脂肪酸在醇酸树脂中的体积分数为30%～60%，甚至更高。根据脂肪酸的含量，醇酸树脂可分为短油度、中油度、长油度三类。短油度醇酸树脂含有30%～45%油，一般需经烘烤才形成硬的漆膜，中油度（46%～55%油）和长油度（56%～75%油）品种，加入金属干燥剂（如萘酸钴）可以室温固化。

醇酸树脂主要用于涂料行业，可以用作多种自干或烘干磁漆、底漆、面漆和清漆，广泛用于桥梁等建筑物以及机械、车辆、船舶、飞机、仪表等涂装。具有端羧基的醇酸树脂可以和硝化纤维素的羟基反应改性，生产硝基漆。氨基树脂改性的醇酸树脂可以用作烤漆涂装金属橱柜、家用电器等。氯化橡胶和醇酸树脂共混后可以用作油漆，用于高速公路标志、混凝土和游泳池等。

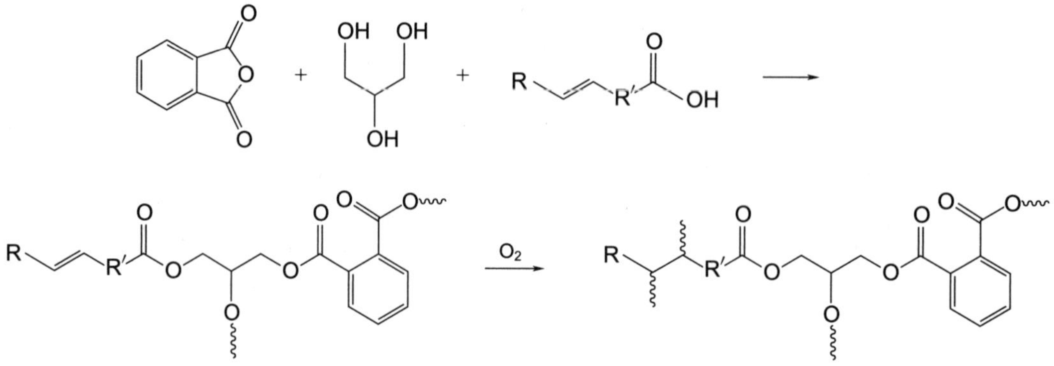

醇酸树脂有两种生产方法。

脂肪酸法：在 200～240℃，将脂肪酸、邻苯二甲酸与甘油一起酯化，可以不加溶剂，由惰性气流脱除水分和未反应物质，也可加入少量溶剂（如二甲苯），共沸蒸馏，帮助脱水。

醇解法：第一阶段，将干性油与脂肪酸甘油酯（如植物油）在 240℃下共热，在酯交换碱催化剂的作用下醇解和酯交换，形成甘油单酯或二元醇，也可能部分形成二酯或一元醇。第一阶段结束后，加入邻苯二甲酸酐（或其他二元酸），进行共缩聚酯化，条件与第一阶段相同。

(4) 生物可降解聚酯

① 聚羟基脂肪酸酯

聚羟基脂肪酸酯（polyhydroxyalkanoates，PHA）是微生物体内一类由 3-羟基脂肪酸组成的线形聚酯，其基本分子结构如下所示，单体的羧基与相邻单体的羟基形成酯键，单体皆为 R 构型。不同的 PHA 主要区别于 C3 位上的侧链基团不同，以侧链为甲基的 3-羟基丁酸酯［poly(3-hydroxybutyrate)，P(3HB)］和 3-羟基戊酸酯［poly(3-hydroxyvalerate)，P(3HV)］最为常见。

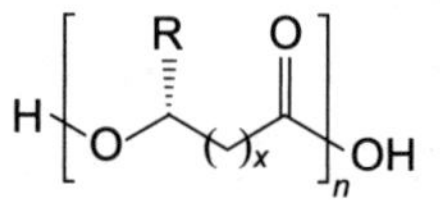

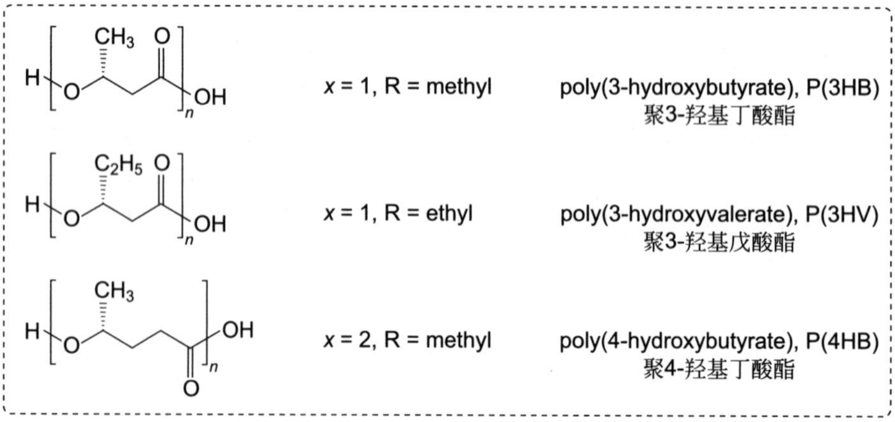

PHA 由微生物大规模发酵生产，迄今为止，聚 3-羟基丁酸酯（PHB）、3-羟基丁酸和 3-羟基戊酸共聚物（PHBV）、3-羟基丁酸和 4-羟基丁酸共聚物（P3HB4HB）均实现大规模生产。PHA 具有 150 多种单体结构，根据 PHA 的单体结构，可将其大致分成 2 类：单体组成在 3～5 个 C 原子的 PHA 称为短链 PHA（scl-PHA，short chain length PHA）；单体组成在 6～16 个 C 原子的 PHA 称为中长链 PHA（mcl-PHA，medium chain length PHA）。

PHA 依结构单元组成的不同，呈现出从硬的晶体到软的弹性体等一系列不同聚合物的性质。大多数短链 PHA 有比较高的结晶度，表现出强而硬的塑料特性；而中长链 PHA 由于结晶度很低，表现出软而韧的弹性体特征。PHA 主要用于一次性塑料用品，例如剃面刀、器皿、化妆品容器和杯子等。通过向 PHB 结构中掺入不同摩尔分数的 4HB、3HV 或者中长链单体获得的共聚 PHA，其韧性和弹性可获大幅提高（表 2-4），具有广阔的应用前景。与传统塑料相比，PHA 的优点是其生物降解性和生物相容性好，PHA 的环境降解主要在微生物分泌的胞外酶作用下进行，PHA 自身的分子结构、聚集态结构、添加组分等会影响其降解性能。

大多生物可降解高分子材料的水气透过性很高，不利于食品保鲜，而 PHA 的透气阻隔性可和 PET、PP 等产品相媲美，能够用于保存期较长的食品包装。与其他生物高分子材料相比，PHA 结构多元化，可以通过改变菌种、给料、发酵过程等调控 PHA 组成及结构，在性能多样化和可调控性方面具有明显优势。此外，PHA 与其他聚烯烃类、聚芳烃类聚合物相比，抗紫外线稳定性能更好。早期研究表明，PHB 和 PHBV 两种聚合物植入体内时，会引起免疫反应。最近发现，PHB 泡沫材料在体内未见明显免疫排斥反应，并且具有良好的生物降解性，大约 3 个月可完全降解。如表 2-4 所示，PHB 和 PHBV 两种聚合物的力学性能、熔点温度等参数与 PP 和 HDPE 十分相似，其发展潜力大，应用空间广。

表 2-4 PHA、PBS、PLA 及其他通用塑料的材料性能

	PHB	PHBV①	PLA	PBS	PP	HDPE
T_g/℃	4	−1	55	−32	−10	−120
T_m/℃	180	145	175	114	175	135
拉伸强度/MPa	40	20	66	34	33	28
断裂伸长率/%	6	50	4	560	415	700
冲击强度/MPa	—	—	29	300	21	40
结晶度/%	—	—	—	35～45	56	69

① P[3HB-co-20%（摩尔分数）3HV]。

② 聚丁二酸丁二醇酯

聚丁二酸丁二醇酯[poly(butylene succinate)，PBS]可在泥土中完全降解，降解产物无毒，并且具有与 PE 和 PP 相近的力学性能（表 2-4），可注塑、挤出和吹塑加工成型。合成 PBS 的原材料丁二酸，可由纤维素等天然产物经生物降解制备，来源广泛且价格低廉。因此，PBS 聚酯是一种可完全生态循环的生物基材料。

PBS 聚酯可采用生物发酵法合成，但目前工业生产效率低，产品价格昂贵，产量较少。化学合成法是当今制备 PBS 聚酯的主要途径。PBS 的聚合方法有缩合聚合法和开环聚合法等，缩合聚合法占主要地位。合成 PBS 的缩合聚合法有三种途径。

- 直接酯化法：丁二酸和 1,4-丁二醇在较低的反应温度下脱水形成羟基封端的低聚物，

然后在高温、高真空和催化剂存在下继续缩合反应,即可得到较高分子量的 PBS。

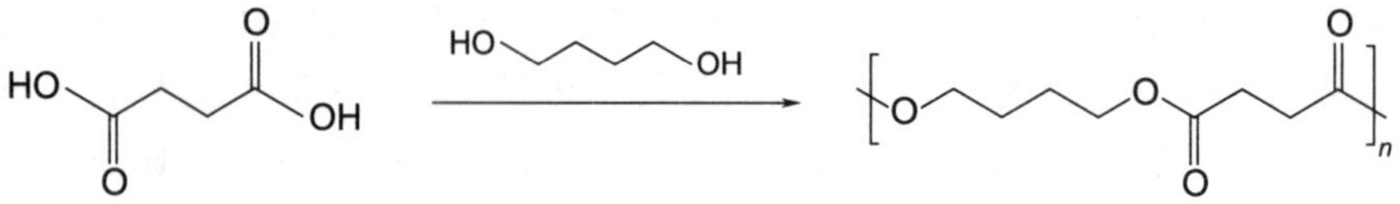

- 酯交换法:以丁二酸二甲酯或二乙酯和 1,4-丁二醇为原料,在催化剂存在下,经高温、高真空脱醇得到聚酯。

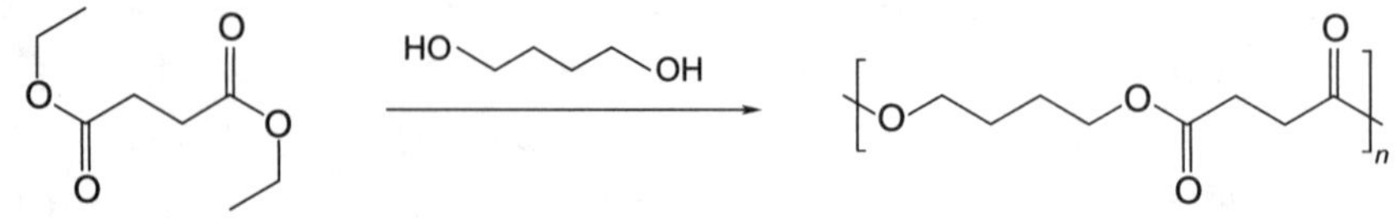

- 环状碳酸酯法:丁二酸和等物质的量的环状碳酸酯在催化剂作用下脱除 CO_2 后得丁二酸单丁二醇酯,进一步在高温、高真空下脱水得到聚酯。

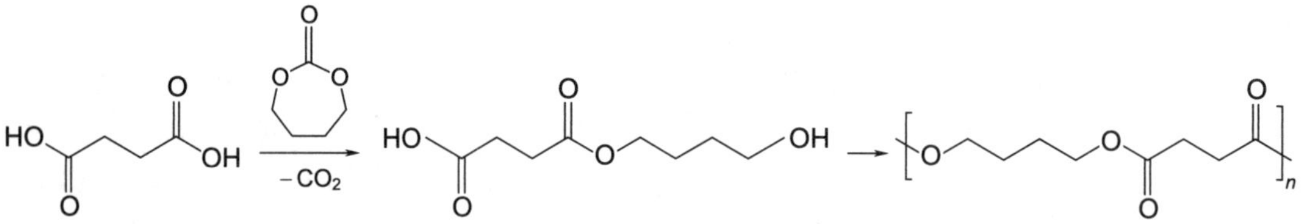

③ 聚乳酸

聚乳酸具有良好的生物降解性和生物相容性,在自然环境中可彻底降解为二氧化碳和水,在人体内可分解为参与新陈代谢的乳酸。乳酸单体可以和多种单体共聚,得到性能多样的乳酸共聚物。目前,可降解的聚乳酸广泛用于塑料、纤维、手术缝线、骨钉等多种商业用途。乳酸分子中有一个手性碳原子,因此有 L-乳酸和 D-乳酸两个光学异构体。L-乳酸由生物发酵制备,D-乳酸以及外消旋乳酸一般通过化学合成制得。

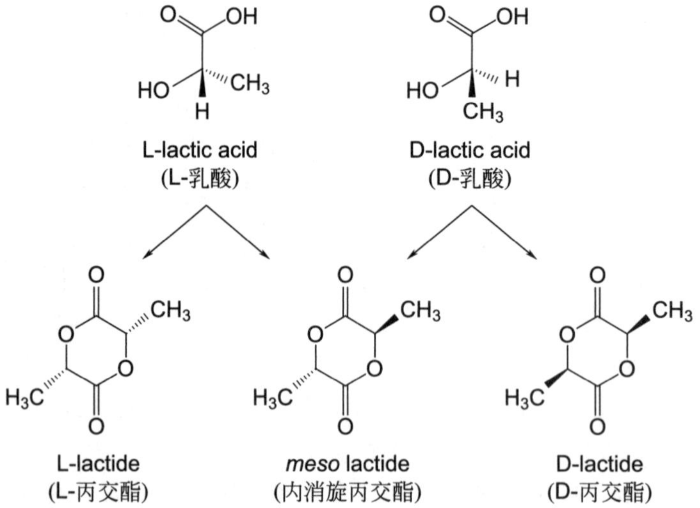

L-lactic acid
(L-乳酸)

D-lactic acid
(D-乳酸)

L-lactide
(L-丙交酯)

meso lactide
(内消旋丙交酯)

D-lactide
(D-丙交酯)

聚乳酸的聚合有丙交酯开环聚合法(也称两步法)和乳酸直接缩聚法(也称一步法)两条路线。丙交酯开环聚合法首先将乳酸脱水生成低聚物(分子量 1000～5000),然后解聚生成丙交酯,经开环聚合生成聚乳酸,如图 2-3 所示。丙交酯的开环聚合反应可以得到数十万的聚乳酸。直接缩聚法是指乳酸分子之间的羟基和羧基发生直接脱水缩合反应生成聚乳酸的一种合成工艺(图 2-3)。与丙交酯开环聚合法相比,直接缩聚不需要丙交酯及纯化,成本较低，但是由于聚合过程中水等副产物较难除去，较难得到高分子量的聚合物。聚合温度高于180℃时聚合产物带黄色。最近结合二异氰酸酯等扩链剂偶联聚乳酸低聚物的合成方法能够获得较高分子量，成为比较被期待的合成路线。

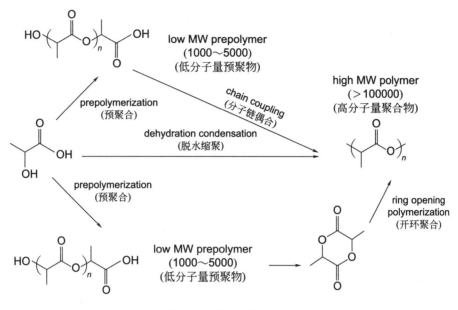

图 2-3　聚乳酸聚合路线

2.3.2　聚碳酸酯

由双酚 A 和光气合成的聚碳酸酯（PC）具有良好的透明性和抗冲击性。PC 常用作大型灯罩、防护玻璃、照相器材、眼镜片、飞机座舱玻璃等光学相关产品。PC 的刚性和耐热性较好，可在 15～130℃内保持良好的力学性能，是重要的工程塑料，但容易受热水解，生成双酚 A。双酚 A 在人体内不易被代谢，具有一定的毒性。

尽管 PC 能够结晶（$T_m = 265～270$℃），但是大多数 PC 材料是非晶态的（$T_g = 150$℃），因此具有较好的光学透明性。

PC 的聚合目前有两种工业化生产路线：酯交换法和光气直接法（图 2-4）。

- 酯交换法：双酚 A 与碳酸二苯酯熔融缩聚，进行酯交换，在高温减压条件下排除苯酚，提高反应程度，获得高分子聚碳酸酯。在 PC 聚合过程中，因苯酚较难从高黏熔体中脱除，PC 的分子量一般较难超过 3 万。
- 光气直接法：光气化学活性高，可以与双酚 A 直接酯化。工业上多采用界面缩聚技术，双酚 A 和氢氧化钠配成双酚钠水溶液作为水相，光气的有机溶液为另一相，在催化剂存在 50℃下反应。光气直接法比酯交换法经济，所得分子量也较高，但是毒性很强的光气的使用需要严密的安全保障。

2.3.3　聚酰胺

聚酰胺（PA）是主链中含有酰胺基团的杂链聚合物，可以分为脂肪族和芳香族两类。

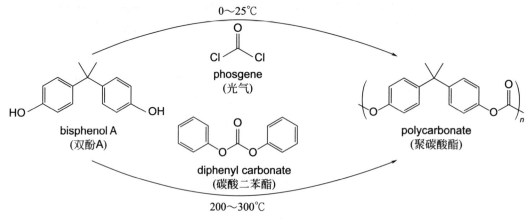

图 2-4 双酚 A 型聚碳酸酯的合成路线

聚酰胺通常由二元胺和二元酸通过熔融缩聚法合成，典型的产品是聚酰胺-66（尼龙-66）。尼龙-66 由己二酸和己二胺缩聚而成。一般是在水溶液中配制浓度为 50% 的铵盐，以保证羧酸和氨基官能团数相等。由己二酸和己二胺制备的铵盐通常称为 66 盐或尼龙盐。聚合中，先将 66 盐水溶液加热至 100℃ 以上浓缩，然后在密闭系统内加压（1.4～1.7MPa）升温至 200～215℃ 进行预缩聚，使体系反应程度达 0.8～0.9。然后慢慢升温至 270～275℃，保持恒温完成最终缩聚反应（图 2-5）。因反应最后阶段的温度在尼龙-66 的 T_m（265℃）以上，属于熔融缩聚。分子量的控制通过在尼龙盐的浆液中添加乙酸来实现。

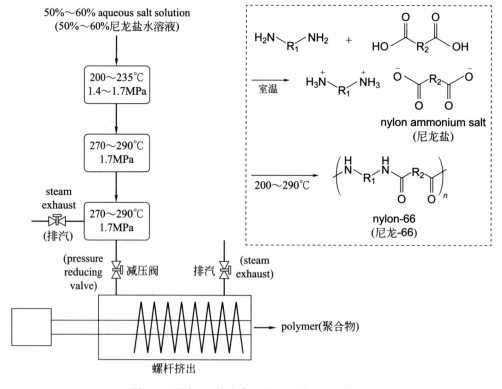

图 2-5 尼龙-66 的聚合反应及工艺流程示意图

尼龙-66结晶度中等，熔点高，能溶于甲酸、苯酚等有机溶剂，具有拉伸强度高、柔韧性好、蠕变低、耐溶剂等优点，是排在 PET 之后的第二大类合成纤维。大约有 60% 以上的尼龙产品用于纤维，包括服装、轮胎补强、降落伞、渔网等。目前，尼龙是产量最大的工程塑料（engineering plastic），其应用遍及汽车制造、电子工业、机械制造、体育用品及玩具生产等工业领域。

在聚酰胺主链中引入苯环可提高高分子材料的耐热性和刚性。半芳族聚酰胺由芳族二元酰氯或酸与脂肪族二元胺缩聚而成。例如，对苯二甲酰氯与己二胺聚合得到的聚合物的商品名为尼龙-6T，其熔点为 370℃，185℃下受热 5h 强度几乎不变。

全芳聚酰胺主要由芳二酰氯与芳二胺缩聚而成。由于苯环的共轭作用，大多芳胺反应活性较低，但可以和酰氯在适当溶剂中于室温下进行缩聚。

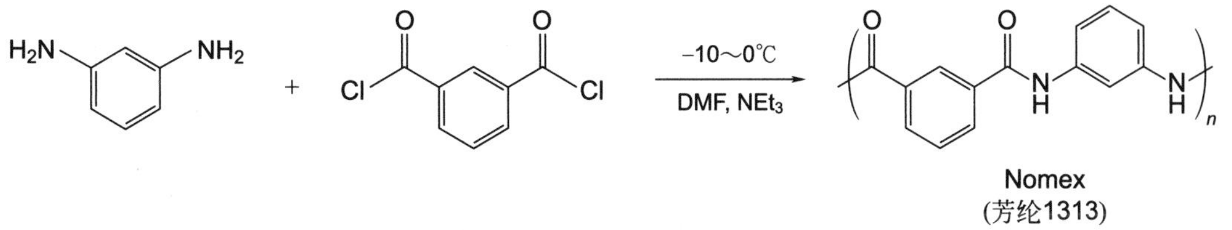

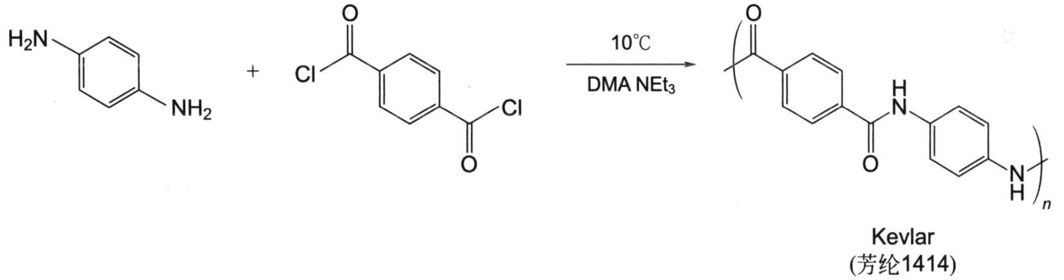

例如杜邦公司开发的聚间苯二甲酰间苯二胺（商品名 Nomex 或芳纶1313）是在 N,N-二甲基甲酰胺（DMF）溶液中，添加三乙胺中和副产物酸，在 $-10\sim0℃$ 的较低温度下进行缩聚而得。Nomex 具有一定的阻燃性，熔点为 370℃，可溶于含 LiCl（5%）的 N,N-二甲基乙酰胺（DMA）有机溶剂中，能够湿法纺丝。另外一个大规模生产的全芳聚酰胺是聚对苯二甲酰对苯二胺（PPD-T），商品名为 Kevlar（国内称作芳纶1414），可加工成纤维。PPD-T 结构单元中有刚性苯环和强极性酰胺键，分子链结构规整，可制成高性能纤维。具有强度高（2.4~3.0GPa），模量高（62~14GPa），耐高温（$T_g=375℃$，$T_m=530℃$）等优良性能，适用于航天、军事装备、轮胎帘子线等方面。

对苯二甲酰氯活性高，能与对苯二胺在低温（10℃）溶液中快速聚合。可用溶剂有 N,N-二甲基乙酰胺、N-甲基-2-吡咯烷酮（NMP）等，氯化锂、氯化钙等为助溶盐，以吡啶或叔胺类有机物作酸吸收剂，可得淡黄色微细粉末状的 PPD-T 产品。

2.3.4 聚酰亚胺

由 4,4'-二氨基二苯基醚和均苯四甲酸酐合成的聚合物是由杜邦公司最早推出的商业化聚酰亚胺产品（PI，商品名：Kapton，Vespel），具有良好的自润滑性、热稳定性、耐溶剂性以及抗蠕变性能。Vespel 薄膜一般用作电动机、导弹、飞机中导线的绝缘材料，还可以替代陶瓷和金属制造汽车和飞机的发动机零部件，如套管、密封件、活塞环和轴承等。

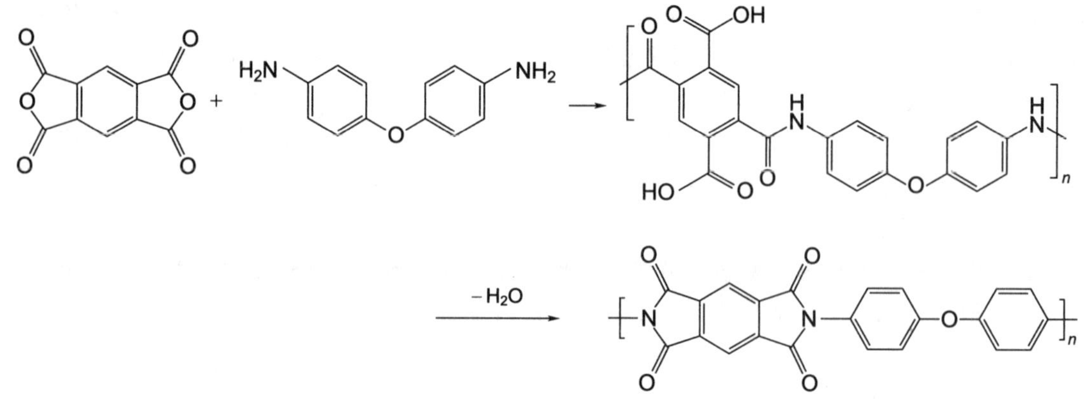

polyimide (Vespel)

因为大多数 PI 难以溶解和熔融，一般采用两步法合成。第一步在有机溶剂（二甲基甲酰胺、二甲基乙酰胺、二甲基亚砜、N-甲基-2-吡咯烷酮等）中进行酰胺化缩聚，生成高分子量的聚酰胺酸。为了将环化反应控制在较低程度，保证聚酰胺酸处于溶解状态，第一步反应通常在低于 30℃ 的温度下进行。第二步是将聚酰胺酸加热到 150℃ 以上，进行脱水环化反应。因得到的 PI 产品难以溶解和熔融，加工困难，加工成型一般要在环化反应前完成。

为了提高 PI 产品在全酰亚胺状态下的加工性能，通过选用适当的酸酐和有机胺原料降低 PI 分子链的有序程度，已经得到了较多 PI 新产品。其中聚醚酰亚胺（PEI）和聚酰胺酰亚胺（PAI）的加工性能和材料性能较为均衡，被广泛应用。

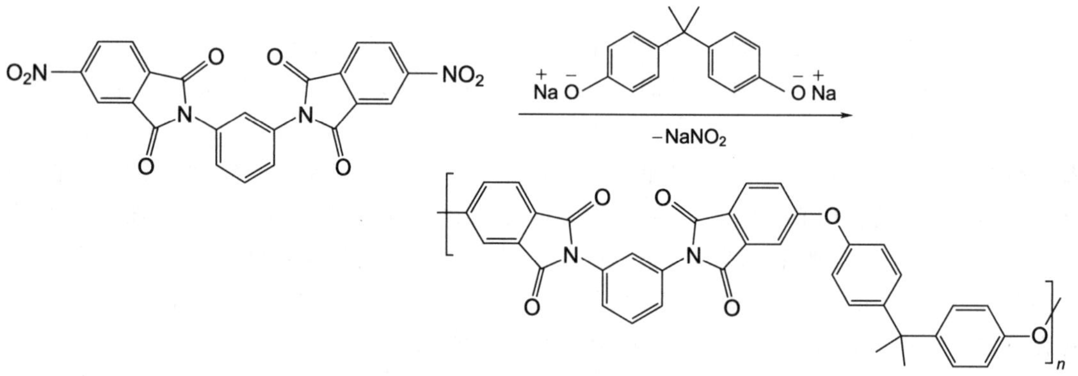

PEI

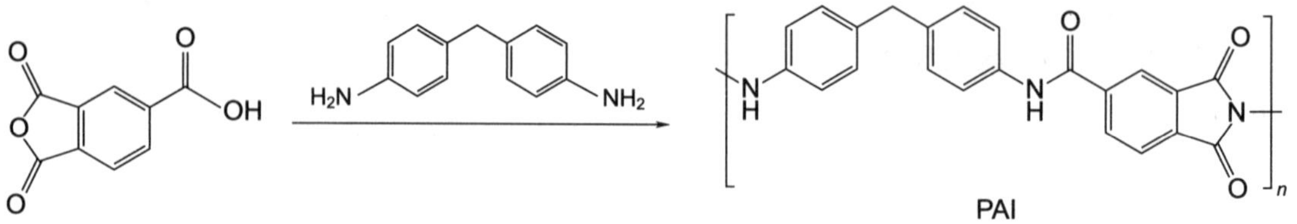

PAI

PEI 的 T_g 为 215℃，长期使用温度为 170~180℃，耐溶剂性和耐化学性能与 PI 相当，但是可以溶解于部分有机溶剂中，可利用传统方法进行加工成型。PAI 的 T_g 为 270~285℃，长期使用温度为 220~230℃，耐溶剂性和耐化学性能良好。

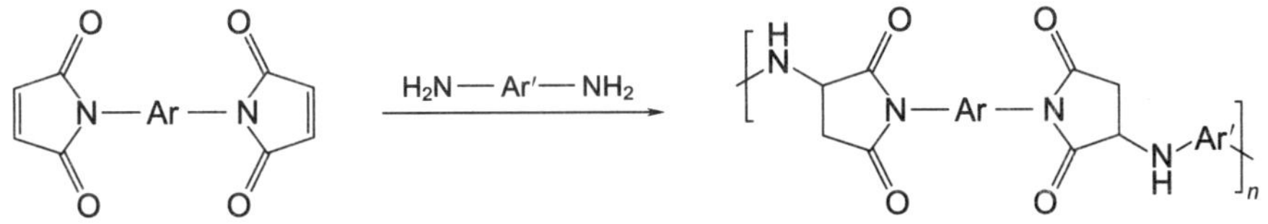

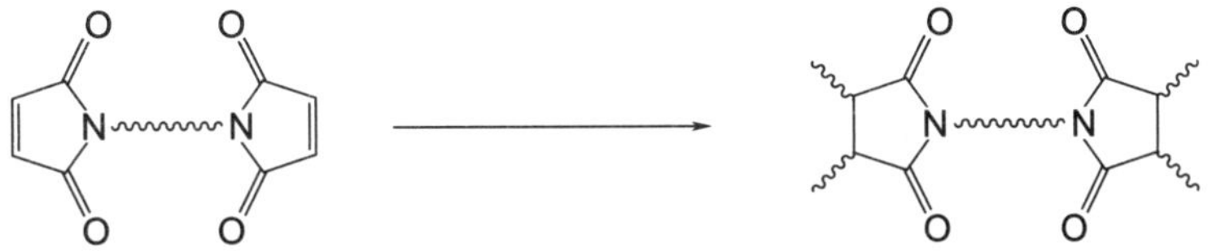

双马来酰亚胺（BMI）属热固性聚酰亚胺树脂，由马来酰亚胺和二元胺加成聚合制备。反应体系中马来酰亚胺过量，得到分子量较低、具有流动性的低聚物，加热到180℃或更高温度，马来酰亚胺分子结构中双键进一步聚合交联，形成热固性树脂。BMI的T_g大于260℃，可在220～230℃下长期使用。BMI的流动性与环氧树脂相近，加工方法也类似。因其耐热性好于环氧树脂，常用作150℃以上温度下工作的复合材料及黏结剂。

2.3.5 聚苯并咪唑类聚合物

聚苯并咪唑（polybenimidazoles，PBI）是一种耐高温高分子，熔点超过400℃，其薄膜和纤维达300℃仍能保持良好的力学性能。PBI一般由芳香族四元胺和芳香族二元酸或酯分两步缩聚而成。

第一步在250℃形成可溶性氨基-酰胺预聚物，第二步再在350～400℃成环固化。用苯酯替代羧酸，可以解决高温脱羧问题。

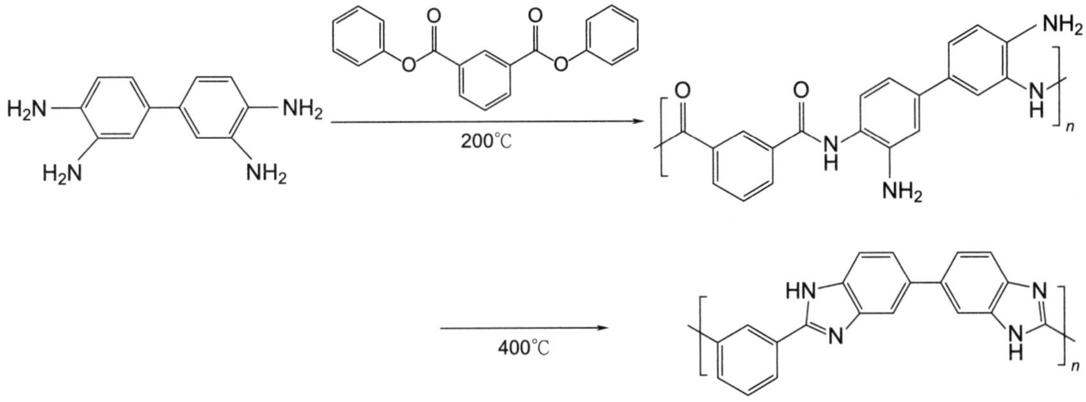

二羟基联苯胺与间苯二甲酰氯或二巯基联苯胺与间苯二甲酸苯酯进行缩聚，可以制备聚苯并噁唑（PBO）、聚苯并噻唑（PBT）等多种聚合物，它们均是耐高温的高性能高分子材料。

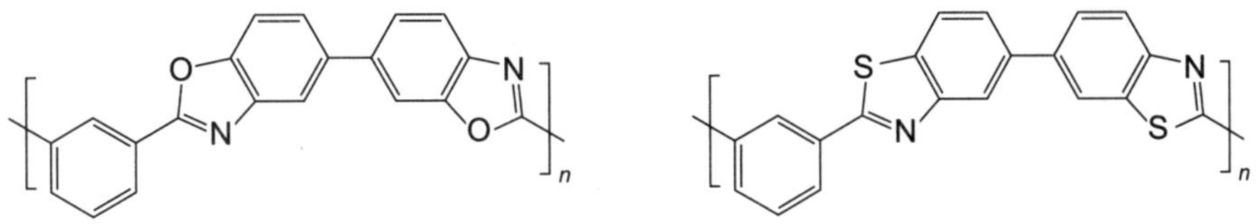

PBI可制作宇宙飞船中的绳索、耐烧蚀热屏蔽材料、减速用阻力降落伞及宇航员的加压安全服等。PBI曾经用作阿波罗号和空间实验室宇航员的航天服和内衣。在一般工业中可作石棉替代品，包括耐高温手套、高温防护服、传送带等，使用温度常为250～300℃，能在500℃下短时间使用。

2.3.6 聚氨酯

聚氨酯（PU）是带有—NH—COO—特征基团的杂链聚合物，是逐步聚合物中非常重要的一员。聚氨酯制品用作黏合剂、涂料、（弹性）纤维、弹性体、软硬泡沫塑料、人造革等。

TDI　　MDI　　IPDI

HDI　　PAPI　　NDI

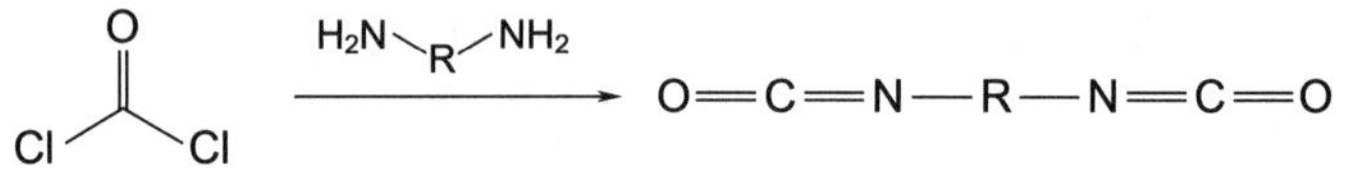

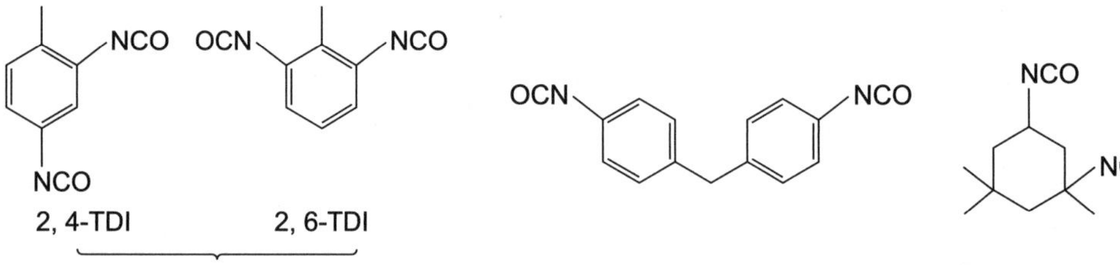

工业上大多选用异氰酸酯合成聚氨酯。异氰酸酯一般由光气和二元胺反应生成。异氰酸基是很活泼的基团，能与醇、胺、脲等反应，与水、羧酸等也很容易反应，同时释放出二氧化碳，可以用来制备聚氨酯泡沫塑料。

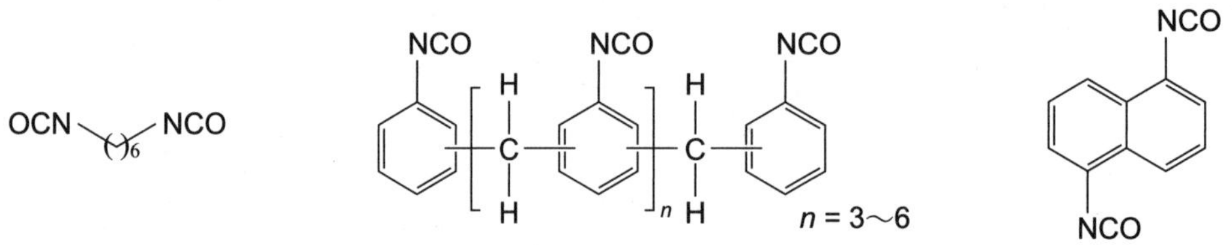

甲苯二异氰酸酯（TDI）和六亚甲基二异氰酸酯（HDI）是常用的异氰酸酯工业化原料。TDI的英文名为toluene diisocynate，有2,4-TDI和2,6-TDI两种异构体。TDI工业品以2,4-TDI和2,6-TDI质量比80∶20（简称TDI-80）的混合物为主，也有纯2,4-TDI（TDI-100）和两种异构体质量比为65∶35（TDI-65）的产品。TDI的工业化生产主要由甲苯经硝化生成二硝基甲苯，再经催化加氢生成二氨基甲苯，与光气反应制得。

二苯基亚甲基二异氰酸酯（4,4′-diphenylmethane diisocyanate，MDI）由苯为原料制得。苯经硝酸硝化为硝基苯，经加氢制得苯胺，与甲醛缩合得二氨基二苯甲烷，再与光气反应得MDI。

多亚甲基多苯基异氰酸酯（polyaryl polyisocyanate，PAPI）常温下为中低黏度液体，其生产方法与 MDI 类似，只是苯胺的用量略小。

聚氨酯的另一种原料是多元醇，起软段的作用。用于聚氨酯合成的多元醇种类繁多，合理选择多元醇是调整聚氨酯材料性能的主要手段。较多使用的是分子量从几百到几千的聚醚多元醇和聚酯多元醇，其中聚醚多元醇占聚氨酯产业中多元醇总量的 90% 以上。

合成聚氨酯时常常使用催化剂来缩短反应时间，抑制副反应。聚氨酯产业中常用的催化剂主要有叔胺和有机金属化合物两类。使用较多的催化剂如图 2-6 所示，有三亚乙基二胺、N-乙基吗啉、双(2-甲基氨基乙基)醚、二丁基锡二月桂酸酯、辛酸亚锡等。

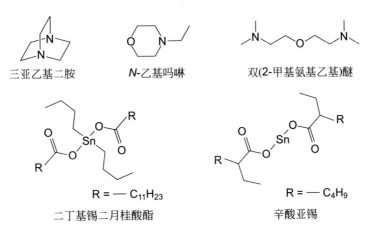

图 2-6 聚氨酯合成用催化剂

一般而言，聚氨酯的 T_g 为 80℃ 左右，具有良好的耐磨性能。此外，聚氨酯分子链的刚性和弹性容易通过选择适当的多元醇来控制，可以制成弹性较好的纤维、发泡材料、弹性体、绝热材料等。

由于聚氨酯所用原材料基本上为液体材料，可以采用不同于普通橡胶、塑料等高分子材料的加工方法进行加工，制成泡沫、弹性体、黏合剂、涂料等各种制品。

聚氨酯发泡体具有软质聚氨酯泡沫塑料和硬质聚氨酯泡沫塑料两类。前者产量最大，一般具有密度低、弹性回复好、吸声、透气、保温等性能，主要用作家具垫材、交通工具座椅垫材、各种软性衬垫层压复合材料。后者具有绝热效果好、质轻、比强度大、耐化学品优良以及隔声效果好等特点。其主要应用有：冰箱、冷柜、冷藏集装箱、冷库等的保温层材料，石油输送管道及热水输送管道保温层，建筑墙壁及屋顶保温层、保温夹心板等。聚氨酯弹性体具有缓冲性能好、质轻、耐磨、防滑等特点，加工性能好，广泛用于制造运动鞋的鞋底、鞋跟、鞋头等。聚氨酯黏合剂广泛用于制鞋、纺织、皮革、薄膜等柔软材料的黏接，同时还能广泛用于建筑业、汽车制造业等高强度结构黏合领域以及超低温特种黏合领域。

聚脲是碳酸的聚酰胺，因脲基团极性大，可以形成更多的氢键，因此聚脲的熔点比相应的聚酰胺要高，韧性也大，适于纺制纤维。

合成聚脲最好的方法是参照聚氨酯的合成方法，即二元胺与二异氰酸酯反应，反应伴有猛烈放热，一般在数秒内完成。因为是逐步加成反应，无小分子副产物生成。

2.3.7 环氧树脂

环氧树脂（Epoxy resin）泛指含有环氧官能团（epoxy group）的高分子预聚物。三元环状的环氧基团因环张力具有较高的反应活性，可以与各种固化剂进行交联反应，也可通过开环聚合形成线形或体形的聚合物。环氧树脂具有优良的力学性能、电绝缘性能、耐化学腐蚀性能、耐热及黏结性能，广泛用于机械电子、化学化工、航空航天、汽车、船舶、建筑等工业领域。是应用领域最为广泛的一类热固性树脂。商用的环氧树脂品种较多，其中由双酚A和环氧氯丙烷缩聚而成的双酚A二缩水甘油醚（diglycidyl ether of bis phenol A，DGEBA）所占比例最大。

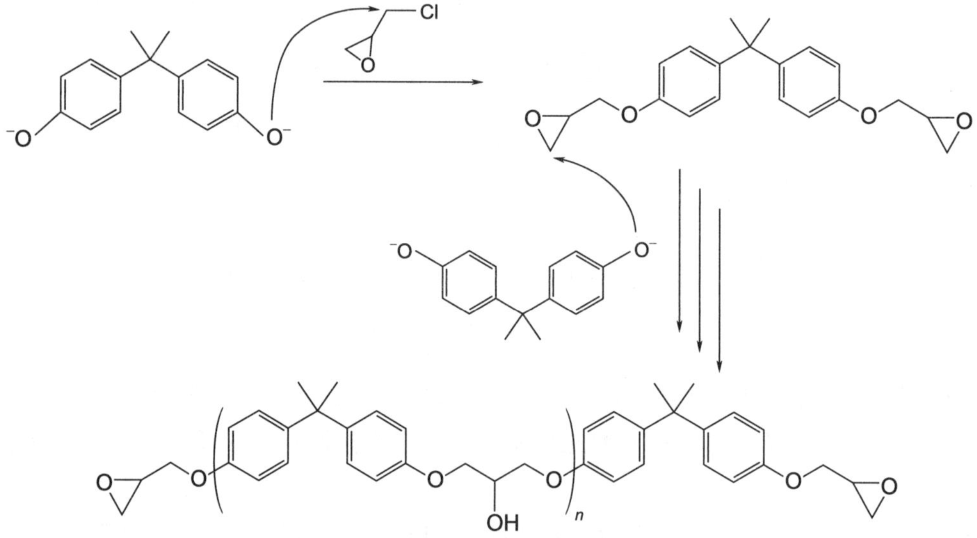

在碱催化条件下，双酚A钠盐和环氧氯丙烷先缩合成低分子中间体，再进一步缩聚成分子量约为340～7000的低聚物，也就是E型环氧树脂产品。市售E型环氧树脂室温下大多呈黄色黏稠液体，其黏度因其分子量而异。上式中的 n 值在2～19范围内，由原料配比、加料次序、操作条件来控制。在实际使用环氧树脂时，常常使用便于配方计算的环氧值来表示分子量大小。

环氧值指每100g树脂中所含环氧官能团的物质的量。例如E-51环氧树脂，其中E表示双酚A类别，51表示其环氧值为0.51mol/100g树脂。

固化剂种类多、固化方便是环氧树脂的一大特点。其固化剂按固化温度可分为：低温和超低温固化剂（＜0℃），室温固化剂（0～25℃），中温固化剂（80～120℃），高温固化剂（160～180℃）。图2-7列出了可与环氧官能团进行固化反应的官能团。

一般而言，低温固化剂以巯基化合物为主，室温固化剂多为脂肪胺，芳香胺类常用作中温固化剂，酸酐类化合物需要较高温度才能固化。

固化环氧树脂中含大量羟基和醚键，除聚四氟乙烯、聚乙烯、聚丙烯以外，对大部分物质表面有高黏结力，有"万能胶"之称。它与许多金属和非金属材料（如玻璃、陶瓷、木

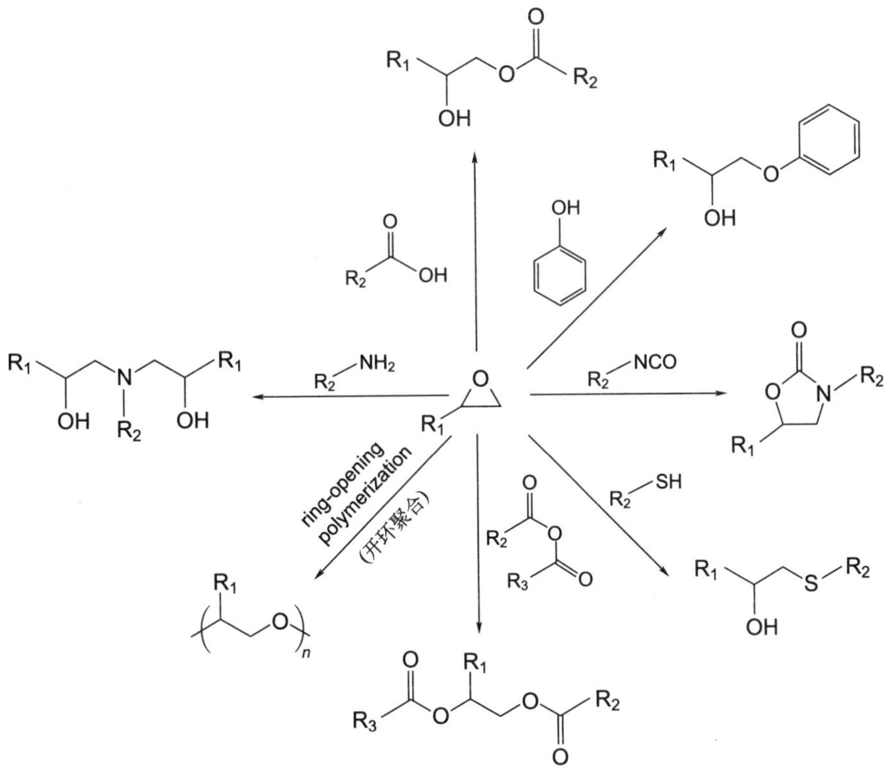

图 2-7 环氧官能团常见反应

材）的黏结强度往往超过材料本身的强度，是结构胶黏剂的主要品种。作为胶黏剂，环氧树脂固化过程的收缩率低，一般为 1%～2%，其固化制品的内应力小，不易开裂，同时还具有耐酸碱、吸水率低、电绝缘性优良、加工方便等特点。

也正是由于上述性能优点，环氧树脂的应用非常广泛，在建筑、汽车制造、航空航天、电器电子、机械加工等领域均有多种多样的应用，尤其是在复合材料的生产中占有重要地位。

2.3.8 酚醛树脂

酚醛树脂由苯酚和甲醛缩聚而成，甲醛官能度为 2，苯酚的邻、对位氢参与聚合反应，官能度为 3。酚醛聚合反应使用两类催化剂，聚合得到相应的两类预聚物产品：碱催化酚醛预聚物（Resoles）和酸催化酚醛预聚物（Novolacs）。两种预聚物需要进一步固化，得到性能稳定的酚醛树脂。碱性酚醛树脂主要用作黏结剂，生产层压板；酸性酚醛树脂则多用于模塑粉。

（1）酸催化酚醛预聚物

盐酸、硫酸、磷酸等无机酸都可以催化酚醛缩聚反应，但草酸腐蚀性较小，优先选用。酸催化条件下的聚合反应机理如下：

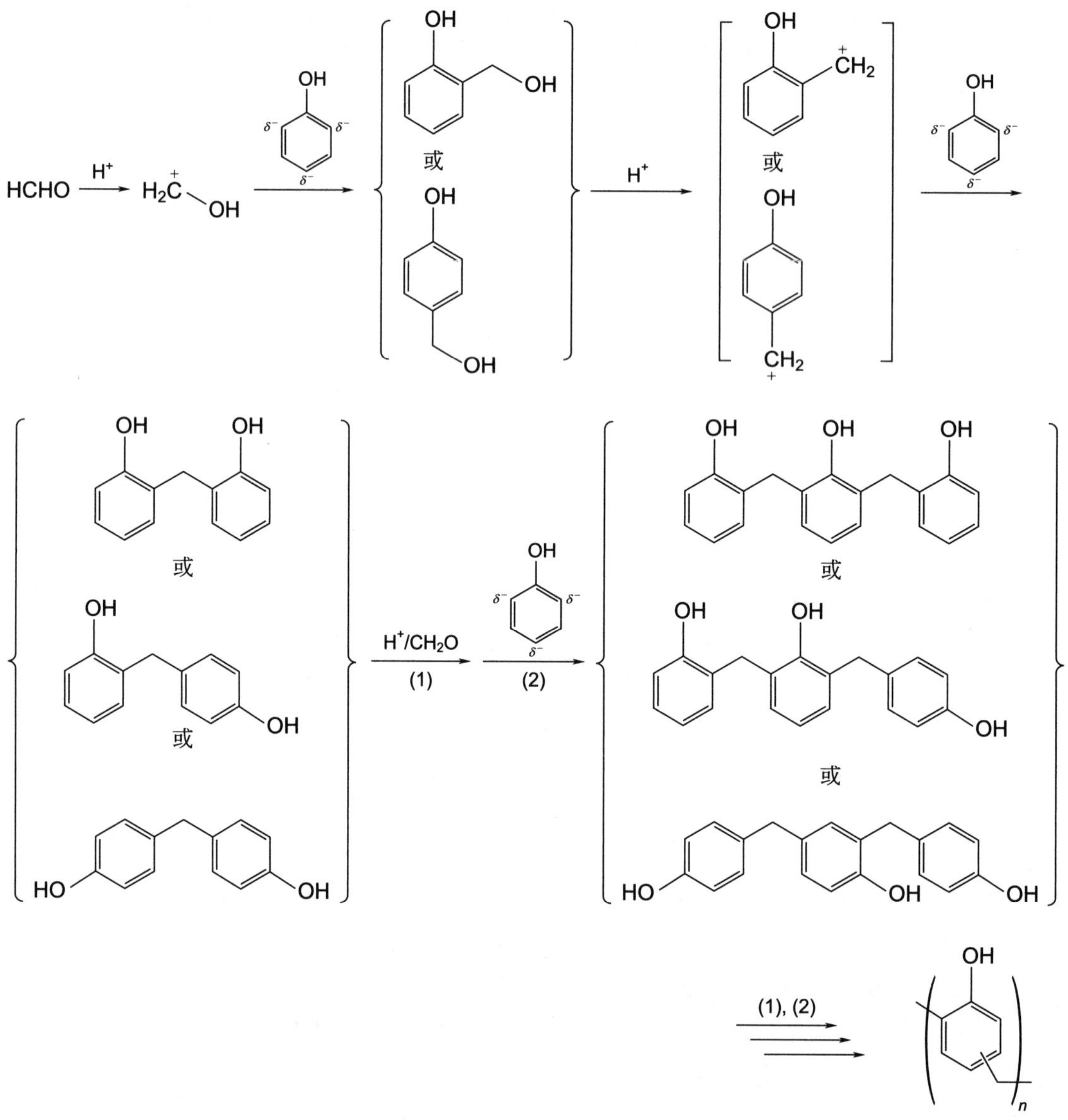

首先，甲醛的羰基质子化，而后作为亲电试剂进攻苯酚的邻、对位，通过芳香亲电取代反应形成邻或对羟甲基酚。芳香环上的羟甲基脱水缩合形成亚甲基桥，形成苯酚的邻-邻、对-对或邻-对随机连接。式中（1）（2）反应重复进行，即可获得 Novolacs 树脂。

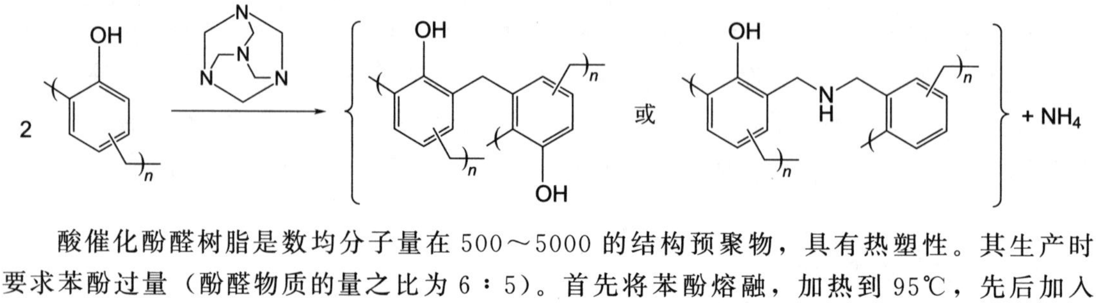

酸催化酚醛树脂是数均分子量在 500～5000 的结构预聚物，具有热塑性。其生产时要求苯酚过量（酚醛物质的量之比为 6∶5）。首先将苯酚熔融，加热到 95℃，先后加入草酸（苯酚的 1%～2%）和甲醛水溶液，回流下反应 2～4h，甲醛即可耗尽。因甲醛量不足，树脂结构中无羟甲基，加热也无交联危险，因此可称为热塑性酚醛树脂。酚醛树脂从水中沉析出来，经常压和减压蒸出水分和未反应的苯酚，直至160℃，冷却，破碎，即成酚醛树脂粉末。

酸催化树脂粉末由于羟甲基不足，加热不能固化，需要使用六亚甲基四胺等进行交联固化。六亚甲基四胺分解，提供交联所需的亚甲基，其作用与甲醛相当，也有可能部分形成苄胺桥。

(2) 碱催化酚醛预聚物

碱催化并且甲醛过量条件下（酚醛物质的量比6∶7），苯酚处于共振稳定的阴离子状态，邻、对位阴离子与甲醛进行亲核加成，先形成邻、对位羟甲基酚。它们进一步相互缩合，形成由亚甲基桥连接的多元酚醇，进而形成二、三环结构较多的低分子量（200~500）酚醛树脂（Resoles）。

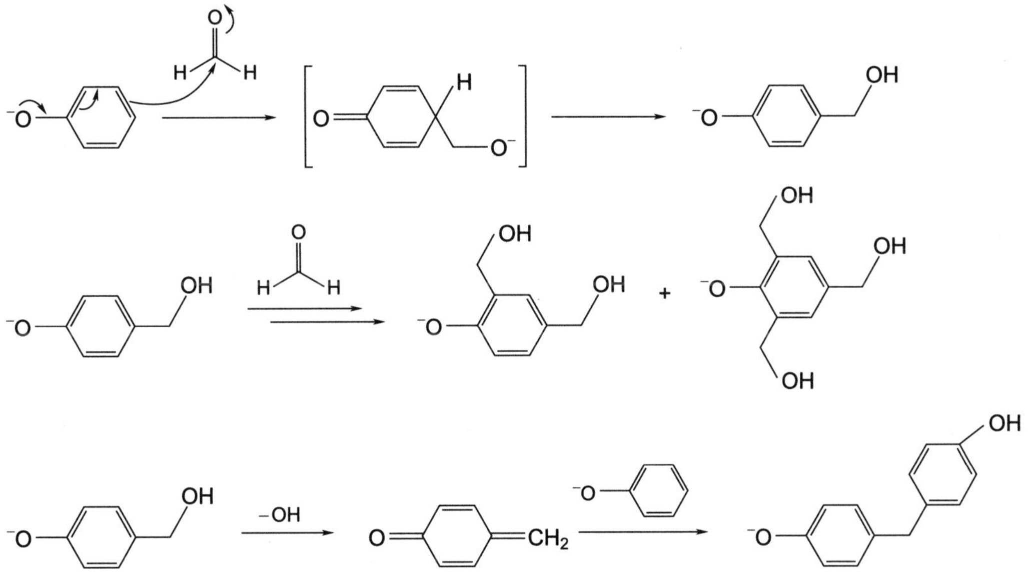

例如，将苯酚、40%甲醛水溶液，氢氧化钠或氨（苯酚量的1%）等混合，回流1~2h，即可达到预聚要求。反应结束前，中和成微酸性，减压脱水，冷却，即得酚醛预聚物。

碱催化酚醛树脂加热即可固化，无需固化剂。其反应机理是在较高温度下，酚醇在两苯环间形成亚甲基桥或二苄基醚键交联结构。交联固化时会有水和甲醛等小分子生成。如果后期干燥过程不彻底，会造成甲醛残留，甚至影响人体健康。

(3) 改良酚醛树脂及苯并噁嗪树脂

其他酚类和醛类化合物可用来改性酚醛树脂。例如，糠醛取代甲醛可以增加酚醛树脂的流动性和耐热性。利用间苯二酚取代苯酚，可以提高酚醛树脂预聚物的固化速度，甚至可以室温固化。单取代苯酚，如邻、对位甲酚或氯代苯酚，是二官能度单体，与苯酚混用时，可以适当降低酚醛树脂的交联密度。

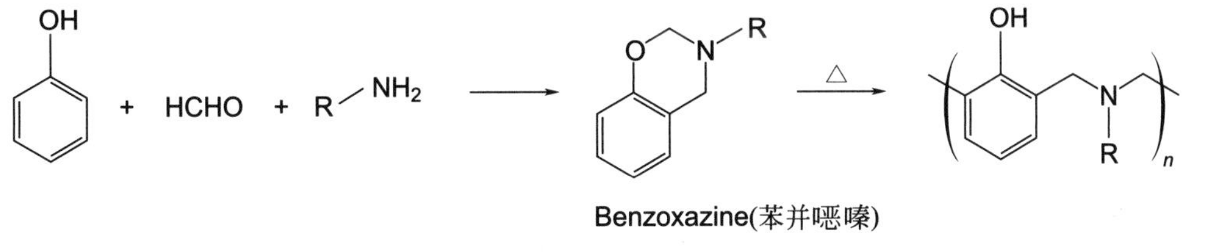

Benzoxazine(苯并噁嗪)

> 通用塑料（general purpose plastics）：长期使用温度低于100℃，产量大、价格较低、性能一般，主要品种有聚乙烯（PE）、聚丙烯（PP）、聚苯乙烯（PS）、聚氯乙烯（PVC）、丙烯腈-丁二烯-苯乙烯共聚物（ABS）等，占热塑性塑料的90%。主要用途有农膜、喷灌管、养殖箱、电子仪表壳体、防腐容器、塑料门窗、上下水管道、包装薄膜、塑料玩具、牙刷等生活用品。
>
> 工程塑料（engineering-plastics）：长期使用温度在100～150℃之间，具有更好的力学性能，能经受较宽的温度变化范围和较苛刻的使用条件，可用于工程中用作力学构件。工程塑料主要包括聚碳酸酯（PC）、聚酰胺（PA）、聚甲醛（polyoxymethylene，POM）、聚苯醚（polyphenylene oxide，PPO）、聚苯硫醚（polyphenylene sulfide，PPS）等。工程塑料在力学性能、耐久性、耐腐蚀性、耐热性等方面均优于通用塑料，可用于汽车保险杠、燃油箱、车灯罩以及发动机相关零部件等，也可制作轴承、齿轮等机械零件。
>
> 特种工程塑料（high performance plastics or super engineering plastics）：热变形温度在200℃以上，长期使用温度在150℃以上，且具有优良力学性能的聚合物，主要品种有聚苯硫醚（PPS）、聚砜（PSF）、聚醚醚酮（PEEK）、聚酰亚胺（PI）、聚醚酰亚胺（PAI）等。特种工程塑料种类多，性能优异，价格昂贵，主要用于高新技术或国防军事等特定用途。

苯并噁嗪（benzoxazine）可由酚、醛和伯胺类化合物合成。该单体加热通过开环聚合得到体形聚合物，较酚醛树脂具有更为优良的热力学性能，是酚醛树脂的一个衍生分枝。由于胺和酚的适用范围广，且合成容易，分子结构设计的变化性较强，最近受到广泛关注。尤其是大多苯并噁嗪加热固化的热收缩率接近零，可有效降低产品的内应力。这是大多数树脂所不具备的特点。

2.3.9 氨基树脂

含氨基有机物与醛类化合物缩聚而成的一类树脂统称为氨基树脂（amino formaldehyde resin，AF），主要的氨基树脂有尿素甲醛（脲醛）树脂（urea formaldehyde resin，UF）和三聚氰胺甲醛（蜜胺）树脂（melamine formaldehyde resin，MF）。

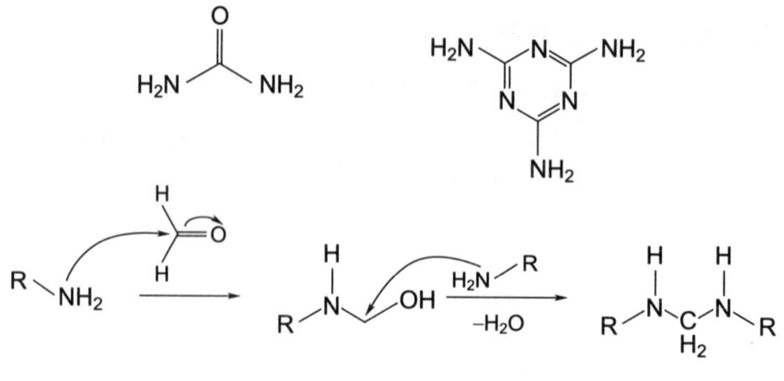

(1) 脲醛树脂

尿素呈碱性，分子中的一个羰基不足以平衡两个氨基，与甲醛反应时，先亲核加成，形成羟甲基衍生物，构成预聚物。

经加热固化所得的脲醛树脂色浅或无色，比酚醛树脂硬，可用作涂料、黏结剂、层压材料和模塑品。脲醛树脂也可用作木粉、碎木的黏结剂，制作木屑板和合成板。

（2）三聚氰胺甲醛树脂

三聚氰胺甲醛树脂的合成和用途与脲醛树脂相似。在微碱性条件下，三聚氰胺与甲醛亲核加成，形成羟甲基衍生物，原则上每个氨基可以形成两个羟甲基，1 分子就可能有 6 个羟甲基，但实际上也有不少单羟甲基衍生物存在。加热条件下，三聚氰胺甲醛树脂能交联固化，其中羟基和氨基缩合，可形成亚甲基或亚甲基醚桥交联网状结构。蜜胺树脂的硬度和耐水性均比脲醛树脂好，最大的用途是用来制作色彩鲜艳的餐具，也可制作电器制品。

2.3.10 聚苯醚

聚苯醚，简称 PPO（polyphenylene oxide）或 PPE（polypheylene ether），又称聚(2,6-二甲基-1,4-苯醚)。工业上，聚苯醚以 2,6-二甲基苯酚为单体，以亚铜盐-三级胺类（吡啶）为催化剂，在有机溶剂中，经氧化偶合反应而成。例如，反应器中加入铜氨络合催化剂，鼓泡通入氧气，逐步加入 2,6-二甲基苯酚和乙醇溶液，进行氧化偶联聚合，产品聚合物经离心分离，用含 30% 硫酸的乙醇溶液洗涤，再用稀碱溶液浸泡、水洗、干燥、造粒，即得聚苯醚的粒状树脂。聚苯醚的分子量可达 30000。

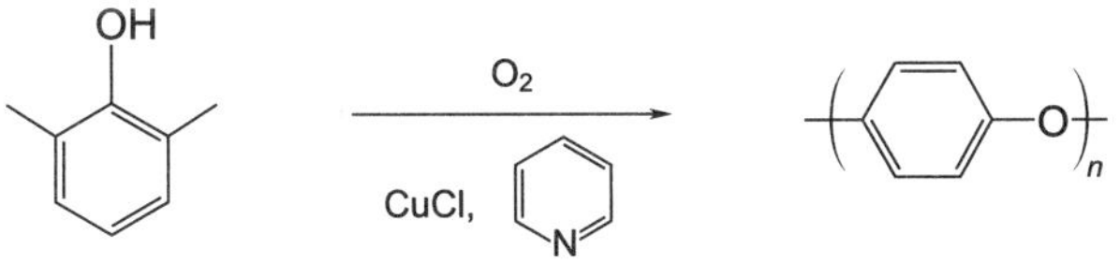

聚苯醚是耐高温塑料，可在 190℃ 下长期使用，其耐热性、耐水解、力学性能、耐蠕变都比聚甲醛、聚酰胺、聚碳酸酯、聚砜等工程塑料好，可用来制作耐热机械零部件。聚苯醚与（抗冲）聚苯乙烯是一对相容性好的聚合物，两者共混（约 1:1~1:2），可提高加工性能和抗冲击性能。添加 5% 磷酸三苯酯后，能够提高阻燃性能。

2,6-二苯基苯酚也可以氧化偶合成相应的聚苯醚，其 $T_g=235℃$，$T_m=480℃$，空气中 175℃ 下稳定，经干纺和高温拉伸，可成晶态纤维。其短纤维加工成纸，可用作超高压电缆的绝缘材料。

在 Lewis 酸催化剂和 $CuCl_2$、$FeCl_3$ 或 $MoCl_5$ 等氧化剂共同作用下，苯经偶合反应，可制成聚亚苯基。聚亚苯基在 500~600℃ 下仍很稳定，但难加工。苯与联苯、二苯、二苯基苯共聚，引入侧苯基，适当破坏其规整性，则可成为可溶性聚合物，在 300~400℃ 时能熔融，但分子量低于 3000，使其应用受到限制。

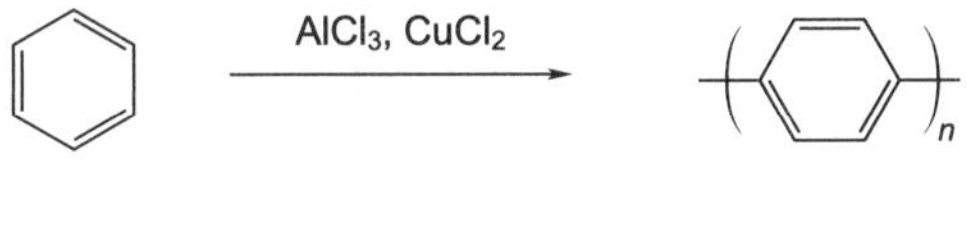

2.3.11 聚砜

聚砜是主链上含有砜基（—SO_2—）的杂链聚合物。比较重要的聚砜是芳香族聚砜，多称作聚芳醚砜，或简称聚芳砜。商业上最常用的聚砜由双酚 A 钠盐和 4,4′-二氯二苯砜缩聚而成。

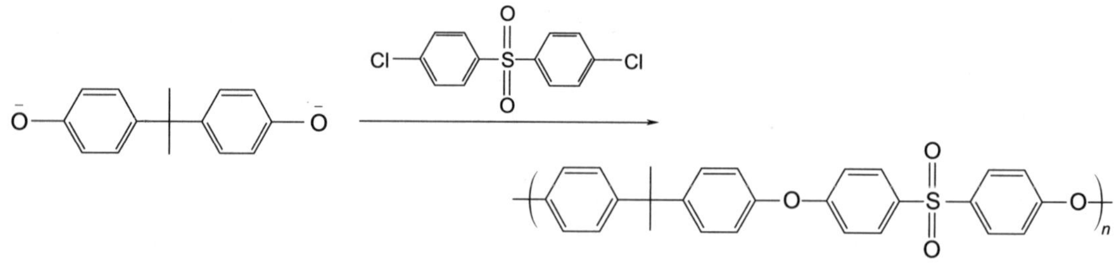

聚砜聚合时，混合双酚 A 和氢氧化钠浓溶液，形成双酚钠盐，所产生的水分经二甲苯蒸馏带走，温度约 160℃，除净水分，防止水解，这是获得高分子量聚砜的关键。以二甲基亚砜为溶剂，用惰性气体保护，使双酚钠与二氯二苯砜进行亲核取代反应，即成聚砜。商品聚砜分子量约为 20000～40000。

苯环的引入，可以提高聚合物的刚性、强度和玻璃化温度，引入砜基可提高耐氧性。与苯环共振赋予砜基热稳定性，醚氧键则赋予大分子链以柔性，异亚丙基对柔性也有一定贡献，这些都改善了其加工性能。上述诸多结构的综合，使双酚 A 聚砜成为高性能的工程塑料。

双酚 A 聚芳砜为无定形线形聚合物，玻璃化温度 195℃，能在 -180～150℃ 温度范围长期使用，耐热性和力学性能都比聚碳酸酯和聚甲醛好，并有良好的耐氧化性能。

无异丙基聚苯醚砜耐氧化性能和耐热性更好，玻璃化温度 180～220℃，在空气中 500℃ 下稳定，可模塑。在 150～200℃ 下，能保持良好的力学性能，在水中有很好的抗碱和抗氧化性。无异丙基聚苯醚砜可以用 $FeCl_3$、$SbCl_5$、$InCl_3$ 作催化剂，通过 Friedel-Crafts 反应制得。

2.3.12 聚芳醚酮

聚芳醚酮是指大分子主链重复单元中亚苯基环通过醚键和酮基连接而成的聚合物。与聚芳砜相似，聚芳醚酮也是性能良好的工程塑料。

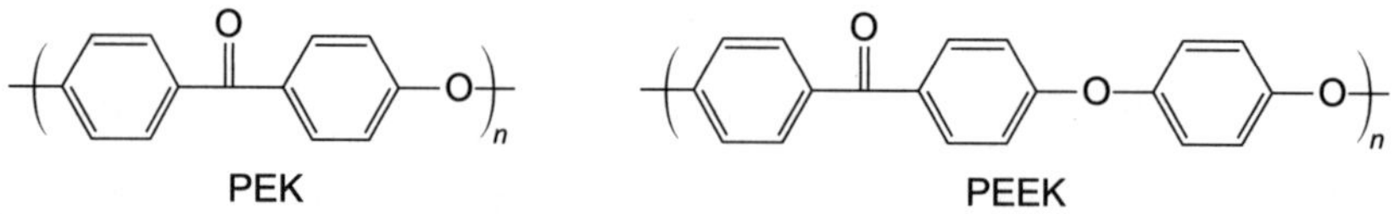

单醚键的聚醚酮（polyether ketone，PEK）和双醚键的聚醚醚酮（polyether ether ketone，PEEK）都耐高温，其玻璃化温度分别为 165℃ 和 143℃，熔融温度为 365℃ 和 334℃，可在 240～280℃ 下连续使用，在水和有机溶剂中使用性能优良。PEEK 热变形温度在 260℃ 左右，有相当好的热稳定性，在 200℃ 下使用寿命可达 $5×10^4$ h。PEEK 具有优良的长期耐蠕变性能和疲劳特性，它在高交变外力作用下经几万次循环受力形变仍保持完好。PEEK 有优良的化学稳定性，除一些如浓硫酸、氯磺酸等强酸外，在常温下可以耐受绝大多数化学试剂。PEEK 具有优良的耐 X 射线、β 射线和 γ 射线性能，能承受高剂量的辐射而不明显地改变其性能，并具有优良的电绝缘性能。PEEK 具有良好的阻燃性，其氧指数较高。

聚芳醚酮可用来制作机械零部件，如汽车轴承、汽缸活塞、泵和压缩机的阀、飞机构件等，使用于条件比较苛刻的环境。

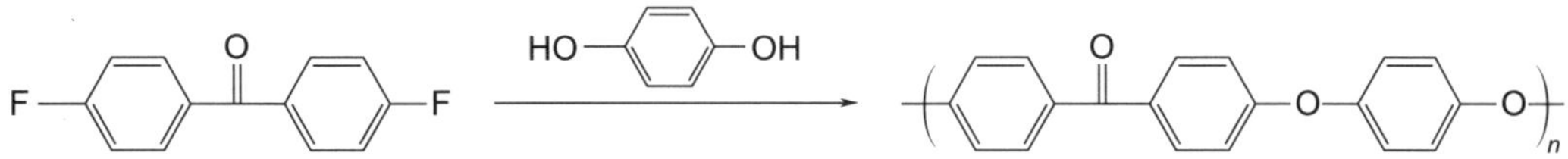

$$F-\phi-CO-\phi-F + HO-\phi-OH \longrightarrow +\phi-CO-\phi-O-\phi-O+_n$$

聚芳醚酮中最重要的是聚醚醚酮，是由4,4′-二氟二苯甲酮、对苯二酚、碳酸钾或碳酸钠在二苯砜溶剂中合成制得的。

2.3.13 聚苯硫醚

聚苯硫醚（polyphenylene sulfide，PPS）为第六大工程塑料和第一大特种工程塑料，属热塑性结晶树脂。其分子主链由苯环和硫原子交替而成，$T_g=85℃$，$T_m=285℃$，耐溶剂，可在220℃以上长期使用。缺点是韧性不够，有一定的脆性。与无机填料、增强纤维以及其他高分子材料复合，可制得各种PPS工程塑料及合金。

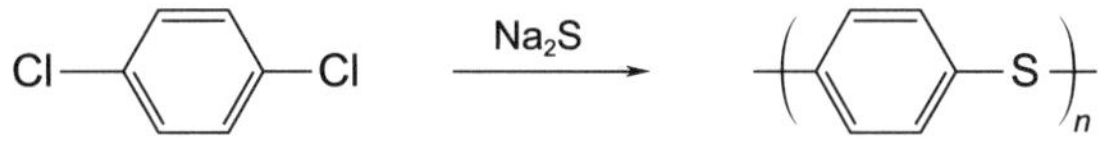

$$Cl-\phi-Cl \xrightarrow{Na_2S} +\phi-S+_n$$

聚苯硫醚由对二氯苯和硫化钠为原料制备，反应属离子机理，但具逐步特性。

在聚苯硫醚分子结构中，刚性苯环与柔顺性硫醚键交替连接而成，分子链有很大的刚性和规整性，因而PPS为结晶性聚合物，具有耐热、阻燃、耐化学药品等诸多优异性能。硫原子上的孤对电子使得PPS树脂与玻璃纤维、无机填料及金属具有良好的亲和性，这样就易于制成各类增强复合材料及合金。聚苯硫醚熔体流动性比较好，容易与其他聚合物共混。例如，和聚酰亚胺共混使用可改善聚酰亚胺的加工性。聚苯硫醚的耐化学药品性优异，200℃以下几乎不溶于任何有机溶剂。除强氧化性酸（如发烟硝酸、氯磺酸、氟酸等）外，可经受各类强酸、碱、盐的侵蚀，是一种耐腐蚀性优异的材料。高温下经各种化学药品浸泡后，强度保持率仍较高。

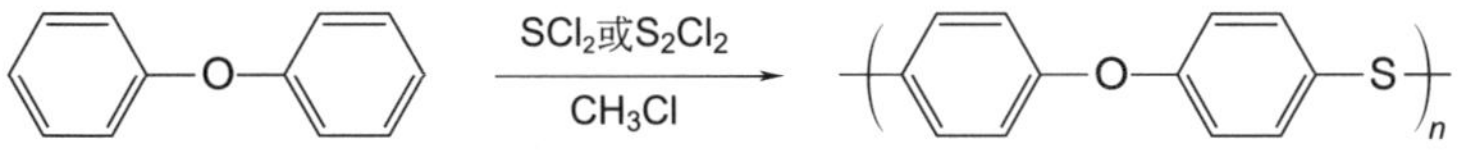

$$\phi-O-\phi \xrightarrow[CH_3Cl]{SCl_2或S_2Cl_2} +\phi-O-\phi-S+_n$$

联苯醚与SCl_2或S_2Cl_2在氯仿溶液中反应，可以制得同时含有醚键和硫键的聚合物，除耐化学药品和耐热外，柔性和加工性能也得到改善。

苯硫醚受热时变化比较复杂，包括氧化、交联和断链。从315℃加热到415℃，物理形态发生多种变化，从熔融、增稠、凝胶化，最后甚至变成不能再熔融的深色固体。聚苯硫醚具有许多加工方法和应用途径，可以从涂料（粉末或浆料）到模塑（注塑、模压、烧结）。

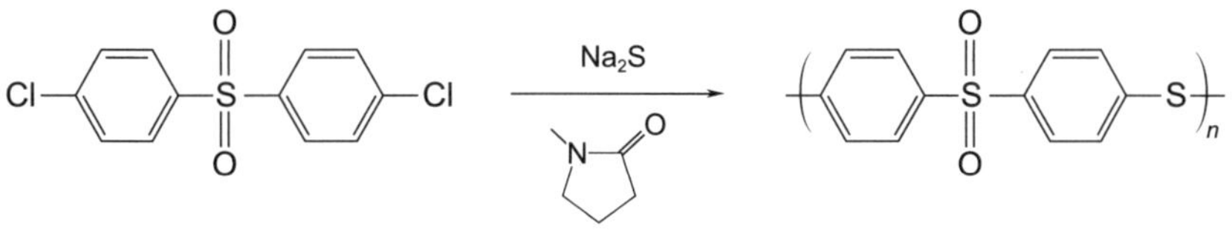

$$Cl-\phi-SO_2-\phi-Cl \xrightarrow{Na_2S, NMP} +\phi-SO_2-\phi-S+_n$$

以N-甲基吡咯烷酮为溶剂，200~220℃下，二氯二苯砜与硫化钠（配比为0.98~1.02）聚合2~8h，还可制得聚苯硫醚砜（PPSS）。聚苯硫醚砜为非结晶性固体，$T_g=215℃$，耐腐蚀、耐辐射、阻燃，热稳定性、抗冲击性能比聚苯硫醚好，可以单独使用，也可以与聚苯硫醚共混改性使用。

英语读译资料

(1) Polymerization

It was Dr W. H. Carothers, an American chemist, who classified polymerization into two groups known as addition polymerization and condensation polymerization.

Addition polymerization involves chain reactions, in which the chain carrier may be a free radical or an ion. A free radical is usually formed by the decomposition of a relatively unstable substance called an initiator, which then initiates chain building. This process occurs very rapidly, often of the order of a second. Addition polymerization is particularly common with double-bonded compounds, such as ethane and its derivatives.

Condensation polymerization is regarded as analogous to condensation reactions in low relative molecular mass materials. In such polymerizations, reaction occurs between two polyfunctional molecules to give one larger polyfunctional molecule accompanied by the elimination of a small molecule, such as water. For example, consider the reaction between the bifunctional monomers ethylene glycol and terephthalic acid.

Note that the product is still bifunctional and so further reaction can now occur, producing a linear polymer, until one of the reactants is used up. It should also be noted that both branched and cross-linked polymers can be formed by both types of polymerization.

The term copolymer refers to a polymer made from polymerizing two or more different suitable monomers together. Different types of copolymers exist. In a random copolymer, the different repeating units are arranged randomly in a chain; in an alternating copolymer, the different units alternate in a particular chain; in a block copolymer, blocks of one type of repeating unit alternate with blocks of another type; finally, in a graft copolymer, blocks of one type of repeating unit are attached or grafted to the backbone of a linear polymer containing the other type of repeating unit.

(2) PET

The most important commercial polyester is poly(ethylene terephthalate), often referred to as PET. Two processes are used for the synthesis of PET, one based on dimethyl terephthalate (DMT) and the other on terephthalic acid (TA). The DMT process was the first to be commercialized because DMT was available in the required purity, but TA was not. That is no longer the case, pure TA is available, and both processes are used.

The DMT process is a two-stage ester interchange process between DMT and ethylene glycol. The first stage is an ester interchange to produce bis(2-hydroxyethyl) terephthalate along with small amounts of larger-sized oligomers. The reactants are heated at temperatures increasing from 150 to 210℃ and the methanol is continuously distilled off.

In the second-stage the temperature is raised to 270-280℃ and polymerization proceeds with the removal of ethylene glycol being facilitated by using a partial vacuum of 66Pa. The first stage of the polymerization is a solution polymerization. The second stage is a melt polymerization since the reaction temperature is above the crystalline melting temperature of the polymer.

The TA process is a modification of the DMT process. Terephthalic acid and an excess of ethylene glycol are used to produce the bis(2-hydroxyethyl)terephthalate, which is then polymerized as described above. The TA process has grown to exceed the DMT process.

Because of its high crystalline melting temperature and stiff polymer chains, PET has good mechanical strength, toughness, and fatigue resistance up to 150-175℃ as well as good chemical, hydrolytic, and solvent resistance. Fiber applications account for about 45% of the total PET production.

(3) Nylon-66

The synthesis of polyamides follows a different route from that of polyesters. Although several different polymerization reactions are possible, polyamides are usually produced either by direct amidation of a diacid with a diamine or the self-amidation of an amino acid. The polymerization of amino acids is not as useful because of a greater tendency toward cyclization. Nylon 66, is synthesized from hexamethylene diamine and adipic acid. A stoichiometric balance of amine and carboxyl groups is readily obtained by the preliminary formation of a 1∶1 ammonium salt in aqueous solution at a concentration of 50%. The salt is often referred to as a nylon salt. Stoichiometric balance can be controlled by adjusting the pH of the solution by appropriate addition of either diamine or diacid. The aqueous salt solution is concentrated to a slurry of approximately 60% or higher salt content by heating above 100℃. Polymerization is carried out by raising the temperature to about 210℃. Reaction proceeds under a steam pressure of 1.7MPa, which effectively excludes oxygen. The pressure also prevents salt precipitation and subsequent polymerization on heat-transfer surfaces.

Unlike polyester synthesis, polyamidation is carried out without an external strong acid, since the reaction rate is sufficiently high without it. Further, amidation may not be an acid-catalyzed reaction. The equilibrium for polyamidation is much more favorable than that for the synthesis of polyesters. For this reason the amidation is carried out without concern for shifting the equilibrium until the last stages of reaction. Steam is released to maintain the pressure at about 1.7MPa, while the temperature is continuously increased to 275℃. When 275℃ is reached the pressure is slowly reduced to atmospheric pressure and heating continued to drive the equilibrium to the right. The later stage of the reaction is a melt polymerization since the reaction temperature is above the T_m.

(4) Epoxy resins

Epoxy resins are a class of versatile polymer materials characterised by the presence of two or more oxirane ring or epoxy groups within their molecular structure. Like other thermosets they also form a network on curing with a variety of curing agents such as amines, anhydrides, thiols etc. Amine curing agents are most widely used because of the better understanding of epoxy-amine reactions.

Undoubtedly, there exists more publications based on the basic and applied research on epoxy resins than that for any other commercially available thermosetting resins. The broad interest in epoxy resins originates from the versatility of epoxy group towards a wide variety of chemical reactions and the useful properties of the network polymers such as high strength, very low creep, excellent corrosion and weather resistance, elevated temperature service capability and adequate electrical properties. Epoxy resins are unique among all the thermosetting resins due to several factors namely, minimum pressure is needed for fabrication of products normally used for thermosetting resins:

① Shrinkage is much lower and hence there is lower residual stress in the cured product than that encountered in the vinyl polymerisation used to cure unsaturated polyester resins.

② Use of a wide range of temperature by judicious selection of curing agent with good control over the degree of crosslinking.

③ Availability of the resin ranging from low viscous liquid to tack free solid, etc.

Because of these unique characteristics and useful properties of network polymers. Epoxy resins are widely used in structural adhesives, surface coatings, engineering composites, and electrical laminates.

(5) Why polycarbonate?

Essentially all CDs, and related audio and video storing devices, have similar components. Here, we will focus on the composition of purchased CDs already containing the desired information and CDs that can be recorded on, CD/Rs. The major material of all of these storage devices is a polycarbonate base. Thus, these devices are polycarbonate, laid over with thin layers of other materials. Of the less than 20g CD weight, over 95% is polycarbonate.

The aromatic rings contribute to the polycarbonate's high glass transition temperature andstiffness. The aliphatic groups temper this tendency giving polycarbonate a decent solubility. The two methyl groups also contribute to the stiffness because they take up space, somewhat hindering free rotation around the aliphatic central carbon moiety. Factors contributing to polycarbonate chain association are interaction between the aromatic rings of different parts of the same or different polycarbonate chain segments and the permanent dipole present within the carbonyl group. The lack of "hydrogen-bonding" hydrogen on polycarbonate means that this type of association is not present. The associations between polycarbonate segments contribute to a general lack of mobility of individual chains. This results in polycarbonate having a relatively high viscosity, which ultimately leads to a low melt flow during processing. The moderate inflexibility, lack of ready mobility, and nonlinear structure contribute to polycarbonate having a relatively long time constant for crystallization. Cooling is allowed to be relatively rapid so that most polycarbonate products possess a large degree of amorphous nature, and it accounts for polycarbonate having a high impact strength that is important in its use to blunt high impacts and important to CDs to provide a semirigid disc that can be dropped and not readily shattered. Thus, control of the rate of flow and cooling is an important factor in producing CD-quality polycarbonate material. A high degree of amorphous nature also contributes to the needed optical transparency, with amorphous polycarbonate having a transparency near that of window glass.

While polycarbonate has the desirable qualities as the basic material for information storage, it also has some debits. First, polycarbonate is relatively expensive in comparison with many polymers. Its superior combination of properties and ability for a large cost markup allows it to be an economically feasible material for specific commercial uses. Second, the polar backbone is susceptible to long-term hydrolysis so that water must be ruthlessly purged. The drying process, generally 4h, is often achieved by placement of polycarbonate chips in an oven at 120℃ with a dew point of －18℃.

The polycarbonate utilized for information storage has strict requirements including high purity, greater than 87% spectral light transmission based on a 4 mm thick sample, yellowness index less than 2, and light scattering less than 0.3cd/(m^2-lx). The two main sources of polycarbonate are virgin and recycled. Virgin polycarbonate has a yellowness index of 1.8 but the first reground polycarbonate has a yellow index of about 3.5. Thus, CDs employ only virgin polycarbonate.

Requirements for CD-quality material are polycarbonate with low levels of chemical impurities, low particle levels, thermal stability, excellent mold release, excellent clarity, as well as constant flow and constant mechanical behavior (for reproducibility). There exists a time/cost balance. High molecular weight polycarbonate offers a little increase in physical property but the flow rate is slow, making rapid production of CDs difficult. The molecular weight where good mechanical strength and reasonable flow occurs, and that allows for short cycles, is in the range of 16000-28000Da.

Injection molding requires the barrel temperature to be about 350℃ with a barrel pressure in excess of 138MPa. The mold is maintained at 110℃ to ensure uniform flow and high definition, and to discourage an uneven index of refraction, birefringence. The CD is about four one-hundredths of an inch (0.5mm) thick. For prerecorded CDs, the PC is compression-molded on a stamper imprinted with the recorder information. This takes about 4 sec. Once the clear piece of polycarbonate is formed, a thin, reflective aluminum layer is sputtered onto the disc. Then, a thin acrylic layer is sprayed over the aluminum to protect it. The label is then printed onto the acrylic surface and the CD is complete.

(6) The microbial polyhydroxyalkanoates

Bacterial polyhydroxyalkanoates (PHAs) are a unique family of polymers that act as a carbon/energy store for more than 300 species of Gram-positive and Gramnegative bacteria as well as a wide range of archaea. Synthesized intracellularly as insoluble cytoplasmic inclusions in the presence of excess carbon when other essential nutrients such as oxygen, phosphorous or nitrogen are limited, these polymeric materials are able to be stored at high concentrations within the cell since they do not substantially alter its osmotic state. The resulting polymers are piezoelectric and perfectly isotactic/optically active (having only the (R)-configuration). They are hydrophobic, water-insoluble, inert and indefinitely stable in air, and are also thermoplastic and/or elastomeric, nontoxic and have very high purity within the cell. PHA has a much better resistance to UV degradation than polypropylene but is less solvent resistant. Most importantly, these biopolymers are completely biodegradable. To date more than 150 different monomer units have been incorporated into biological PHA, and the polymer properties available within this family are as a result very broad. In general, PHAs can be divided into two main groups, these being the short-chain-length PHAs (scl-PHAs) that contain monomer units with 3-5 carbon atoms, and the medium-chain-length PHAs (mcl-PHAs) that contain monomer units of 6-18 carbon atoms. The most common PHAs are poly (3-hydroxybutyrate) (P3HB) and poly(3-hydroxybutyrate-co-3-hydroxyvalerate) (P(3HB-co-3HV)). These materials have mechanical properties that are comparable to those of polypropylene and polyethylene, although they have much lower elongation to break and are more brittle.

The French bacteriologist Lemoigne first isolated and characterized poly-3-hydroxybutyrate (P(3HB)) from bacteria between 1923 and 1927, and showed that this extract could be cast into a transparent film. Although it was some time before this discovery was turned into a practical outcome, PHA production using pure cultures still has a long history. Stanier and Wilkinson and their co-workers were responsible for some of the initial fundamental research into the mechanisms of PHA biosynthesis, beginning in 1957. This was followed in 1959 by the first attempted commercialization, when W. R. Grace and Company patented a P(3HB) production process using bacteria, although production inefficiencies, poor thermal stabilities and a lack of available extraction technologies limited this application. In 1970, Imperial Chemical Industries Ltd. commercialized the production of P(3HB-co-3HV) under the trade name of BiopolTM, with the technology since being sold to Monsanto and then to Metabolix. Since that time, there have been a range of new technologies developed, and recent focus within pure culture production has been on the synthesis of alternative copolymers such as P(3HB-co-3HHx), a wide range of functionalized PHAs (incorporating monomer units with novel and active functionalities on the side chains) and also on the use of metabolic engineering to reengineer the central metabolism of PHA producers for more efficient PHA production. In its simplest form, PHA production using pure cultures adopts a two-stage batch production process, with an inoculum of bacteria being introduced into a sterile solution of trace metal nutrients and a suitable carbon source and nutrients in the first (growth) stage. In the second stage, an essential nutrient (such as N, P or O_2) is deliberately limited and PHA accumulation takes place. The properties of the final polymer depends on the mix of carbon feed stocks fed during accumulation, the metabolic pathways that the bacteria use for the following conversion into precursors, and the substrate specificities of the enzymes involved.

习 题

1. 简述线形缩聚和体形缩聚的关系和区别。
2. 简述线形缩聚的逐步机理，转化率和反应程度的关系。
3. 简述官能团等活性理论的适用性和局限性。
4. 简述缩聚反应中聚合度与平衡常数、副产物残留量之间的关系。

5. 影响线形缩聚物聚合度有哪些因素？如何控制聚合度？

6. 简单比较熔融缩聚和固相缩聚、溶液缩聚和界面缩聚的特征。

7. 聚酯化和聚酰胺化的平衡常数有何差别，对缩聚条件有何影响？

8. 简述不饱和聚酯的配方原则和固化原理。

9. 简述涤纶聚酯合成的两条技术路线及其选用原则，聚合度控制方法。

10. 比较聚碳酸酯两条合成路线以及聚合度控制方法。

11. 比较聚酰胺-66和聚酰胺-6的合成方法。

12. 为什么合成全芳聚酰胺的条件比脂肪族聚酰胺苛刻？

13. 合成聚酰亚胺时，为什么要采用两步法？

14. 聚氨酯合成为什么多采用异氰酸酯路线？写出两种二异氰酸酯和两种多元醇及其聚氨酯的合成反应式。

15. 软、硬聚氨酯泡沫塑料的发泡原理有何差异？

16. 简述环氧树脂的合成原理和固化原理。

17. 简述聚芳砜的合成原理。

18. 比较聚苯醚和聚苯硫醚的结构、主要性能和合成方法。

19. 从原料配比、预聚物结构、预聚条件、固化特性等方面来比较碱催化和酸催化酚醛树脂。

20. 简述合成脲醛树脂的工艺条件。

21. 简述并解释以下聚酰胺的熔点和分子结构的关系。

结构	T_m/℃	结构	T_m/℃
聚酰胺-66	265	间苯二甲酰间苯二胺	371
对苯二甲酰己二胺	370	对苯二甲酰对苯二胺	530
对苯二甲酰戊二胺	430		
间苯二甲酰戊二胺	250		

22. α-羟基酸进行线形缩聚，测得聚合物的重均分子量为18400，试计算：(1) 羧基已经酯化的百分比；(2) 数均聚合度；(3) 结构单元数。

23. 等物质的量的二元醇和二元酸经外加酸催化缩聚，试证明从开始到 $p=0.98$ 所需的时间与 p 从0.98到0.99的时间相近。计算自催化和外加酸催化聚酯化反应时不同反应程度 p 下 X_n、$[c]/[c]_0$ 与时间 t 的关系，列表作图来说明。

24. 由1mol丁二醇和1mol己二酸合成 $M_n=5000$ 的聚酯，试计算：

(1) 终止缩聚时的反应程度 p；

(2) 缩聚过程中如果有 0.5%（摩尔分数）丁二醇脱水成乙烯而损失，求到达同一反应程度时的聚合度；

(3) 如何补偿丁二醇的脱水损失才能获得 $M_n=5000$ 的聚酯？

(4) 假定原始混合物中羧基的总浓度为 2mol，其中 1.0% 为乙酸，无其他因素影响两基团数量比，求获得同一数均聚合度所需的反应程度 p。

25. 等物质的量的乙二醇和对苯二甲酸在 280℃下封管内进行缩聚，平衡常数 $K=4$，求最终 X_n。另外在排除副产物水的条件下缩聚，欲得 $X_n=100$，问体系中残留水分有多少？

26. 等物质的量的二元醇和二元酸缩聚，另加乙酸 1.5%（摩尔分数），$p=0.995$ 或 0.999 时，求聚合度。

27. 己二胺和己二酸合成聚酰胺，反应程度 $p=0.995$，分子量 15000，试计算原料比。

28. 邻苯二甲酸酐与甘油或季戊四醇缩聚，两种基团数相等，试求：（1）平均官能度；（2）按 Carothers 法求凝胶点。

29. 按 Carthers 法计算下列混合物的凝胶点：（1）邻苯二甲酸酐和甘油物质的量比为 1.50∶0.98；（2）邻苯二甲酸酐、甘油、乙二醇物质的量比为 1.50∶0.99∶0.002。

30. 用二亚乙基三胺使 1000g 环氧树脂（环氧值 0.2）固化，计算固化剂用量。

31. 2.5mol 邻苯二甲酸酐、1.0mol 乙二醇、1.0mol 丙三醇体系进行缩聚，为控制凝胶点需要，在聚合过程中定期测定树脂的熔点、酸值 [mg(KOH)/g 试样]、溶解性能，试计算反应酸值多少时会出现凝胶。

32. 醇酸树脂配方为：1.21mol 季戊四醇，0.50mol 邻苯二甲酸酐，0.49mol 丙三羧酸，能否不产生凝胶而反应完全？

英语习题

1. The repeat unit structures of three polymers, labelled (a)-(c), are shown below. For each polymer, write down the chemical structures of the monomers from which the polymer could be prepared.

(a) $\left[O-\overset{O}{\underset{\|}{C}}-NH-\underset{}{\bigcirc}-CH_2-\underset{}{\bigcirc}-NH-\overset{O}{\underset{\|}{C}}-O-CH_2-CH_2-CH_2-CH_2 \right]_n$

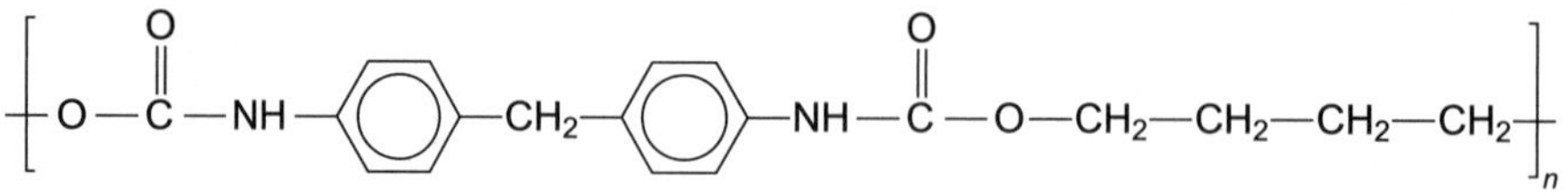

(b) $\left[O-\overset{O}{\underset{\|}{C}}-\underset{}{\bigcirc}-\overset{O}{\underset{\|}{C}}-O-CH_2-CH_2-CH_2-CH_2-CH_2 \right]_n$

(c) $\left[\overset{O}{\underset{\|}{C}}-NH-(CH_2)_7-NH-\overset{O}{\underset{\|}{C}}-(CH_2)_5 \right]_n$

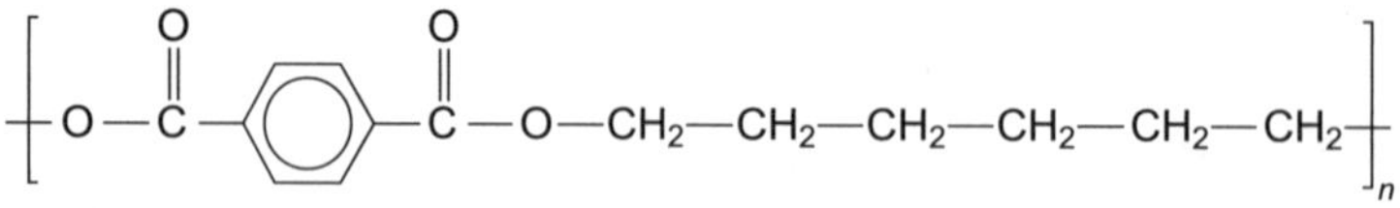

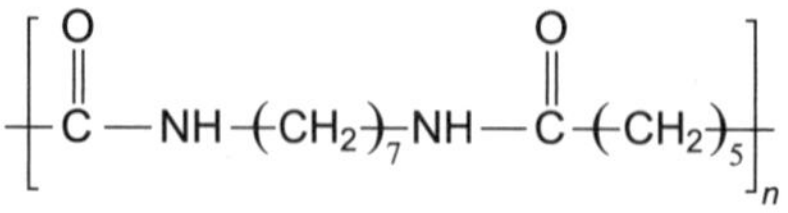

2. Neglecting the contribution of end groups to polymer molar mass, calculate the percentage conversion of functional groups required to obtain a polyester with a number-average

molar mass of 24000g/mol from the monomer $HO(CH_2)_{14}COOH$.

3. A polyamide was prepared by bulk polymerization of hexamethylene diamine (9.22g) with adipic acid (11.68g) at 280℃. Analysis of the whole reaction product showed that it contained 2.6×10^{-3} mol of carboxylic acid groups. Evaluate the number-average molar mass M_n of the polyamide, and also estimate its weight-average molar mass M_w by assuming that it has the most probable distribution of molar mass.

4. One kilogram of a polyester with a number-average molar mass $M_n = 10000$g/mol is mixed with one kilogram of another polyester with a $M_n = 30000$g/mol. The mixture then is heated to a temperature at which ester interchange occurs and the heating continued until full equilibration is achieved. Assuming that the two original polyester samples and the new polyester, produced by the ester interchange reaction, have the most probable distribution of molar mass, calculate M_n and weight-average molar mass M_w for the mixture before and after the ester interchange reaction.

5. Consider 1 ∶ 1 stoichiometric polymerization of 1,4-diaminobutane with sebacoyl chloride.

(a) Write down a balanced chemical equation for the polymerization.

(b) Name the polymer formed by this polymerization.

(c) Using the Carothers equation, calculate the extent of reaction required to produce this polymer with a number-average molar mass $M_n = 25.0$kg/mol. You should neglect the effects of end groups on molar mass.

(d) Explain briefly the consequences of including 1,1,4,4-tetraminobutane in the polymerization with an exact stoichiometric ratio of 2 ∶ 1 ∶ 4 of 1,4-diaminobutane to 1,1,4,4-tetraminobutane to sebacoyl chloride.

6. A polycondensation reaction takes place between 1.2mol of a dicarboxylic acid, 0.4mol of glycerol (a triol) and 0.6mol of ethylene glycol (a diol).

(a) Calculate the critical extents of reaction for gelation using the Carothers theory.

(b) Comment on the observation that the measured value of the critical extent of reaction is 0.866.

7. Poly(hexamethylene adipamide) (Nylon-66) was synthesized by condensation polymerization of hexamethylenediamine and adipic acid in 1 ∶ 1 mole ratio. Calculate the acid equivalent of the polymer whose average DP is 440.

8. Name the polamides made from the following monomers and draw their structural formulas (one repeating unit).

(a) caprolactam; (b) ω-aminoundecanoic acid; (c) dodecyl lactam; (d) ethylene diamine and sebacic acid, and (e) ethylene diamine and decanedioic acid.

9. Represent, by showing a repeating unit, the structure of the polymer which would be obtained by polymerization of the following monomers:

(a) ω-aminolauric acid; (b) lauryl lactam; (c) ethylene oxide; (d) ethylene glycol and terephthalic acid; (e) hexamethylene diamine and sebacic acid; (f) m-phenylene diamine and isophthaloyl chloride.

10. Explain the fact that conversion of the amide groups —CONH— in nylon to methylol groups —CON(CH$_2$OH)— by reaction with formaldehyde, followed by methylation to ethergroups —CON(CH$_2$OCH$_3$)— results in the transformation of the fiber to a rubbery product with low modulus and high elasticity.

11. Show that the time required to go from $p=0.98$ to $p=0.99$ is very close to the time to reach $p=0.98$ from the start of polymerization for the external acid-catalyzed polymerization of an equimolar mixture of a diol and diacid.

12. The polymerization between equimolar amounts of a diol and diacid proceeds with an equilibrium constant of 200. What will be the expected degree of polymerization and extent of reaction if the reaction is carried out in a closed system without removal of the by-product water? To what level must [H$_2$O] be lowered in order to obtain a degree of polymerization of 200 if the initial concentration of carboxyl groups is 2mol/L?

13. A heat-resistant polymer Nomex has a number-average molecular weight of 24116. Hydrolysis of the polymer yields 39.31% by weight m-aminoaniline, 59.81% terephthalic acid, and 0.88% benzoic acid. Write the formula for this polymer. Calculate the degree of polymerization and the extent of reaction. Calculate the effect on the degree of polymerization if the polymerization had been carried out with twice the amount of benzoic acid.

14. Calculate the number-average degree of polymerization of an equimolar mixture of adipic acid and hexamethylene diamine for extents of reaction 0.500, 0.800, 0.900, 0.950, 0.970, 0.990, 0.995.

15. Calculate the feed ratio of adipic acid and hexamethylene diamine that should be employed to obtain a polyamide of approximately 15000 molecular weight at 99.5% conversion. What is the identity of the end groups of this product? Do the same calculation for a 19000-molecular-weight polymer.

16. What proportion of benzoic acid should be used with an equimolar mixture of adipic acid and hexamethylene diamine to produce a polymer of 10000 molecular weight at 99.5% conversion? Do the same calculation for 19000 and 28000 molecular weight products.

17. Show by equations the polymerization of melamine and formaldehyde to form a crosslinked structure.

18. Calculate the extent of reaction at which gelation occurs for the following mixtures:

(1) Phthalic anhydride and glycerol in stoichiometric amounts.

(2) Phthalic anhydride and glycerol in the molar ratio 1.500 : 0.980.

(3) Phthalic anhydride, glycerol, and ethylene glycol in the molar ratio 1.500 : 0.990 : 0.002.

(4) Phthalic anhydride, glycerol, and ethylene glycol in the molar ratio 1.500 : 0.500 : 0.700.

Compare the gel points calculated from the Carothers equation (and its modifications) with those using the statistical approach. Describe the effect of unequal functional groups reactivity (e.g., for the hydroxyl groups in glycerol) on the extent of reaction at the gel point.

第 3 章
烯烃类单体的连锁聚合

3.1 自由基聚合

自由基聚合　活性/可控自由基聚合

碳纤维自行车架从竞赛用车逐渐走向普通市场。它的质量非常轻，只有 1.2kg 左右，但是能够比合金车架更好地吸收震动，并保持刚性。用于增强的碳纤维布由聚丙烯腈纤维面料经碳化处理而得，拉伸强度可达 3.0GPa，拉伸弹性模量达 200GPa。聚丙烯腈采用自由基聚合工艺进行工业化生产。

本节目录

3.1.1　加成聚合和连锁聚合 / 85
3.1.2　烯烃类单体的结构及聚合特性 / 85
3.1.3　自由基活性及化学反应 / 86
3.1.4　自由基聚合机理 / 87
3.1.5　链引发反应和引发剂 / 88
3.1.6　聚合速率 / 94
3.1.7　动力学链长和聚合度 / 96
3.1.8　阻聚作用和阻聚剂 / 97
3.1.9　活性自由基聚合 / 99
3.1.10　聚合方法 / 101
3.1.11　自由基聚合重要聚合物 / 108

重点要点

加成聚合，连锁聚合，自由基聚合，自由基聚合机理，动力学链长，链转移，偶合终止，歧化终止，阻聚剂，可控"活性"自由基聚合，ATRP，RAFT，本体聚合，溶液聚合，悬浮聚合，乳液聚合。

3.1.1 加成聚合和连锁聚合

加成聚合（addition polymerization）大多按连锁聚合（chain polymerization）机理进行，一般用于含有烯烃官能团（viny group）的单体聚合（图 3-1）。

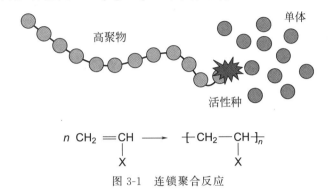

图 3-1 连锁聚合反应

连锁聚合反应依靠高活性的活性种与单体进行加成反应，活性种可由不稳定的引发剂或通过外加能量诱导共价键断裂产生。

如图 3-2 所示，共价键可以按均裂和异裂两种方式分解。均裂时，形成各带 1 个单电子的 2 个中性自由基（free radical）。异裂时，共价键上一对电子全属于某一基团，形成阴（负）离子；另一个基团就成为缺电子的阳（正）离子。

$$R : R \longrightarrow 2R^\cdot$$

$$A : B \longrightarrow A^+ + B^-$$

图 3-2 可引发连锁聚合反应的活性种

自由基、阴离子、阳离子都可能成为活性种，打开烯类的 π 键，从而引发聚合。烯烃类单体的连锁聚合形式大致有自由基聚合（free radical polymerization）、阳离子聚合（cationic polymerization）、阴离子聚合（anionic polymerization）和配位聚合（coordination polymerization）等。配位聚合的反应机理以及反应动力学特征与阴离子聚合相似，可以归为阴离子聚合。

3.1.2 烯烃类单体的结构及聚合特性

单体对聚合机理的选择与其分子结构有关，其中共轭效应和诱导效应的影响较大，位阻效应对聚合速率有一定的影响，但不会影响聚合机理的选择。

一般而言，如果烯烃具有两个取代基，并位于烯键两侧（CHX=CHY），该单体因位阻效应较大而难以聚合。取代基位于烯键一侧的 CH_2=CXY 和 CH_2=CHX 这两种结构一般都可聚合。烯类单体分子结构中的供电子取代基团（electron-donating group），如烷氧基、烷基、氨基等，可使 C=C 双键电子云密度增加，有利于阳离子的进攻，倾向于阳离子聚合。

烯类单体中的吸电子基团（electron-withdrawing group），如氰基、羰基（醛、酮、酸、酯等），使双键 π 电子云密度降低，并使阴离子活性种共振稳定，有利于阴离子聚合，聚合

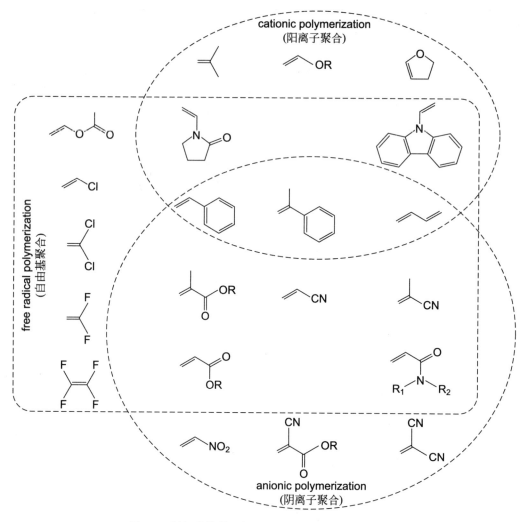

图 3-3 烯烃类单体机构和连锁聚合反应机理的关系

反应机理的关系见图 3-3。

大部分烯类单体都能按自由基聚合机理聚合。当取代基团的吸电性或供电性过强，就只能进行离子聚合。

带有共轭体系的烯类，如苯乙烯、α-甲基苯乙烯、丁二烯、异戊二烯等，电子流动性较大，易诱导极化，能按阳离子、阴离子、自由基三种机理进行聚合。

3.1.3 自由基活性及化学反应

碳自由基的碳原子基团有 7 个电子，其中有一个单电子在 sp^2 杂化轨道，共轭效应对自由基的反应活性具有较大影响。

就稳定性而言，叔碳自由基最稳定，甲基自由基最活泼。

$$H_3C-\underset{CH_3}{\overset{CH_3}{C}}\cdot \; > \; H_3C-\underset{H}{\overset{CH_3}{C}}\cdot \; > \; H-\underset{H}{\overset{CH_3}{C}}\cdot \; > \; H-\underset{H}{\overset{H}{C}}\cdot$$

碳自由基如和共轭结构相连，稳定性增强。与3个苯环共轭的碳自由基非常稳定，无引发能力，是一种阻聚剂。

自由基可以和其他有机物按取代（substitution）、加成（addition）、消除（elimination）三种反应机理进行反应。分别对应自由基聚合中的链转移、链增长、歧化链终止反应。

$$\text{取代} \quad X\cdot \; Y-Z \longrightarrow X-Y + Z\cdot$$

$$\text{加成} \quad X\cdot \; Y=Z \longrightarrow X-Y-Z\cdot$$

$$\text{消除} \quad \cdot X-Y-Z \longrightarrow X=Y + Z\cdot$$

3.1.4 自由基聚合机理

自由基聚合反应由链引发（initiation）、链增长（propagation）、链终止（termination）、链转移（chain transfer）等四个基本反应构成。

链引发

$$R-R \longrightarrow R\cdot + R\cdot \tag{1}$$

$$R\cdot + H_2C=\underset{X}{CH} \longrightarrow R-\underset{H}{\overset{H}{C}}-\overset{\cdot}{C}\underset{X}{H} \tag{2}$$

链增长

$$\sim\sim\underset{H}{\overset{H}{C}}-\overset{\cdot}{C}\underset{X}{H} + H_2C=\underset{X}{CH} \longrightarrow \sim\sim\underset{H}{\overset{H}{C}}-\underset{X}{\overset{H}{C}}-\underset{H}{\overset{H}{C}}-\overset{\cdot}{C}\underset{X}{H} \tag{3}$$

链终止

偶合终止

$$\sim\sim\underset{H}{\overset{H}{C}}-\overset{H}{C}H + H\overset{H}{C}-\underset{H}{\overset{H}{C}}\sim\sim \longrightarrow \sim\sim\underset{H}{\overset{H}{C}}-\underset{X}{\overset{H}{C}}-\underset{X}{\overset{H}{C}}-\underset{H}{\overset{H}{C}}\sim\sim \tag{4}$$

歧化终止

$$\sim\sim\underset{H}{\overset{H}{C}}-\overset{H}{C}H + H\overset{H}{C}-\underset{H}{\overset{H}{C}}\sim\sim \longrightarrow \sim\sim\underset{H}{\overset{H}{C}}-\underset{X}{\overset{H}{C}}-H + \underset{X}{\overset{H}{C}}=\underset{H}{\overset{H}{C}}\sim\sim \tag{5}$$

链转移

$$\sim\sim\underset{H}{\overset{H}{C}}-\overset{\cdot}{C}H + A-Y \longrightarrow \sim\sim\underset{H}{\overset{H}{C}}-\underset{X}{\overset{Y}{C}}H + A\cdot \tag{6}$$

(1) 链引发

链引发是形成单体自由基的反应，由下列两步反应组成。第一步引发剂分解，形成初级自由基 R·。第二步初级自由基与单体加成，形成单体自由基。

第一步引发剂分解是吸热反应，活化能高，约 105～150kJ/mol，反应速率小，分解速率常数仅为 10^{-4}～10^{-6} s^{-1}。

第二步是放热反应，活化能低，反应速率大。其活化能和反应速率均与链增长反应相当。

(2) 链增长

单体自由基与烯类单体分子进行加成反应，形成新自由基并继续与单体连锁加成。通过链增长反应，聚合产物的分子链不断增长。

链增长反应具有强放热特征，一般烯类单体聚合热为 55～95kJ/mol。链增长反应活化能低，20～34kJ/mol，反应速率极快，在 10^{-1}～10^{1} s 内就可使聚合度达到 10^3～10^4。

在链增长反应中，两结构单元的键接以"头-尾"为主，偶尔会间有"头-头"（或"尾-尾"）键接。例如，苯乙烯聚合容易头-尾键接。因为头-尾键接时，苯基与单电子接在同一碳原子上，形成共轭体系，对自由基有稳定作用。另一方面，亚甲基一端的位阻较小，也有利于头-尾键接。两种键接方式的活化能差达 34～42kJ/mol。

(3) 链终止

两个自由基相互作用而终止活性，有偶合和歧化两种方式。

偶合终止是两自由基的单电子结合成共价键的终止方式，聚合形成的高分子链的聚合度是聚合链自由基结构单元数的两倍，大分子两端均为引发剂残基 R。

歧化终止是某自由基夺取另一自由基的氢原子或其他原子而终止的方式。歧化终止所得到的高分子的聚合度与聚合链自由基的结构单元数相同，只有一端是引发剂残基，另一端为饱和或不饱和，两者各半。

链终止反应活化能很低，仅为 8～21kJ/mol。终止反应速率常数极高，为 10^6～10^8 L/(mol·s)。但由于在聚合反应后期，体系黏度较高，双基终止受扩散控制。

(4) 链转移

链自由基还有可能从单体、引发剂、溶剂或大分子上夺取一个原子而终止，同时将电子转移给相应分子而成为新自由基，继续新链的增长。自由基向低分子转移往往导致聚合物分子量降低。链自由基向大分子转移，大多形成支链。

3.1.5 链引发反应和引发剂

自由基聚合引发剂可在聚合温度以适当的分解速率分解，生成自由基，引发单体聚合。链引发是控制聚合反应速率和分子量的主要因素。

(1) 引发剂的种类

常用于自由基聚合的引发剂有偶氮类和过氧化物两大类化合物。过氧化物还可以和适当的还原剂搭配组合，按氧化还原引发体系使用。引发剂也可根据溶解性分为油溶性和水溶性两类。

① 偶氮类引发剂。偶氮二异丁腈（AIBN）是代表性的偶氮类引发剂，其热分解反应式如下：

$$\underset{CN}{\underset{|}{H_3C-\overset{CH_3}{\overset{|}{C}}}}-N=N-\underset{CN}{\underset{|}{\overset{CH_3}{\overset{|}{C}}-CH_3}} \xrightarrow{\triangle} 2\ \underset{CN}{\underset{|}{H_3C-\overset{CH_3}{\overset{|}{C}}\cdot}} + N_2$$

AIBN 多在 45~80℃使用，其分解反应的特点是呈一级反应，无诱导分解，只产生一种自由基，因此广泛用于聚合动力学研究。AIBN 虽然在低温下比较稳定，但 80~90℃下会剧烈分解，甚至产生爆炸。

偶氮类引发剂分解时有氮气产生，可利用氮气的放出速率来测定其分解速率，计算半衰期。工业上还可用作泡沫塑料的发泡剂和光聚合的光引发剂。

② 过氧化合物。过氧化二苯甲酰（BPO）是常用的过氧类引发剂，其活性与 AIBN 相当。BPO 中 O—O 键的电子云密度大而相互排斥，容易断裂，用于 60~80℃下的聚合反应。

BPO 按两步分解。第一步均裂成苯甲酸基自由基，可引发单体聚合。在无单体时，容易进一步分解成苯基自由基，并析出二氧化碳气体。

$$Ph-\overset{O}{\overset{\|}{C}}-O-O-\overset{O}{\overset{\|}{C}}-Ph \xrightarrow{\triangle} 2\ Ph-\overset{O}{\overset{\|}{C}}-O\cdot \longrightarrow 2\ Ph\cdot + 2CO_2$$

过氧类引发剂种类很多，活性差别很大，可适用于不同温度下的聚合反应。其中也有活性较高可在较低温度下使用的引发剂，例如过氧化二碳酸二乙基己酯（EHP）。

过硫酸盐类引发剂为水溶性，多用于乳液聚合和水溶液聚合。过硫酸钾和过硫酸铵是这类引发剂的代表。有效分解温度在 60℃以上，在酸性（pH<3）介质中，分解加速。

$$^-O-\overset{O}{\underset{O}{\overset{\|}{S}}}-O-O-\overset{O}{\underset{O}{\overset{\|}{S}}}-O^- \longrightarrow 2\ ^-O-\overset{O}{\underset{O}{\overset{\|}{S}}}-O\cdot$$

③ 氧化还原体系。过氧化物和还原剂组成氧化还原引发体系使用时，可在室温下产生自由基，引发单体聚合。氧化还原体系的引发反应活化能低，在较低温度下也能获得较快的速率。

a. 水溶性氧化还原引发体系　该体系的氧化剂组分有过氧化氢、过硫酸盐、氢过氧化物等，还原剂则有无机还原剂（Fe^{2+}、Cu^+、亚硫酸盐、硫代硫酸盐等）和水溶性的有机还原剂（醇、胺、草酸、葡萄糖等）。过氧化氢、过硫酸钾、异丙苯过氧化氢单独热分解的活化能分别为 220kJ/mol、140kJ/mol、125kJ/mol，与亚铁盐构成氧化还原体系后，活化能却降为 40kJ/mol、50kJ/mol、50kJ/mol，能够在 5℃下引发聚合并达到较高聚合速率。

$$^-O-\overset{O}{\underset{O}{\overset{\|}{S}}}-O-O-\overset{O}{\underset{O}{\overset{\|}{S}}}-O^- + Fe^{2+} \longrightarrow\ ^-O-\overset{O}{\underset{O}{\overset{\|}{S}}}-O\cdot +\ ^-O-\overset{O}{\underset{O}{\overset{\|}{S}}}-O^- + Fe^{3+}$$

$$HO-OH + Cu^+ \longrightarrow OH^- + \cdot OH + Cu^{2+}$$

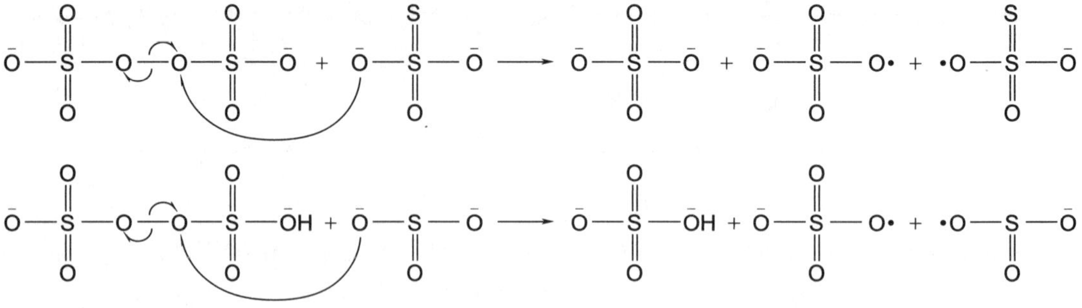

$$\text{示意反应式(过硫酸盐分解,略)}$$

四价铈盐和醇、醛、酮、胺等也可以组成氧化还原体系,有效地引发烯类单体聚合或接枝聚合。在淀粉接枝丙烯腈制备水溶性高分子时,常采用这一引发体系,葡萄糖单元中的醇羟基或醛基参与氧化还原反应。

b. 油溶性氧化还原体系 该体系的氧化剂有氢过氧化物、过氧化二烷基、过氧化二酰基等,还原剂有叔胺、环烷酸盐、硫醇、有机金属化合物(如三乙基铝、三乙基硼等)。过氧化二苯甲酰/N,N-二甲基苯胺是常用体系,可用来引发甲基丙烯酸甲酯共聚合,制备牙科自凝树脂和骨水泥。

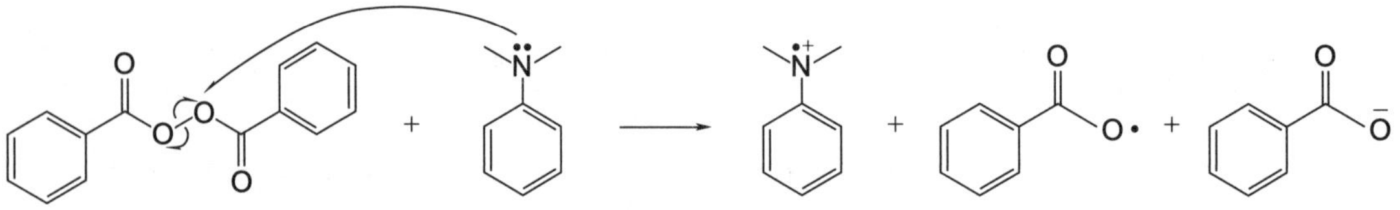

$$\text{BPO} + N,N\text{-二甲基苯胺} \longrightarrow \text{产物}$$

90℃下,BPO 在苯乙烯中的分解速率常数 $k_d = 1.33 \times 10^{-4} \, \text{s}^{-1}$,而 BPO/$N,N$-二甲基苯胺体系在 60℃时的 k_d 可达 $1.25 \times 10^{-2} \, \text{L/(mol·s)}$,30℃时 k_d 也有 $2.29 \times 10^{-3} \, \text{L/(mol·s)}$,表明 BPO/$N,N$-二甲基苯胺体系的引发活性更高,可在室温下使用。

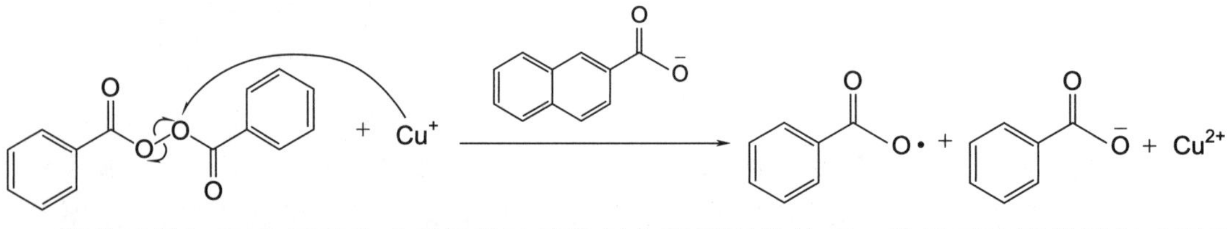

$$\text{BPO} + \text{Cu}^+ + \text{萘酸根} \longrightarrow \text{PhCOO·} + \text{PhCOO}^- + \text{Cu}^{2+}$$

萘酸亚铜与 BPO 可以构成高活性油溶性氧化还原引发体系,常用于油漆催干剂或不饱和树脂的室温自由基聚合引发剂。

(2) 引发剂分解动力学

自由基聚合反应中,链引发是最慢的一步,是影响速率和分子量的关键因素。

引发剂分解速率 R_d 与引发剂浓度 [I] 成正比:

$$\text{I} \longrightarrow 2\text{R·}$$

$$R_d = -\frac{\text{d}[I]}{\text{d}t} = k_d[I]$$

式中,负号代表引发剂浓度随时间增加而减少;k_d 是分解速率常数,s^{-1}。

上式经积分得:

$$\ln \frac{[I]}{[I]_0} = -k_d t$$

式中,$[I]_0$、$[I]$ 分别代表起始($t=0$)和时间为 t 时的引发剂浓度,mol/L;$[I]/[I]_0$

代表引发剂残留率，随时间呈指数关系而衰减。

固定温度下，以 $\ln([I]/[I]_0)$ 对 t 作图，由直线斜率即可求得引发剂分解速率常数 k_d。对于偶氮类引发剂，可以测定分解时析出的氮气量来计算引发剂分解量；对于过氧类引发剂，则多用碘量法测定残留的引发剂浓度。

工业上常用半衰期 $t_{1/2}$ 来衡量一级反应速率的大小。所谓半衰期是指引发剂分解至起始浓度一半时所需的时间。当 $[I]=[I]_0/2$，半衰期与分解速率常数有如下关系：

$$t_{1/2}=\frac{\ln 2}{k_d}=\frac{0.683}{k_d}$$

引发剂分解速率常数与温度的关系遵循 Arrhenius 经验式：

$$k_d=A_d e^{-E_d/RT}$$

$$\ln k_d=\ln A_d-E_d/RT$$

在不同温度下，测定某一引发剂的分解速率常数，作 $\ln k_d$-$1/T$ 图，呈直线关系。由斜率求出分解活化能 E_d。常用引发剂的 k_d 为 $10^{-4}\sim 10^{-6} s^{-1}$，$E_d$ 为 $105\sim 140 kJ/mol$。

(3) 引发剂效率

引发剂分解后，并不能全部转化为链增长活性种。能够转化为链增长活性种的引发剂占引发剂消耗总量的比例称作引发剂效率（f）。引发剂的无意义消耗的原因有诱导分解和笼蔽效应两种作用。

① 诱导分解。诱导分解实际上是自由基向引发剂的转移反应。诱导效应和聚合体系中的引发剂和单体的分子结构有关。氢过氧化物容易诱导分解，它们的引发剂效率一般不高于0.5。丙烯腈、苯乙烯等活性较高的单体容易被自由基所引发，自由基向引发剂转移的概率较低，引发剂效率较高。乙酸乙烯等活性较低的单体，引发剂效率也较低。聚合体系内如果有阻聚剂成分存在，也会使诱导分解时间增长。

② 笼蔽效应。初级自由基的寿命只有 $10^{-11}\sim 10^{-9} s$，同时，引发剂一般浓度很低，引发剂分子可能被溶剂分子包围。初级自由基只有在短时间内扩散，与单体碰撞反应，才能引发笼外单体聚合。否则，只能向溶剂和引发剂转移，降低引发剂效率。因此，引发剂效率与单体、溶剂、引发剂、温度、体系黏度等因素有关，在 $0.1\sim 0.8$ 之间波动。

(4) 引发剂的选择

引发剂的选择往往是设计自由基聚合反应最为关键的一步。一般需从聚合方法、聚合温度、聚合物性能等多方面来考虑。

首先根据聚合方法选择引发剂种类：本体、溶液、悬浮聚合法选用油溶性引发剂，乳液聚合和水溶液聚合则用水溶性引发剂。过氧类引发剂具有氧化性，易使聚合物着色，偶氮类含有氰基，具有毒性，需考虑这些对聚合物性能的影响。储存时应避免高温或撞击，以防爆炸。

引发剂活性差别很大，应根据聚合温度来选用（表 3-1）。引发剂和温度是影响聚合速率和分子量的两大因素，应该综合考虑它们的影响。

(5) 其他引发聚合的方法

① 热引发聚合。有一些单体仅靠加热就能聚合，如苯乙烯，这可能与单体活性高有关。仅凭热能打开乙烯基单体的双键使其成为自由基，此法需 $210 kJ/mol$ 以上的能量。

表 3-1　常用自由基引发剂的适用温度及分解活化能

分解活化能 /(kJ/mol)	适用温度范围 /℃	类别	典型引发剂
140~190	>100	高温引发剂	异丙苯过氧化氢、叔丁基过氧化氢、过氧化二异丙苯、过氧化二叔丁基
110~140	40~100	中温引发剂	过氧化二苯甲酰（BPO）、过氧化十二酰、偶氮二异丁腈（AIBN）、过硫酸盐
60~110	−10~40	低温引发剂	氧化还原体系：过氧化氢/亚铁盐、BPO/N,N-二甲基苯胺、过硫酸盐/亚硫酸氢盐
<60	<−10	超低温引发剂	氧/三乙基硼、过氧化氢/三乙基铝

根据苯乙烯的聚合速率与单体浓度的 2.5 次方成正比的实验，推论出热引发反应属于三级反应，普遍认可的机理是：

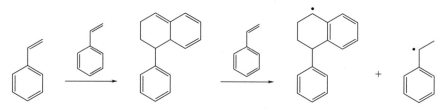

两分子苯乙烯先经 Diels-Alder 加成形成二聚体，再与一分子苯乙烯进行氢原子转移反应，生成两个自由基，然后引发单体聚合。

② 光引发聚合。工业上的光引发聚合大多在室温条件下进行，而且单体和光引发剂混合后在黑暗处可长时间保存，在光照后迅速开始反应。这在热分解引发剂体系中较难得到。因此，光引发聚合在印刷、集成电路、3D 打印领域均有大量应用。

光引发聚合有光直接引发、光引发剂引发和光敏剂间接引发三种。

光直接引发：如果选用波长较短的紫外光，其能量大于单体的化学键能，就可能直接引发聚合。单体吸收一定波长的光量子后，先形成激发态 M*，再分解成自由基，引发聚合。比较容易直接光引发聚合的单体有丙烯酰胺、丙烯腈、丙烯酸、丙烯酸酯等。

光引发剂引发：光引发剂吸收光后，分解成自由基，然后引发烯类单体聚合。

光敏剂间接引发：二苯甲酮和荧光素、曙红等染料，吸收光能后，将光能传递给单体或引发剂，然后引发聚合。

③ 辐射引发聚合。核能放射线引发的聚合，称作辐射聚合。

不稳定元素衰变时，辐射出来的放射线主要有 α、β、γ 三种。其中 γ 射线波长最短，能量最高，穿透性最强。钴-60（^{60}Co）γ 射线是一种常见的 γ 射线，其能量为 1.17~1.33MeV，远高于有机化合物的共价键键能（2.5~4.0eV）和电离能（9~11eV），所以可以打断共价键，使被照射体系产生自由基、阳离子以及阴离子等连锁聚合活性中心。但是由于自由基之间、阴阳离子之间均可迅速复合，只有小部分能够转化为链增长活性中心。其中自由基转化为链增长活性点的概率要大得多，所以绝大部分辐射聚合体系都以自由基聚合为主。

吸收剂量（D）是用来表示吸收辐射能量的单位，其定义是被照射物质所吸收的辐射平

均能量 E 对被照射物质质量的微商。吸收剂量的单位是戈瑞（Gy），$1Gy=1J/kg$。

单体的电离辐射与热引发聚合的主要区别在于链引发阶段。辐射聚合不需要加引发剂，电离辐射能通过电离和激发过程生成活性粒子（自由基、离子）引发聚合反应。在链增长和链终止阶段，则无明显区别。

相比普通自由基聚合，γ射线辐射聚合的特殊的优势有：

a. 无需引发剂即可获得交联聚合，产物比较纯净；
b. γ射线穿透力强，聚合物均匀，且无尺寸限制；
c. 调节辐射剂量和辐射总剂量可控制反应的速率和程度，工艺简单；
d. 引发活化能极低，可在室温或低温下进行聚合。

④ 等离子体引发聚合。等离子体是部分电离的气体，由电子、离子（正、负离子数相等）、自由基，以及原子、分子等高能中性粒子组成。

等离子体可以粗分为高温（热）和低温两类。用于有机反应的是低温等离子体，多由 13.56MHz 的射频低气压辉光放电产生，其能量约为 2~5eV，恰好与有机化合物的键能（2.5~5eV）相当。

等离子体可以直接引发烯类单体进行自由基聚合，或使杂环开环聚合，与传统聚合机理相同。但其特征是在气相中引发，在液、固凝聚态中（尤其在表面）增长和终止。因等离子体聚合在凝聚相中增长和终止，链自由基易被包埋，寿命很长。经等离子体辐照几分钟至十几分钟后，停止辐照，几天至十几天后，仍能继续聚合。

许多有机物，例如饱和烷烃（如甲烷、乙烷）和环烃（如苯、环己烷）乃至很稳定的六甲基二硅氧烷和饱和碳氟化合物，经等离子体作用，都可能解离、重排，再结合成高分子，往往交联。反应机理复杂，有自由基产生，却不能用自由基聚合的基元反应来表述，无法检出和写出结构单元，也谈不上明确的单体。

等离子体也常用于高分子材料的表面处理。高能态的等离子体粒子轰击高分子表面，使链断裂，产生长寿自由基（可达 10 天），发生交联、化学反应、刻蚀等。例如，以聚乙烯、聚丙烯、聚酯或聚四氟乙烯为基材，经在电场中加速的 Ar、He 等离子体处理，可使表面刻蚀和粗面化，提高黏结性；用等离子体处理，与空气接触，引入极性基团，可提高亲水性；再经化学反应，还可引入目标基团；以 NF_3、BF_3、SiF_4 等离子体处理，可使表面氟化，提高防水-防油性和光学特性；经氨、氧、氢等离子体前处理，产生长寿自由基，再与丙烯酰胺、丙烯酸等接枝聚合，改善抗静电性和吸湿性。

⑤ 微波引发聚合。微波是频率 $3\times10^2\sim3\times10^5$ MHz（相当于波长为 1m~1mm）的电磁波，属于无线电中波长最短的波段，亦称超高频。微波最常用的频率为 $2450MHz\pm50MHz$（相当于波长 120mm），进入分米波段，该频率与化学基团的旋转振动频率接近，可以活化基团，促进化学反应。

微波具有热效应和非热效应双重作用。热效应是交变电场中介质的偶极子诱导转动滞后于频率变化而产生的，因分子转动摩擦而内加热，加热速率快，受热均匀。微波可以加速化学反应，速率提高十到千倍不等。这不局限于热效应的影响，非热效应（电特性）起着更重要的作用。在微波作用下，苯乙烯、（甲基）丙烯酸酯类、丙烯酸、丙烯酰胺，甚至马来酸酐都曾（共）聚合成功，也可用于接枝共聚。无引发剂时，可激发聚合；有引发剂时，则加速聚合，还可以降低引发剂浓度和/或聚合温度。

3.1.6 聚合速率

(1) 聚合速率方程

根据自由基聚合机理和质量作用定律，可以写出各基元反应的速率方程。

① 链引发速率

$$I \xrightarrow{k_d} 2R\cdot$$

$$R\cdot + M \xrightarrow{k_i} RM\cdot$$

链引发由两步反应构成，一个是引发剂分子分解成两个初级自由基。但并不是所有初级自由基都能够引发单体聚合，需引入引发剂效率 f，则链引发速率方程可写成下式：

$$R_i = fk_d[I]$$

式中，I、M、R、k 分别代表引发剂、单体、初级自由基、速率常数；下标 d 和 i 分别代表分解和引发反应。

$$RM\cdot + M \xrightarrow{k_{p1}} RM_2\cdot$$

$$RM_2\cdot + M \xrightarrow{k_{p2}} RM_3\cdot$$

$$\vdots$$

$$RM_x\cdot + M \xrightarrow{k_{px}} RM_{x+1}\cdot$$

② 链增长速率

链增长是单体自由基连续加聚单体的连锁反应。因聚合链一端的自由基活性与聚合链的链长基本无关，因此可作等活性假定，即体系内所有链增长反应的速率常数相等。

用 [M·] 代表各种聚合度的活泼自由基浓度的总和，则链增长速率方程可写成：

$$R_p = -\frac{d[M]}{dt} = k_p[M][M\cdot]$$

③ 链终止速率

链终止速率以自由基消失速率表示，两个活性自由基能迅速反应形成新的共价键，反应方式分为偶合终止和歧化终止两种形式。

$$RM_x\cdot + RM_y\cdot \xrightarrow{k_{tc}} RM_{x+y}R$$

$$RM_x\cdot + RM_y\cdot \xrightarrow{k_{td}} RM_x + RM_y$$

链终止反应及其速率方程可写成下式：

$$R_t = -\frac{d[M\cdot]}{dt} = k_{tc}[M\cdot][M\cdot] + k_{td}[M\cdot][M\cdot] = k_t[M\cdot]^2$$

式中，下标 p、t、tc、td 分别代表链增长、链终止、偶合终止和歧化终止。

链增长和链终止的速率方程中都出现自由基浓度 [M·]，但由于自由基活泼，寿命很短，浓度极低，可作"稳态"假定，设法消去 [M·]。所谓"稳态"假定是指经过一段聚合时间，引发速率与终止速率相等，体系内自由基浓度保持恒定不变，也就是，$R_t = R_i$。

$$[M\cdot] = \left(\frac{R_i}{k_t}\right)^{1/2}$$

一般而言，自由基聚合产物的聚合度很大，链引发过程远少于链增长过程所消耗的单体，因此可假设单体仅消耗于链增长反应，聚合体系中的聚合总速率等于链增长速率。即：

$$R=-\frac{d[M]}{dt}=R_p$$

将自由基浓度和链引发速率代入，即得总聚合速率的普适方程：

$$R=k_p\left(\frac{fk_d}{k_t}\right)^{1/2}[I]^{1/2}[M] \tag{3-1}$$

由式(3-1)可知聚合速率与引发剂浓度的平方根、单体浓度成正比。

(2) 温度对聚合速率的影响

可以利用 Arrhenius 式来分析聚合速率常数 k 与温度的关系。

$$k=Ae^{-E/RT}$$

将聚合速率的普适方程中的速率常数设定为 k，则：

$$k=k_p\left(\frac{k_d}{k_t}\right)^{1/2}$$

由此可得聚合反应的各基元反应的活化能为：

$$E=\left(E_p-\frac{E_t}{2}\right)+\frac{E_d}{2}$$

选取 $E_p=29kJ/mol$，$E_t=17kJ/mol$，$E_d=125kJ/mol$，则 $E=85kJ/mol$，总活化能为正值。温度升高，将使聚合速率常数增大，例如，温度从 50℃ 升高到 60℃，聚合速率将增为 2.5 倍。

热引发聚合活化能约为 80～90kJ/mol，与引发剂引发时相当或稍大，温度对聚合速率的影响很大。而光和辐射引发聚合时，无 E_d 项，聚合活化能很低，约 20kJ/mol，温度对聚合速率的影响较小，甚至在 0℃ 以下也能聚合。

(3) 凝胶效应和宏观聚合动力学

自由基聚合过程中，单体和引发剂浓度随转化率上升而降低，聚合总速率应该降低。但实际上，当转化率达 15%～20% 时，聚合速率会快速上升，这种现象称为自动加速现象。自动加速现象主要是体系黏度增加所引起的，因此又称为凝胶效应。

图 3-4 为甲基丙烯酸甲酯本体聚合以及在苯溶液中进行自由基聚合的转化率-时间曲线，可以较好地说明自由基聚合反应中的凝胶效应。40% 浓度以下的 MMA 溶液聚合时，无自动加速现象；60% 以上才出现加速。MMA 本体聚合时，10% 转化率以下，加速现象尚不明显；转化率在 10%～50% 时，加速显著。

凝胶效应也可用扩散控制理论和自由基双基终止机理来解释。聚合初期，体系黏度较低，聚合正常。随转化率增加，黏度上升，具有较高分子量的链自由基运动受阻，双基终止困难，k_t 下降。但由于单体分子量较小，在体系中的扩散速率受影响不大，k_p 几乎保持不变。

据实验测定，当转化率达 40%～50% 时，k_t 下降达上百倍。因此 $k_p/k_t^{1/2}$ 增加近 10 倍，活性链寿命也增长 10 多倍，导致聚合速率和分子量大幅度上升。转化率继续上升后，黏度增大至单体的运动受阻，则 k_t 和 k_p 都下降，聚合总速率下降，最后甚至停止反应。

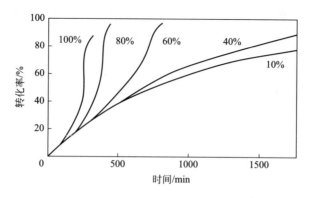

图 3-4　甲基丙烯酸甲酯自由基聚合转化率-时间曲线
BPO 引发剂，苯溶剂，反应温度 50℃

3.1.7　动力学链长和聚合度

（1）动力学链长

聚合度是表征聚合物的重要指标，在自由基聚合反应的聚合速率普适公式的基础上，可建立数学模型来解释引发剂浓度、单体浓度、温度等因素对聚合度的影响。

将一个活性种从引发开始到链终止所消耗的单体分子数定义为动力学链长 ν，无链转移时，相当于每一链自由基所连接的单体单元数，可由链增长速率和链引发速率之比求得。稳态时，引发速率等于终止速率，因此动力学链长的定义表达式为：

$$\nu = \frac{R_p}{R_i} = \frac{R_p}{R_t} = \frac{k_p[M]}{2k_t[M\cdot]}$$

将 $[M\cdot]$ 及 R_i 关系式代入上式，得动力学链长公式：

$$\nu = \frac{k_p}{2(fk_dk_t)^{1/2}} \cdot \frac{[M]}{[I]^{1/2}}$$

上式表明，动力学链长与引发剂浓度的平方根成反比。由此看来，增加引发剂浓度来提高聚合速率的措施，往往使聚合度降低。

聚合物平均聚合度 $\overline{X}_n$ 与动力学链长的关系与终止方式有关：偶合终止，$\overline{X}_n = 2\nu$；歧化终止，$\overline{X}_n = \nu$；兼有两种终止方式，则可按下式计算：

$$\overline{X}_n = \frac{R_p}{R_{td} + \frac{R_{tc}}{2}} = \frac{\nu}{\frac{C}{2} + D} \tag{3-2}$$

式中，C、D 分别代表偶合终止和歧化终止的分数。

升温使自由基聚合速率增加，却使聚合产物的聚合度降低。应用 Arrhenius 公式：

$$k' = A'e^{-E/RT} \quad 和 \quad k' = \frac{k_p}{(k_dk_t)^{\frac{1}{2}}}$$

得：

$$E' = \left(E_p - \frac{E_t}{2}\right) - \frac{E_d}{2}$$

E' 是影响聚合度的综合活化能。取 $E_p = 29 \text{kJ/mol}$，$E_t = 17 \text{kJ/mol}$，$E_d = 125 \text{kJ/mol}$，则 $E' = -42 \text{kJ/mol}$，表明温度升高，聚合度将降低。

热引发聚合时，温度对聚合度的影响，与引发剂引发时相似。光和辐射引发时，E 是很小的正值，表明温度对聚合度和速率的影响甚微。

(2) 链转移反应和聚合度

在自由基聚合中，除了链引发、增长、终止基元反应外，往往伴有链转移反应。活性链向单体、引发剂、溶剂等低分子链转移的反应式和速率方程如下：

$$M_x \cdot + M \xrightarrow{k_{tr,M}} M_x + M \cdot \qquad R_{tr,M} = k_{tr,M}[M_x \cdot][M]$$

$$M_x \cdot + I \xrightarrow{k_{tr,I}} M_x R + R \cdot \qquad R_{tr,I} = k_{tr,I}[M_x \cdot][I]$$

$$M_x \cdot + YS \xrightarrow{k_{tr,S}} M_x Y + S \cdot \qquad R_{tr,S} = k_{tr,S}[M_x \cdot][S]$$

式中，下标 tr、M、I、S 分别代表链转移、单体、引发剂、溶剂，例如 $k_{tr,M}$ 代表向单体的链转移速率常数。

按定义，动力学链长是每个活性中心自链引发到链终止所消耗的单体分子数，这在无链转移情况下是很明确的。但有链转移反应时，转移后，动力学链尚未真正终止，仍在继续引发增长。因此，动力学链长应该考虑自初级自由基链引发开始，包括历次链转移以及最后双基终止所消耗的单体总数。而聚合度则等于动力学链长除以链转移次数和双基终止之和。假设双基终止全部为歧化终止，平均聚合度就是增长速率与形成大分子的所有链终止（包括链转移）速率之比。

$$\overline{X}_n = \frac{R_p}{R_t + \sum R_{tr}} = \frac{R_p}{R_t + (R_{tr,M} + R_{tr,I} + R_{tr,S})}$$

转成倒数，化简得：

$$\frac{1}{\overline{X}_n} = \frac{2k_t R_p}{k_p^2 [M]^2} + \frac{k_{tr,M}}{k_p} + \frac{k_{tr,I}[I]}{k_p[M]} + \frac{k_{tr,S}[S]}{k_p[M]}$$

令 $k_{tr}/k_p = C$，定名为链转移常数，是链转移速率常数与增长速率常数之比，代表这两个反应的竞争能力。向单体、引发剂、溶剂的链转移常数 C_M、C_I、C_S 的定义如下：

$$C_M = \frac{k_{tr,M}}{k_p}, \quad C_I = \frac{k_{tr,I}}{k_p}, \quad C_S = \frac{k_{tr,S}}{k_p}$$

经简化可得：

$$\frac{1}{\overline{X}_n} = \frac{2k_t R_p}{k_p^2 [M]^2} + C_M + C_I \frac{[I]}{[M]} + C_S \frac{[S]}{[M]} \tag{3-3}$$

式(3-3)描述了链转移反应对平均聚合度的影响，右边四项分别代表正常聚合、向单体转移、向引发剂转移、向溶剂转移对平均聚合度的贡献。

在实际生产中，也经常应用链转移的原理来控制分子量。例如，聚氯乙烯分子量由单体转移决定，可由聚合温度来控制；聚合丁苯橡胶时，可添加十二硫醇来调节分子量等。

3.1.8 阻聚作用和阻聚剂

阻聚剂有分子型和稳定自由基型两大类。分子型阻聚剂有苯醌、硝基化合物、芳胺、酚类、硫和含硫化合物、三氯化铁等；稳定自由基型阻聚剂有 1,1-二苯基-2-三硝基苯肼

(DPPH)、三苯基甲基等。按照阻聚剂与活泼自由基间的反应机理，则有加成型、链转移型和电荷转移型三类，现按这三类介绍阻聚机理。

（1）加成型阻聚剂

苯醌、硝基化合物、氧、硫等属于此类。其中苯醌是最重要的，其阻聚机理比较复杂，苯醌分子上的氧和碳原子都有可能与自由基加成，分别形成醚和酮型，然后偶合或歧化终止。

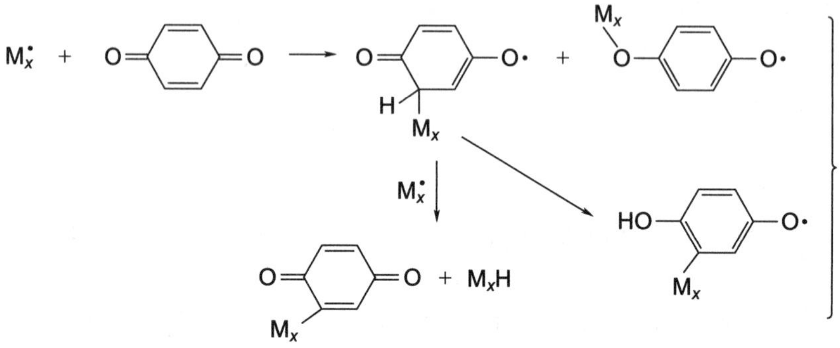

在室温下，氧和自由基反应，先形成不活泼的过氧自由基。过氧自由基本身或与其他自由基歧化或偶合终止。过氧自由基有时也与少量单体加成，形成低分子量的共聚物。因此，氧是阻聚剂，大部分聚合反应需在排除氧的条件下进行。氧有低温阻聚和高温引发的双重作用。低温时稳定的聚合物过氧化合物，高温时却能分解成活泼的自由基，起引发作用。

（2）链转移型阻聚剂

DPPH、芳胺、酚类等属于这类阻聚剂。

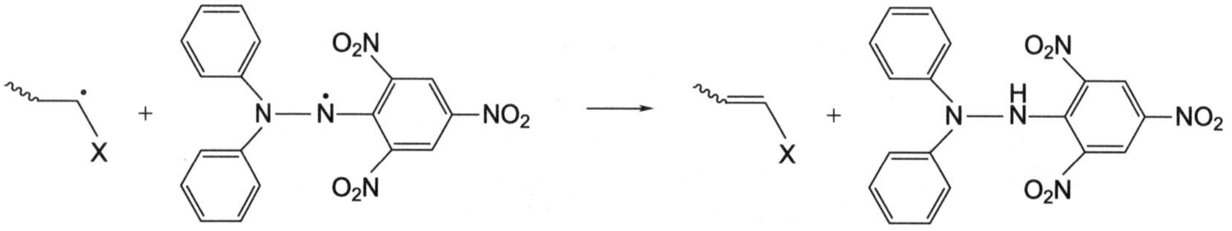

DPPH是自由基型高效阻聚剂，素有"自由基捕捉剂"之称。DPPH原来呈黑色，向自由基转移后，则变成无色，可用比色法定量。据此，可用于引发速率的测定。

苯酚和苯胺缓聚机理是链自由基先夺取酚羟基上的氢原子而终止，同时形成酚氧自由基，再与其他自由基偶合终止。

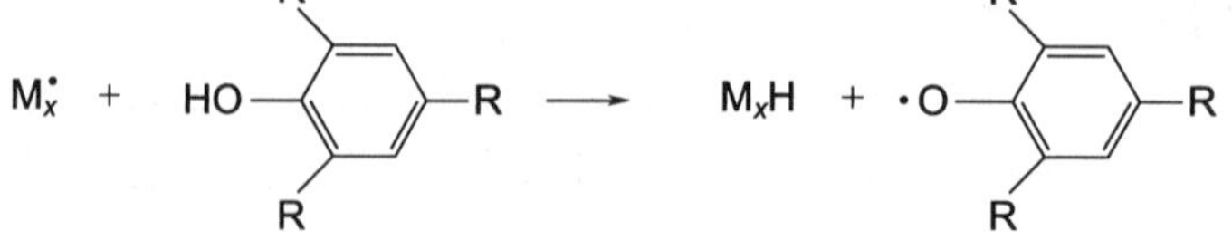

（3）电荷转移型阻聚剂

电荷转移型阻聚剂主要是变价金属的氯化物，如三氯化铁、氯化铜等。三氯化铁阻聚效率很高，能一对一地消灭自由基。亚铁盐也能使自由基终止，但效率较低。

烯丙基自由基的结构特点是自由基p电子与π电子共轭稳定。烯丙基单体中CH_2Y中的 H 活泼，易被链转移成稳定的烯丙基自由基。

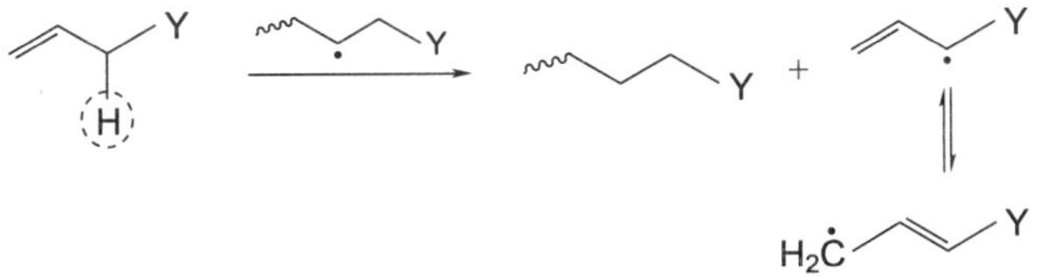

烯丙基自由基因共振而稳定，链引发和链增长都减弱，最后相互或与其他链自由基发生终止反应，显现自阻聚作用。

甲基丙烯酸甲酯、甲基丙烯腈等也有烯丙基的 C—H 键，却不衰减转移，因为酯基和氰基对自由基都有稳定作用，使链转移活性降低，链增长活性增加，能形成高聚物。

3.1.9 活性自由基聚合

自由基聚合的适用单体种类多，反应条件温和，是重要的连锁聚合反应，但有聚合产物分子量分布宽，容易发生支化等缺点。活性聚合（living polymerization）的概念是在 20 世纪 50 年代提出的，它具有以下特征，①聚合物分子量增长与单体转化率呈线性关系；②单体完全转化后加入新单体，链增长继续进行，分子量进一步增加，且仍与单体转化率成正比；③聚合物分子链数目在整个聚合过程中保持恒定；④分子量分布非常窄，接近单分散。

理想的活性聚合具有无终止、无转移等特点，能实现对聚合物结构的控制。将活性聚合概念运用到自由基聚合，形成了一系列可控活性自由基聚合（controlled radical polymerization，CRP）方法。

$$\frac{R_t}{R_p}=\frac{k_t[\mathrm{M}\cdot]}{k_p[\mathrm{M}]}$$

自由基聚合的链增长反应速率 R_p 与自由基浓度和单体浓度呈正比，而链终止反应速率 R_t 与自由基浓度的平方呈正比，二者之比为 R_t/R_p，因已知 k_t/k_p 为 $10^4 \sim 10^5$，单体浓度 [M] 一般为 $1\sim 10\mathrm{mol/L}$，所以，$R_t/R_p \approx 10^4[\mathrm{M}\cdot]$。当自由基浓度 [M·] 足够低，大约为 $10^{-8}\mathrm{mol/L}$ 这个数量级时，仍然能得到满意的链增长速率，但是 $R_t/R_p \approx 10^{-3}\sim 10^{-4}$，$R_t$ 相对于 R_p 可忽略不计。

如果在聚合体系中加入一种化学物质 X，它不直接参与链引发、链增长反应，但是与活性链自由基反应，生成一个不引发单体聚合的"休眠种"——M—X，并构成倾向于休眠种一侧的可逆平衡，这样的话，聚合反应体系中的自由基浓度和链终止速率降低，使自由基聚合有可能变为可控的活性聚合。目前为止，活性可控自由基聚合的主要方法均是添加各种能降低自由基活性的化合物。

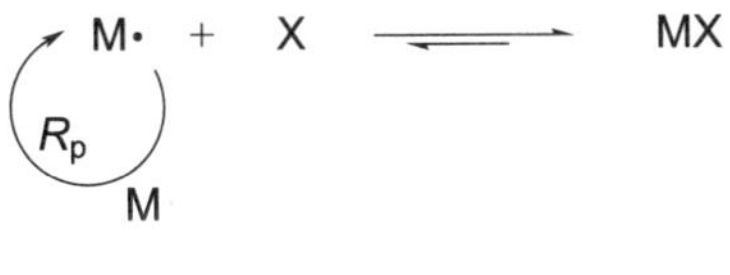

(1) 稳定活性自由基法

2,2,6,6-四甲基哌啶-1-氧基（TEMPO）是氮氧稳定自由基的代表，是一种稳定的自由基，不能引发聚合反应，但可以促进 BPO 等引发剂的分解，并可与处于链增长阶段的链自由基结合，形成休眠种。较高温度（120℃）下，该休眠种又能逆均裂成增长自由基，再参与引发聚合。

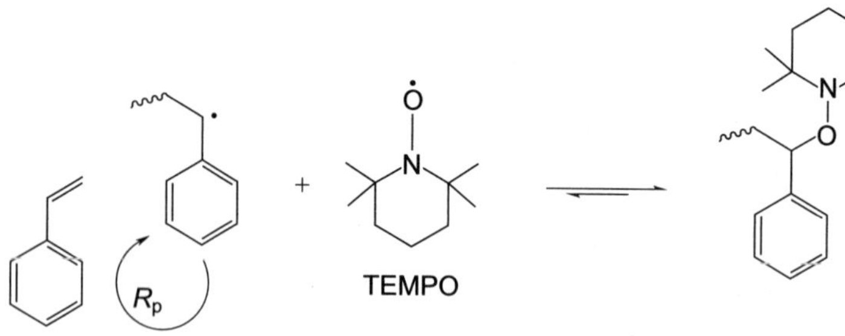

例如，采用 TEMPO/BPO 引发体系，初级自由基引发苯乙烯在 120℃ 以上聚合，增长自由基迅速被 TEMPO 捕捉，偶合成共价休眠种，此后该休眠种又均裂成自由基。所得聚合物分子量随转化率而线性增加，分子量分散指数为 1.15～1.3，显示出活性聚合的特征。

稳定活性自由基法也存在一些缺点，例如只适用于苯乙烯等少数单体，聚合温度较高，聚合速率低，TEMPO 价格昂贵，等等。

（2）原子转移自由基聚合法

原子转移自由基聚合法（ATRP）中运用了过渡金属配位化合物和卤代烷烃的相互作用获得能捕捉自由基的可逆反应体系。以氯化亚铜体系为例，使用有机卤化物 RX（1-氯-1-苯基乙烷）以及氯化亚铜/双吡啶（bpy）的配位化合物构成三元引发体系。

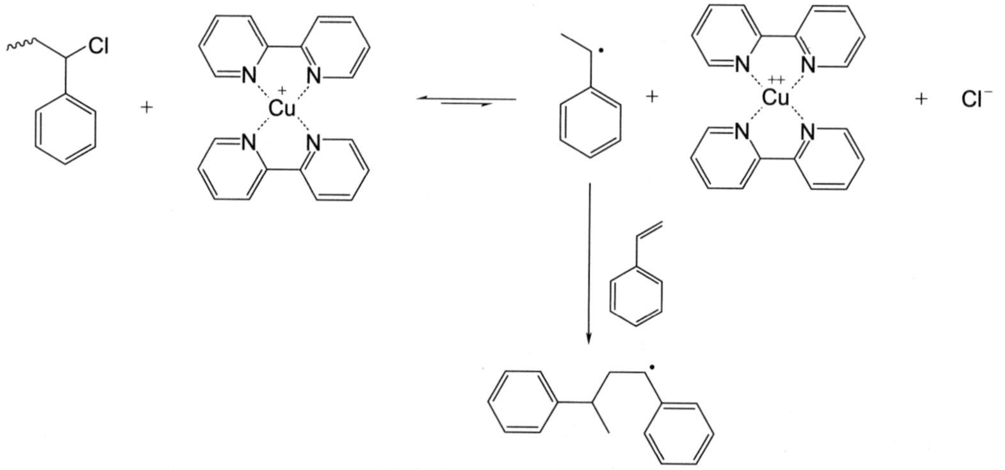

在引发阶段，卤代烃 RX 和亚铜离子反应，卤原子转移氧化亚铜离子为铜离子，同时释放自由基 R·，引发单体生成单体自由基，即活性种。

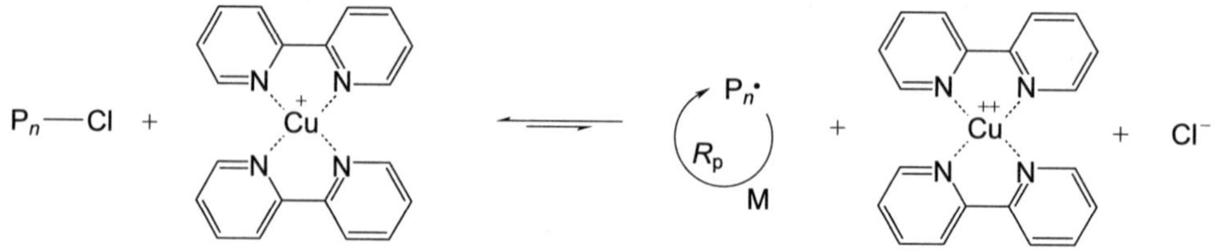

活性种既可以形成链增长反应，也可参与生成休眠种的可逆反应。活性种和休眠种之间形成的快速动态平衡反应可以降低链增长自由基稳态浓度，使体系表现出可控活性聚合特性。由于在引发阶段存在卤原子从卤化物转移到金属络合物，再从金属络合物转移到自由基的原子转移过程，所以称为原子转移自由基聚合。

应用这一方法，苯乙烯、二烯烃、（甲基）丙烯酸酯类等均可聚合成分子量分布较窄（$M_w/M_n=1.05\sim1.5$）的聚合物。原子转移自由基聚合的优点在于适用单体范围广，聚合条件温和，分子设计能力强；但也有需要使用重金属离子，体系毒性较强，不适用于生物医用材料等缺点。

（3）可逆加成-裂解链转移聚合法

可逆加成-裂解链转移聚合法（reversible addition-fragmentation chain transfer，RAFT）的聚合条件温和、单体适应范围广，与 ATRP 法并列，是可控自由基聚合的主要方法，可合成许多窄分子量分布的均聚物和嵌段聚合物。

RAFT 聚合机理如下所示，其中可逆链转移和链平衡两个可逆平衡反应是 RAFT 聚合的关键。这样的平衡反应能生成活性较差的休眠种，使得链增长自由基浓度保持在较低水平，并使聚合反应表现出可控活性聚合的特征。

引发：

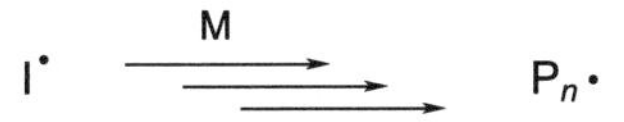

$$I\cdot \xrightarrow{M} P_n\cdot$$

可逆链转移：

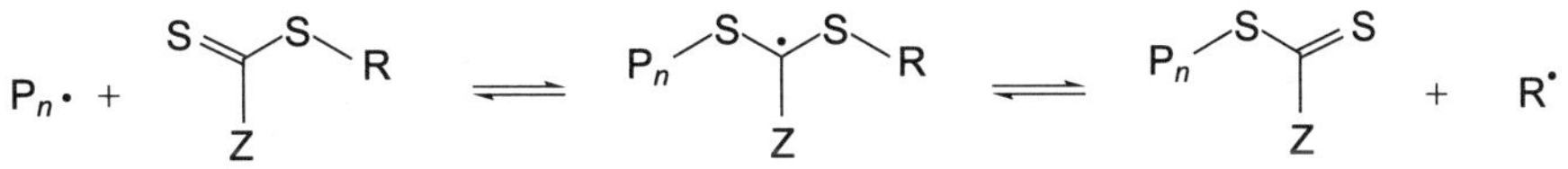

$$P_n\cdot + \underset{Z}{S=C-S-R} \rightleftharpoons P_n-\underset{Z}{S-\overset{\cdot}{C}-S-R} \rightleftharpoons P_n-\underset{Z}{S-C=S} + R\cdot$$

再引发：

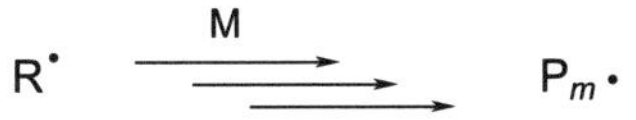

$$R\cdot \xrightarrow{M} P_m\cdot$$

链平衡：

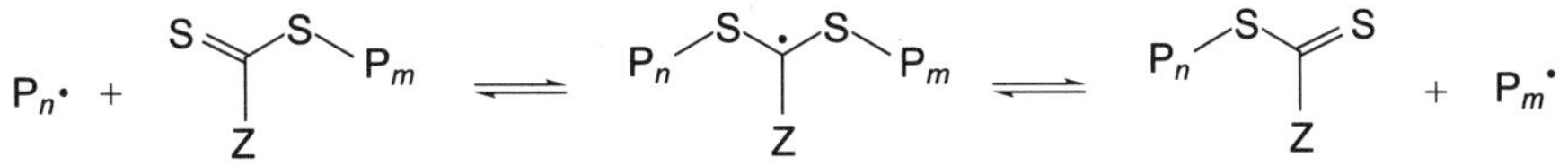

$$P_n\cdot + \underset{Z}{S=C-S-P_m} \rightleftharpoons P_n-\underset{Z}{S-\overset{\cdot}{C}-S-P_m} \rightleftharpoons P_n-\underset{Z}{S-C=S} + P_m\cdot$$

RAFT 聚合与普通的自由基聚合相比，仅仅加入了少量的 RAFT 试剂，在普通自由基聚合的反应条件下就可以实现活性可控聚合。这种简易性是 RAFT 的优点。RAFT 试剂多为双硫酯［ZC(S)S—R］结构，其中 Z 官能团可稳定自由基中间体，多为烷基、苯基等，R 是容易形成活泼自由基的基团，如异丙苯基、氰异丙苯基等。RAFT 法适用的单体包括苯乙烯类、丙烯酸酯类单体，可以在温和条件下调控聚合物结构，合成嵌段、接枝、星形、树枝状、支化及超支化聚合物等。

3.1.10 聚合方法

自由基聚合有本体聚合（bulk polymerization）、溶液聚合（solution polymerization）、悬浮聚合（suspension polymerization）、乳液聚合（emulsion polymerization）四种聚合方法。

本体聚合无需溶剂或分散介质，仅包含单体和少量引发剂，包括熔融聚合和气相聚合。溶液聚合则是单体和引发剂溶于适当溶剂中的聚合。悬浮聚合一般是单体以液滴状悬浮在水中的聚合，体系主要由单体、水、油溶性引发剂、分散剂四部分组成，反应机理与本体聚合相同。乳液聚合则是单体在水中分散成乳液状而进行的聚合，一般体系由单体、水、水溶性引发剂、水溶性乳化剂组成，机理独特。

烯类单体采用上述四种方法进行自由基聚合的配方、机理、生产特征、产物特性等比较如表 3-2 所示。

表 3-2 自由基聚合方法及特征

聚合体系里物质	本体聚合	溶液聚合	悬浮聚合	乳液聚合
聚合体系里物质	单体 引发剂	单体 溶剂可溶引发剂 溶剂	单体 单体可溶引发剂 水 分散剂	单体 水溶性引发剂 水 乳化剂
聚合场所	本体内	溶液内	悬浮液滴内	胶束和乳胶粒内
聚合状态	均匀	均匀	非均匀	非均匀
搅拌	困难	容易	需要	需要
散热	困难	比较容易	容易	容易
聚合速率	快	慢	快	极快
分子量	高	低	高	极高
聚合物分离	易，可直接使用	使用非溶剂沉淀	容易（过滤）	困难
杂质	极少	少	稍多	多

（1）本体聚合

本体聚合无需溶剂，仅由单体和少量（或无）引发剂组成，产物纯净，比较适合学术研究。但是它要求聚合温度下单体必须是液体熔融状态，引发剂需要能够溶于单体，聚合热散热比较困难。工业中，聚苯乙烯、聚甲基丙烯酸甲酯、聚氯乙烯都采用本体聚合的生产工艺。

本体聚合在大规模生产工艺中的关键问题是聚合热的散热。烯类单体的聚合热约为55～95kJ/mol。在转化率小于10%的聚合初期阶段，聚合体系黏度不大，散热无困难。当转化率增加到20%～30%后，凝胶效应出现，聚合速率自动加速。此时如不及时散热，容易爆聚。因此，本体聚合多分段进行，第一阶段一般在转化率达到10%～35%时结束，体系黏度较低，可利用通用聚合釜搅拌散热。此后，在可以应对高黏聚合体系的特殊反应器内聚合。

① 苯乙烯连续本体聚合

图 3-5 所示为苯乙烯连续本体聚合的生产工艺流程。预聚在通用立式搅拌釜内进行，聚合温度控制在 80～90℃，以过氧化物为引发剂，转化率控制在 30%～35%。这时，尚未出现自动加速现象，聚合热不难排除。透明黏稠的预聚物流入聚合塔顶并缓慢流向塔底，温度由上到下由 100℃ 渐增至 200℃，最后达 99% 转化率，自塔底出料，经挤出、冷却、切粒，即成透明粒料产品。

② 甲基丙烯酸甲酯的间歇本体聚合——有机玻璃板的制备

四种聚合方法均可用于聚甲基丙烯酸甲酯的生产，其中，本体聚合可生产有机玻璃板材（图 3-6），悬浮聚合生产模塑粉，乳液聚合生产皮革或织物处理剂。本体聚合生产有机玻璃时，工程上的关键是凝胶效应和体积收缩问题。

预聚合可在通用反应釜内进行，体系含单体、引发剂和增塑剂等，85℃ 加热 10min 后

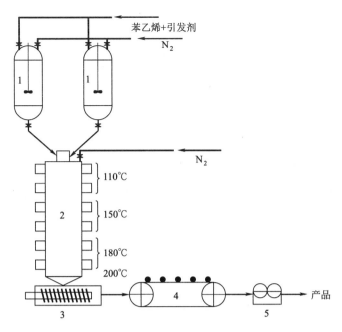

图 3-5　苯乙烯连续本体聚合的生产工艺流程
1—通用反应釜；2—塔式反应器；3—挤出机；4—冷却装置；5—造粒装置

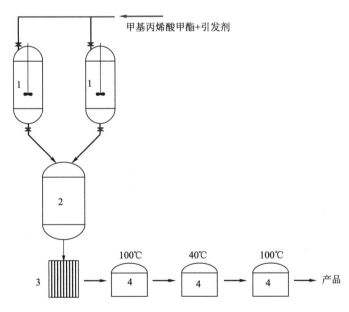

图 3-6　甲基丙烯酸甲酯间歇式本体聚合的生产工艺流程
1—通用反应釜；2—浆料罐；3—灌浆；4—烘房

停止加热，利用聚合热自动升温至 90℃，反应 15min 后，冰水中快速搅拌冷却至 20℃，所得浆液供灌浆使用。预聚合可以部分减弱凝胶效应和体积收缩，黏度增加可减少灌浆时的泄漏。灌浆工段将预聚浆液通过漏斗灌入模具中，根据生产的板材厚度采取不同工艺方法。空气加热法是将灌浆后的模具在 100℃ 烘房进一步聚合后，再先后在 40℃（30h 以上）和 100℃（5h 左右）的烘房聚合，分子量可达百万。

(2) 溶液聚合

溶液聚合体系黏度较低，可以较好地解决本体聚合的散热和凝胶效应问题。其缺点有单体浓度较低，聚合速率较慢，设备生产效率较低，溶剂链转移可导致聚合物分子量降低，溶剂分离回收费用高，聚合物中残留溶剂等。

丙烯腈均聚物中氰基极性强，分子间力大，加热时不熔融，均聚物较难纺丝，纤维性脆不柔软，难染色。一般需要与第二、第三单体形成共聚物，其中丙烯腈约占 90%。丙烯酸甲酯和衣康酸常用作第二单体和第三单体，分别用来增加柔软性和改善染色性能。

丙烯腈均相溶液聚合工艺流程如图 3-7 所示，单体丙烯腈醇、丙烯酸甲酯、衣康酸钠盐溶液、偶氮二异丁腈、二氧化硫脲、硫氰酸钠水溶液分别经计量泵在混合釜内混合，再与从聚合浆液中回收的未反应单体混合，调节 pH 值为 5 左右，在聚合釜中，浆液在两个脱单体塔内真空脱除未聚合的单体后，可直接作为纺丝原液。

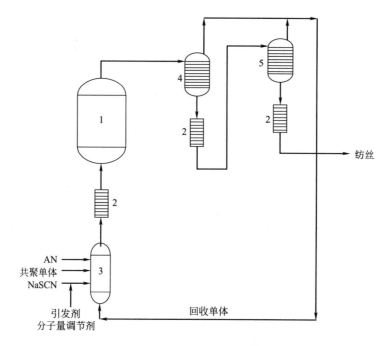

图 3-7 丙烯腈溶液聚合的生产工艺流程

1—聚合釜；2—预热器；3—试剂混合釜；4—第一单体脱出器；5—第二单体脱出器

(3) 悬浮聚合

悬浮聚合是将油性单体和油性引发剂经剧烈搅拌分散成小液滴，呈悬浮液状态进行聚合的方法。每个分散的小液滴就相当于一个小的本体聚合单元。为了防止液滴或聚合物固体粒子凝聚，一般需要添加分散剂，在粒子表面形成保护膜。悬浮聚合体系由单体、油溶性引发剂、水、分散剂四个基本组分构成，实际生产的配方则较复杂。

悬浮聚合的优点包括：聚合热容易经水散除，聚合物分子量及其分布较稳定，分子量较溶液聚合高，杂质含量比乳液聚合少，可以通过过滤法分离聚合物和介质水，产品形状为颗粒，生产成本低。悬浮聚合缺点为含少量分散剂。

悬浮聚合反应机理和动力学与本体聚合相同，仅仅在成粒机理和颗粒控制方面有所差异。油性单体和水混合将分成油水两层。在剧烈搅拌下，油性单体分散成液滴，同时也会凝聚为大液滴，构成动态平衡。当聚合到15%～30%的转化率时，液滴变黏，凝聚速率随之加快。分散剂可在液滴表面形成保护膜，防止凝聚。当转化率达60%～70%时，液滴转变成刚性固体粒子，黏结性减弱，不再凝聚（图3-8）。

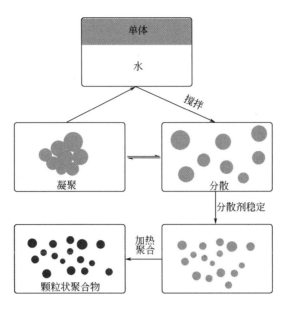

图3-8 悬浮聚合机理

大约80%的聚氯乙烯用悬浮聚合方法生产，聚氯乙烯悬浮聚合的生产流程如图3-9所示。所得颗粒平均粒径为100～160μm。聚氯乙烯悬浮聚合的配方中水和单体之比在1.2∶1～2∶1之间变动。因自由基向单体转移容易发生，所以决定聚氯乙烯聚合度的是温度而非引发剂浓度。聚合温度一般控制在45～70℃之间。水、分散剂、其他助剂、引发剂先后加入聚合釜中，反复几次抽真空和充氮排氧，然后加单体，升温至预定温度聚合。后期压力下降0.1～0.2MPa，即可出料，这时的转化率约为80%～85%。如果降压过多，将不利于疏松颗粒形成。聚合结束后，回收单体，出料，经后处理、离心分离、洗涤、干燥，即得聚氯乙烯树脂成品。

（4）乳液聚合

乳液聚合的优点包括：聚合热散热效率高，聚合速率快，产物分子量高，聚合温度低等。缺点有：需经凝聚、洗涤、脱水、干燥等复杂工序才能获得固体成品，产品中留有乳化剂等杂质且难以去除。乳液聚合的基本组成是单体、水、水溶性引发剂和水溶性乳化剂。传统或经典乳液聚合体系中四组分的比例大致为：单体100份（质量），水约150～250份，乳化剂2～5份，引发剂0.3～0.5份。

单体在水中的溶解度将影响聚合机理和产物性能。苯乙烯、丁二烯难溶于水，乙酸乙烯酯水溶性较大，甲基丙烯酸甲酯介于其间，三者乳液聚合机理和结果各异。水溶性单体需要反相乳液聚合。传统乳液聚合需要水溶性引发剂，一般为过硫酸盐或其氧化还原引发体系。乳液聚合中常用阴离子型乳化剂，首先，用于乳化油性单体为微细液滴（<1μm），更重要的是对引发、聚合和胶束的形成都有决定性的影响。以苯乙烯、过硫酸钾、十二烷基硫酸钠、水组成的经典乳液聚合体系为例，说明乳液聚合的机理。

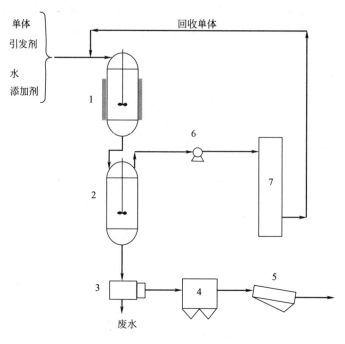

图 3-9 悬浮聚合法生产聚氯乙烯流程
1—聚合釜；2—单体汽提机；3—离心分离机；4—流动干燥机；
5—筛分机；6—真空泵；7—单体精馏塔

① 乳液聚合机理

乳液聚合遵循自由基聚合的一般规律，但其有独特的反应机理和成粒机理，可以使聚合速率和聚合度同时增加。如图 3-10 所示，乳液聚合开始时，单体和乳化剂分别处在水溶液、胶束、液滴三相。微量单体和乳化剂溶于水中，大部分乳化剂以胶束形态存在，每一胶束由 50~150 个乳化剂分子构成，直径为 4~5nm，胶束数约为 $10^{17} \sim 10^{18}$ 个$/cm^3$。小部分单体被分散在胶束内，构成胶束相，大部分单体分散成直径为 1~10μm 的液滴，数量约为

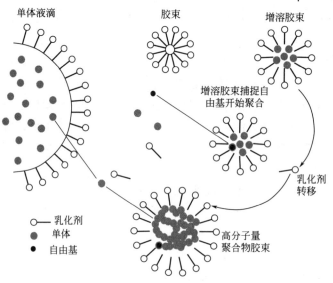

图 3-10 乳液聚合机理

$10^{10} \sim 10^{12}$ 个/cm³，液滴表面吸附有乳化剂，使乳液稳定，构成液滴相。

水溶性引发剂分解成初级自由基，首先引发溶于水中的微量单体，在水相中链增长成短链自由基。聚苯乙烯不溶于水，容易进入到增溶胶束内部，引发其中单体聚合而成核，即所谓的胶束成核。单体液滴由于体积大，比表面积小，数量少，不能捕捉较多的链增长自由基。

胶束成核持续链增长，单体消耗后，由液滴内的单体通过水相扩散来补充。单体液滴消失后，才由胶粒内的残余单体继续聚合至结束，最后成为聚合物胶粒（0.1～0.2μm）。胶粒逐渐增大，造成表面的乳化剂不足，未成核胶束破裂，其乳化剂通过水相扩散来补充。

初期的聚合物胶粒较小（十几纳米），只能容纳 1 个自由基。由于胶粒表面乳化剂的保护作用，包埋在胶粒内的自由基寿命较长（10～100s），允许较长时间的增长，等水相中的第二个自由基扩散进入胶粒内，双基才终止，胶粒内自由基数变为零。乳液聚合的特征就是链引发、链增长、链终止的基元反应在"被隔离"的胶束或胶粒内进行。就是这种"隔离作用"才使乳液聚合兼有高速率和高分子量的特点。

对于亲水性较好的乙酸乙烯酯（溶解度 25g/L）而言，水相中的短链自由基亲水性也较大，聚合度上百后才从水中沉析出来。水相中多条这样较长的短链自由基相互聚集在一起，絮凝成核（原始微粒）。以此为核心，单体不断扩散进入，聚合成胶粒。胶粒形成以后，更有利于吸取水相中的初级和短链自由基，然后在胶粒中引发增长。这就成为水相成核（或均相成核）机理。

② 乳液聚合三阶段

根据胶粒发育情况和相应速率变化，可将经典乳液聚合过程分成三个阶段。

a. 第一阶段——成核期或增速期　增溶胶束捕捉水相中的自由基，引发单体聚合并成核，逐渐转变成单体-聚合物胶粒。这一阶段，胶束减少，胶粒增多，聚合速率相应增加。单体液滴数不变，但体积缩小。到达一定转化率（约 2%～15%），未成核胶束消失，胶粒数趋向恒定（$10^{13} \sim 10^{15}$ 个/cm³），聚合速率恒定，第一阶段结束。

b. 第二阶段——胶粒数恒定期或恒速期　增溶胶束消失，体系内只有胶粒和液滴。单体液滴释放单体经水相不断扩散进入胶粒内，保持胶粒内的单体浓度恒定，因此聚合速率也恒定。胶粒不断长大，直径可达 50～150nm。单体液滴消失或聚合速率开始下降意味第二阶段结束。第二阶段结束的转化率与单体种类有关，随单体水溶性降低。苯乙烯转化率为 40%～50%，甲基丙烯酸甲酯约 25%，乙酸乙烯酯约 15%，聚氯乙烯可达 70%。

c. 第三阶段——降速期　单体液滴消失，仅剩下胶粒一种粒子，胶粒数不变。胶粒内单体继续聚合，浓度不断降低，聚合速率递降。

③ 乳液聚合动力学

乳液聚合总体上遵循自由基聚合机理，具有高聚合速率和高聚合度同时存在的特点。

a. 聚合速率　乳液聚合过程可分为增速、恒速、降速三个阶段。

在自由基聚合中，聚合速率方程可表示为：

$$R_p = k_p[M][M\cdot]$$

该式同样可以适用于乳液聚合，只不过自由基浓度需要使用胶粒的物质的量浓度代替。

设定 N 为胶粒数，N_A 为阿伏伽德罗常数，则 N/N_A 为胶粒的物质的量浓度（mol/L）。n 为胶粒中平均自由基数（理想体系为 0.5），因此乳液聚合中的自由基的物质的量浓度 $[M\cdot]$（mol/L）为：

$$[M\cdot] = \frac{nN}{N_A}$$

代入聚合速率方程，则得乳液聚合第二阶段恒速期的速率表达式：

$$R_p = \frac{k_p[M]nN}{N_A}$$

上式表明聚合速率取决于胶粒数，胶粒数 N 达 10^{14} 个/cm^3，$[M\cdot]$ 可达 10^{-7} mol/L，比一般自由基聚合（10^{-8} mol/L）要大一个数量级。因此乳液聚合速率比较快。

b. 聚合度　自由基聚合物的动力学链长或聚合度可由链增长速率和链终止（或链引发）速率的比值求得。但应考虑1个胶粒内的链增长速率和链引发速率。1个胶粒的链引发速率 r_i 是总链引发速率 R_i 与捕捉自由基的粒子数之比，而捕捉自由基的粒子数是胶粒中平均自由基数 n 与总粒子数 N 的乘积。因此：

$$r_i = \frac{R_i}{nN}$$

1个胶粒的增长速率 r_p 为：

$$r_p = k_p[M]$$

聚合物的平均聚合度为：

$$\overline{X}_n = \frac{r_p}{r_i} = \frac{k_p[M]nN}{R_i}$$

c. 胶粒数　乳液聚合中的胶粒数 N 是决定聚合速率和聚合度的关键因素，且都成正比关系。稳定的胶粒数与体系中的乳化剂总表面积 $a_s S$ 有关。a_s 是一个乳化剂分子所具有的表面积，S 是体系中乳化剂的总浓度。同时，N 也与自由基生成速率 ρ [单位：个/(mL·s)，相当于引发速率 R_i] 直接有关。其定量关系为：

$$N = k\left(\frac{\rho}{\mu}\right)^{2/5} a_s S^{3/5}$$

式中，μ 是胶粒体积增加速率；k 为常数，处于 0.37~0.53 之间，取决于胶束和胶粒捕获自由基的相对效率以及胶粒的几何参数，如半径、表面积或体积等。由于粒子数与粒径有立方根的关系，即胶粒数多，则粒径小。可见，R_p 和 $\overline{X}_n$ 都与 $[S]^{3/5}$ 成正比。自由基生成速率 ρ 影响着胶粒的生成数，进而影响到聚合速率。胶粒数一旦恒定，尽管胶粒内仍进行着引发、增长、终止，但引发速率 ρ 不再影响聚合速率。维持 ρ 恒定，增加乳化剂浓度以增加胶粒数，就可同时提高 R_p 和 $\overline{X}_n$。胶粒数 N 可由乳化剂量来调节，而胶粒内自由基数 n 却无法控制。

d. 温度　乳液聚合中温度升高，将导致 k_p 增加，ρ 增加，N 增加，胶粒中单体浓度 $[M]$ 降低；自由基和单体扩散进入胶粒的速率增加。升高温度除了使聚合速率增加、聚合度降低外，还可能使乳液凝聚和破乳，产生支链和交联（凝胶），并对聚合物微结构和分子量分布产生影响。

3.1.11　自由基聚合重要聚合物

(1) 低密度聚乙烯

早在1933年，英国ICI公司研究人员在高温（200℃）和高压（200MPa）条件下，以微量氧为引发剂，按自由基机理，聚合成功低密度聚乙烯（LDPE）。高温聚合，易发生链转移反应。低密度聚乙烯平均每个高分子链含有1~2个长支链，10~40个短支链。低密度聚乙烯分子链的不规整性影响其结晶程度，其结晶度仅有50%~65%，熔点为105~110℃，密度大约为0.92g/cm³。

低密度聚乙烯加工性能和透明性都很好，低温下也能保持很好的柔软性，电绝缘性和耐溶剂性优良，目前被广泛用于薄膜、包装材料、各种容器等。

（2）聚苯乙烯

工业中大规模生产的聚苯乙烯（polystyrene）大都通过自由基聚合制备，主要有通用聚苯乙烯、发泡聚苯乙烯和抗冲聚苯乙烯（high-impact polystyrene，HIPS）。通用聚苯乙烯是苯乙烯的均聚物，分子量为5万~30万，玻璃化温度为95℃，伸长率为1%~3%，尺寸稳定，电性能好，透明色浅，流动性好，易成型加工。但有性脆、不耐溶剂、不耐紫外光和氧等缺点。

通用聚苯乙烯可以采用悬浮聚合或本体聚合法生产。通用聚苯乙烯聚合分预聚和后聚合两段，分别选用不同反应器。预聚阶段温度为80~100℃，控制转化率在30%~35%，在这一阶段凝胶效应不明显，可在一般搅拌釜内进行。转化率大于35%以后，进入后聚合阶段，黏度增加，自动加速渐增，温度自95℃递增至225℃。温度增加，可降低黏度，扩大温差，有利于传热。聚合结束后，物料进入减压脱挥装置，将残留苯乙烯降低到理想的含量（<0.3%）。

发泡聚苯乙烯采用悬浮聚合技术，将单体、引发剂、石油醚（发泡剂）、分散剂一起在水介质中进行悬浮聚合，聚合物颗粒中溶有石油醚，供以后发泡之用。也可以将常规悬浮聚合所得的聚苯乙烯粒子，在加温加压条件下浸渍石油醚而成可发性树脂。

抗冲聚苯乙烯是将5%~15%的聚丁二烯橡胶、适量引发剂溶于苯乙烯中，成为均相溶液，进行聚合。聚合过程中，部分自由基向橡胶链转移，产生接枝点，苯乙烯在接枝点上聚合，长出支链。其中接枝共聚物处于两种均聚物的界面，起到增溶作用，使聚丁二烯以细小粒子（1~2μm）稳定分散在聚苯乙烯基体中，类似海岛结构。抗冲聚苯乙烯受力时，按银纹机理，吸收了冲击能，而起到抗冲的作用。

（3）聚氯乙烯

大部分聚氯乙烯［poly(vinyl chloride)］用悬浮聚合法生产，也有部分使用乳液聚合和本体聚合的工艺。氯乙烯单体容易进行自由基聚合，很难进行离子聚合。单体结构中碳氢共价键是弱键，自由基容易置换氯原子，发生链转移反应。链转移常数对温度敏感，一般情况下，聚合温度是聚合的主要工艺参数，而引发剂浓度对分子量影响较小，但可以影响聚合速率。

悬浮聚合、乳液聚合和本体聚合的产品呈颗粒状，悬浮聚合和本体聚合的颗粒较大，乳液聚合产品的颗粒较小。为便于成型加工时增塑剂的吸收，要求颗粒结构疏松，这是聚氯乙烯与其他聚合物不同的地方。聚氯乙烯可被氯乙烯单体溶胀，但溶解的聚氯乙烯很少（<0.1%），因此氯乙烯聚合属于沉淀聚合。转化率小于70%时，聚合体系存在单体和聚氯乙烯溶胀粒子两相，转化率高于70%时，单体相消失，为了确保树脂颗粒的疏松结构，一般在转化率达到85%时，结束聚合反应。

聚氯乙烯的玻璃化转变温度大约为80~90℃，结晶度约为10%，添加增塑剂后，软化温度降低，便于加工成型。增塑剂添加50%以上的树脂产品称为软质聚氯乙烯，用于制造雨衣、薄膜、人造革、电缆包皮、软管等产品。硬质聚氯乙烯用于制造各种排水管道、塑料板材、家庭装修用建筑材料等。聚氯乙烯的缺点主要有：产品中的增塑剂容易渗透聚集，使产品性能劣化；燃烧时会释放毒性较强的氯气或盐酸气体；垃圾焚烧时，以金属炉中的金属为催化剂，可能产生毒性极强的二噁英等物质。

（4）聚四氟乙烯

氟是电负性最大的元素，亲电能力强，四氟乙烯分子中有四个氟原子吸电子，使碳碳双键键能大大减弱（398～440kJ/mol），成缺电子状态，是一强电子受体，很容易接受自由基而被引发聚合，聚合活性特高，成为烯类中摩尔聚合热最大的单体。

也是因为碳氟键键能大，自由基聚合中很难向单体或聚合物链转移反应，歧化终止也较少，以偶合终止为主。四氟乙烯沸点低（−76.3℃），常温下为气体，常压下在水中溶解度低（约0.1g/L），只能进行稀水溶液沉淀聚合。一般体系只由四氟乙烯单体、水溶性引发剂、水组成，聚合物一经形成，就从水相中沉析出来，粒子几经聚并，成为悬浮粗粒，工业俗称聚四氟乙烯［poly(tetrafluoroethylene)］悬浮聚合。分散聚合则另加少量全氟辛酸表面活性剂，用来防止粒子聚并，聚合物细粒（<30μm）产物分散在水中，工业俗称聚四氟乙烯分散聚合。因为聚四氟乙烯颗粒中的自由基被包埋，很难终止，能够得到分子量达数百万的聚合物。

因碳氟键键能大，聚四氟乙烯具有优异的化学稳定性，耐溶剂、耐氧化、耐强酸等。又因其结构对称，堆砌紧密，聚合物的结晶度高（93%～97%），熔点达到327℃，玻璃化温度为120℃，可在260℃下长期使用。聚四氟乙烯介电损耗小，电绝缘性能好，可用作高温高频绝缘材料；摩擦系数小，可制自润滑轴承；表面能低，可制超疏水材料。缺点是黏结性能差，加工较难。

聚四氟乙烯即使在较高温度下黏度也很高，例如，在380℃下为10^{10}Pa·s，难以用常规的塑料成型方法加工。一般将粉状产物模压成型，在350℃下定型，再机械加工。也可以将粉末与矿物油（沸点>200℃）捏合成面团状，再经注塑或挤出成形后，烧结定型。

（5）聚乙酸乙烯酯

聚乙酸乙烯酯工业上以过氧化物为引发剂，利用溶液聚合、悬浮聚合或乳液聚合的方法进行聚合。聚乙酸乙烯酯［poly(vinyl acetate)］玻璃化转变温度大约为30℃，为非晶聚合物。乳液聚合的固体含量很难超过50%，添加少量亲水性共单体（如乙烯基磺酸盐），则可提高固体含量。

聚乙酸乙烯酯大多用作木材黏结剂、涂料和水泥混凝土的添加剂（喷雾干燥细粉）。

聚乙酸乙烯酯在碱性条件下水解，可得聚乙烯醇［poly(vinyl alcohol)］。聚乙烯醇在造纸、纺织、黏结剂等工业领域有广泛用途。另外，也可以作为液晶显示器用的偏光薄膜材料，墨水、牙膏、化妆品中的添加剂等。

聚乙烯醇和甲醛反应得到的高分子材料可以纺丝，称之为维尼纶。维尼纶耐磨性优良，是各种作业服、渔网的纤维原料。聚乙烯醇和丁醛反应得到的高分子材料透明性和抗冲击性优良，是一种良好的玻璃黏结剂，可用作汽车防护玻璃。

（6）聚甲基丙烯酸甲酯

聚甲基丙烯酸甲酯［poly(methyl methacrylate)］玻璃化转变温度约为105℃，是一种非晶聚合物。聚甲基丙烯酸甲酯多采用自由基本体聚合，根据产品的不同要求，也可以选用溶液、悬浮、乳液聚合。

本体聚合主要用来生产透明板、棒、型材。由于凝胶效应显著，控制困难，多采用多段聚合技术。先在90℃下聚合至20%转化率的浆料，冷却后加入添加剂，注入平板模中聚合。

聚甲基丙烯酸甲酯透光率达92%，比无机玻璃好且质轻，常用作大型水槽、光学镜片、隐形眼镜镜片（硬质）、室外标牌、光导纤维等材料。

(7) 聚丙烯腈

丙烯腈常温下在水中的溶解度约7.8%，聚合产物不溶于水，很快从水中沉析出来。目前，纺织用腈纶大多与其他单体在水溶液中连续聚合来合成，选用氧化还原引发体系，聚合温度为40~50℃。第二单体有丙烯酸甲酯（7%~10%），可提高溶解性能，添加衣康酸（1%）作第三单体，可改善染色性能。该共聚物溶于二甲基甲酰胺等溶剂，经溶液纺丝，即成腈纶纤维。腈纶是合成纤维的第三大品种，耐光、耐候、强度好，熨烫后（<150℃）能保持良好的形状，手感类似羊毛，有人造羊毛之称，适宜制作针织衫和外套。

聚丙烯腈（polyacrylonitrile）均聚物经纺丝后，在200~300℃时在空气中加热后，再在惰性气体保护下在1000~1800℃继续加热可得碳纤维。若在预热处理和高温处理过程中进行拉伸使石墨网格结构取向，可以得到高强（2.5GPa）、高模（200GPa）碳纤维。碳纤维与热固性树脂复合后，得到的复合材料产品在航空航天、汽车制造、体育用品、土木建筑材料以及军工等领域具有广泛的用途。

(8) 聚丙烯酸

聚丙烯酸［poly(acrylic acid)］及其钠盐都是水溶性聚电解质，因分子量不同，可以形成许多品种。例如，分子量为3000~5000的低聚物是一种胶体保护剂，可在热电厂、中央空调等循环水里起到防水垢的作用，分子量为数百万的聚丙烯酸及其钠盐可以作为污水处理中的絮凝剂。此外，还可以用作分散剂、增稠剂、絮凝剂、涂层剂、吸附剂和胶黏剂，广泛用于涂料、织物整理、水处理、采油采矿冶金、农业等领域。丙烯酸及其钠盐是水溶性单体，可用过硫酸钾作引发剂，进行自由基聚合，主要有水溶液聚合和反相乳液聚合两种实施方法。

丙烯酸钠的水溶液聚合非常简单，丙烯酸钠盐水溶液中加入约1%过硫酸盐，100℃下聚合1h，即可完成。聚合高分子量丙烯酸钠，可在50℃下，使用过硫酸钾/亚硫酸钠氧化还原体系（浓度0.02%），聚合2h后结束。

反相乳液聚合常采用油溶性乳化剂，使50%丙烯酸钠水溶液在二甲苯中乳化，以过氧化二苯甲酰作引发剂，在60℃下聚合18h，经凝聚、过滤、干燥，可得高分子聚丙烯酸钠。

(9) 聚丙烯酰胺

分子量数百万的聚丙烯酰胺的主要用途是生产饮用自来水的絮凝剂，此外还广泛用于造纸、采油、矿冶乃至水敏性凝胶、酶的固定化和生物医学材料等。聚丙烯酰胺的合成可以采用水溶液聚合和反相乳液聚合两种方法。

丙烯酰胺水溶液聚合体系中含单体、水、水溶性引发剂，8%~10%丙烯酰胺水溶液可在搅拌釜内聚合，使用氮气或二氧化碳保护，在温度为20~50℃时，聚合2~4h，转化率可达95%~99%，所形成的水溶胶即为产品。8%~10%聚丙烯酰胺水溶液黏度很高，呈冻胶状，搅拌传热都很困难，分子量受到限制。如欲制备高分子量（10^7）品种，则选用反相乳液聚合法。反相乳液聚合首先将丙烯酰胺、过硫酸钾/亚硫酸钠、水配成水溶液，另将油溶性乳化剂配成有机溶剂溶液，通过高速搅拌形成油包水（W/O）乳液体系，在 70℃下聚合 1h 就可完成反应。

英语读译资料

(1) PAN

Polyacrylonitrile (PAN) forms the basis for a number of fibers and copolymers. As fibers, they are referred to as acrylics or acrylic fibers. The development of acrylic fibers began in the early 1930s in Germany but they were first commercially produced in the United States by DuPont and Monsanto, about in 1950.

Because of the repulsion of the cyanide groups, the polymer backbone assumes a rod-like conformation. The fibers derive their basic properties from this stiff structure of PAN where the nitrile groups are randomly distributed about the backbone rod. Because of strong hydrogen bonding between the chains, they tend to form bundles. Most acrylic fibers actually contain small amounts of other monomers, such as methyl acrylate and methyl methacrylate. Because they are difficult to dye, small amounts of ionic monomers, such as sodium styrene sulfonate, are often added to improve their dye-ability.

Acrylic fibers are used as an alternative to wool for sweaters. PAN is also used in the production of blouses, blankets, rugs, curtains, shirts, craft yarns, and pile fabrics used to simulate fur. At temperatures above 160℃, PAN begins forming cyclic imines that dehydrogenate forming dark-colored heat-resistant fused ring polymers with conjugated C=C and C=N bonding.

Fibers with more than 85% acrylonitrile units are called acrylic fibers but those containing 35%～85% acrylonitrile units are referred to as modacrylic fibers. The remainder of the modacrylic fibers are derived from comonomers such as vinyl chloride or vinylidene chloride, which are specifically added to improve flame resistance.

(2) Atom-transfer radical polymerization

Research into control of cationic polymerizations using transition metals led to the discovery, first reported independently by Sawamoto and Matyjaszewski in 1995, of transition-metal-catalysed radical polymerizations that have the characteristics of living polymerization.

This type of polymerization is now more commonly termed atom-transfer radical polymerization (ATRP). The transition metal is in the form of a halide compound that is complexed by ligands and behaves as a true catalyst, but more usually is called an activator. Organic halides RX (usually bromides or chlorides) are used as initiators. Initiation proceeds via a single-electron transfer from the metal to the halogen atom in the R—X bond, causing homolysis of the bond to yield a free radical, R·, and an increase in the oxidation state of the metal atom which captures the released halogen atom. Sooner rather than later, propagation of the chain radical is intercepted by the reverse process in which the oxidized transition metal complex donates a halogen atom back to the propagating radical resulting in deactivation of the chain radical through formation of a new C—X bond at the chain end and regeneration of the metal. Thus, ATRP has features similar to NMP: (ⅰ) the chain grows through a series of activation-propagation-deactivation cycles; (ⅱ) the equilibrium must be controlled so that it lies far to the left and exchange is rapid, this being achieved for each monomer by careful choice of the initiator, activator complex and temperature; and (ⅲ) most chains exist in the dormant state, so the free radical concentration is very low, thereby massively reducing the probability of bimolecular termination events, but also reducing the rate of polymerization.

(3) Fluorine-containing polymers

Polytetrafluoroethylene (PTFE), better known by its trade name of Teflon, was accidentally discovered by Roy J. Plunkett, a DuPont chemist who had just received his PhD from Ohio State 2 years before. He was part of a group searching for nontoxic refrigerant gases. On April 6, 1938, he and his assistant, Jack Rebok, had filled a tank with tetrafluoroethylene. After some time, they opened the value but no gas came out. The tank weight indicated that there was no weight loss, so what happened to the tetrafluoroethylene. Using a hacksaw, they cut the cylinder in half and found a waxy white powder. He correctly surmised that the tetrafluoroethylene had polymerized.

The waxy white powder had some interesting properties. It was quite inert toward strong acids, bases, and heat and was not soluble in any attempted liquid. It appeared to be quite "slippery." Little was done with this new material until the military, working on the atomic bomb, needed a special material for gaskets that would resist the corrosive gas uranium hexafluoride that was one of the materials being used to make the atomic bomb. General Leslie Groves, responsible for the U.S. Army's part in the atomic bomb project, had learned of DuPont's new inert polymer and had DuPont manufacture it for them.

Teflon was introduced to the public in 1960 when the first Teflon-coated muffin pans and frying pans were sold. Like many new materials, problems were encountered. Bonding to the surfaces was uncertain at best. Eventually, the bonding problem was solved. Teflon is now used for many other applications, including acting as a biomedical material in artificial corneas, substitute bones for nose, skull, hip, nose, and knees; ear parts, heart valves, tendons, sutures, dentures, and artificial tracheas. It has also been used in the nose cones and heat shield for space vehicles and for their fuel tanks.

Over a half million vascular graft replacements are performed every year. Most of these grafts are made of poly(ethyleneterephthalate) (PET) and PTFE. These relatively large-diameter grafts work when blood flow is rapid, but they generally fail for smaller vessels. Polytetrafluoroethylene is produced by the free radical polymerization process. While it has outstanding thermal and corrosive resistance, it is a marginal engineering material because it is not easily machinable. It has low tensile strength, resistance to wear, and it has low-creep resistance.

Molding powders are processed by press and sinter methods used in powder metallurgy. It can also be extruded using ram extruder techniques. PTFE is a crystalline polymer with melting typically occurring above 327℃. Because it is highly crystalline, it does not generally exhibit a noticeable T_g.

The C—F bond is one of the strongest single bonds known with a bond energy of 485kJ/mol. While it is structurally similar to linear polyethylene (PE), it has marked differences. Because of the small size of hydrogen, PE exists as a crank-shaft backbone structure. Fluorine is a little larger (atomic radius of F=71pm and for H=37pm) than hydrogen, causing the Teflon backbone to be helical and forming a complete twist every 13 carbon atoms. The size of the fluorine is sufficient to form a smooth "protective" sheath around the carbon backbone. The concentration of F end groups is low in ultrahigh molecular weight PTFE, contributing to its tendency to form crystals.

The electron density of PE and PTFE are also different. The electronegativity value for C is 2.5, F=4.0, and for H=2.1. Thus, the electron density on the fluorine surface of PTFE is greater than that for PE.

For high molecular weight linear PE, the repeat unit length is about 0.254nm forming crystalline portions with a characteristic thickness of about 10nm. The chain length for tough solids from PE is about 4.5times the crystalline thickness. Thus, tough solids occur at molecular weights greater than 5,000g/mol or chain lengths greater than about 45nm. In comparison, the repeat unit length for PTFE is about 0.259nm. The crystalline thicknesses for PTFE are about 100~200nm or much thicker than for PE. Chain lengths for tough solids are about 4.5times the crystalline thickness. Thus, much greater chain sizes, about 200,000~400,000Da, are required to produce tough solids.

The greater size of the crystalline portions also probably contributes to its higher Tm and greater difficulty in processing. The crystal thickness of PTFE is about 10~20times the crystal thickness found for most other semicrystalline polymers such as PE. At low molecular weights, PTFE is waxy and brittle. To achieve good mechanical properties ultrahigh molecular weights on the order of 10 million dalton is usually needed. These long chains disrupt crystal formation because they are longer than a single crystal. But the long chain lengths connect the crystals together adding to their strength. But these long chains result in extremely high viscosities so that ultrahigh molecular weight PTFE does not flow when melted and is thus, not melt processable. Form restrictive and costly methods are used to produce products from PTFE. While vinyl fluoride was prepared in about 1900, it was believed resistant to typical "vinyl" polymerization.

German scientists prepared vinyl fluoride through reaction of acetylene with hydrogen fluoride in the presence of catalysts in 1933. It was not until 1958 that DuPont scientists announced the polymerization of vinyl fluoride forming poly(vinyl fluoride) (PVF). Polymerization is accomplished using peroxide catalysts in water solutions under high pressure.

In comparison to PTFE, PVF is easily processable using a variety of techniques used for most thermoplastic materials. It offers good flame retardancy, presumably due to the formation of HF that assists in the control of the fire. Thermally induced formation of HF is also a negative factor because of its toxicity. As in the case of PVC, elimination of the hydrogen halide (HF) promotes formation of aromatic polycyclic products that themselves are toxic.

The difference in electronegativity between the adjacent carbons because of the differing electronegativities of H and F results in the C—F bond being particularly polar, resulting in it being susceptible to attack by strong acids. The alternating bond polarities on the PVF chain gives a tight structure, resulting in PVF films having a low permeability. This tight structure also results in good resistance, resistance to cracking, and resistance to fading.

Friction and wear are important related characteristics. If a material has a high friction then it will generally have a shorter wear time because water or other friction event chemicals pass over the material with the higher friction causing greater wear. The friction eventually "wears" away polymer chains layer-by-layer. The engineering laws of sliding friction are simple. According to Amontons' laws, the friction F between a body (rain drop, wind, or board rubbing against the material) and a plane surface (the polymeric material) is proportional to the load L and independent to the area of contact A. The friction of moving bodies is generally less than that of a static body. The kinetic friction is considered independent of the velocity. The coefficient of friction is defined as F/L. Polymers show a wide range of coefficients of friction so that rubbers exhibit relatively high values (BR=0.4~1.5 and SBR=0.5~3.0) whereas some polymers such as PTFE (0.04~0.15) and PVF (0.10~0.30) have low values.

The low coefficient of friction for PVF results in materials coated with it remaining somewhat free of dirt and other typical contaminants, allowing PVF-coated materials to be less frequently cleaned. It is essentially self-cleaning as rain carries away dust and other particulates, including bird droppings, acid rain, and graffiti. The low friction also results in longer lifetimes for materials coated with the PVF and for the PVF coating itself.

PVF has a T_g of about $-20℃$ remaining flexible over a wide-temperature range (from about $-20℃$ to $150℃$), even under cold temperatures. Because of its low coefficient of friction and tightly bound structure, it retains good strength as it weathers. Films, in Florida, retain much of their thickness even after about a decade losing less than 20% of their thickness. To increase their useful lifetimes, relatively thick films, such as 1 mil, are generally employed. The "slickness" also acts to give the material a "natural" mildew resistance.

Unlike PVC that requires plasticizers to be flexible, PVF contains no plasticizers and does not "dry out" like PVC. PVF, because of its higher cost in comparison to PE and PP, is used as a coating and selected "high end" bulk applications such as films. Films are sold by DuPont under the trade name of Tedlar. Tedlar is used in awnings, outdoor signs, roofing, highway sound barriers, commercial building panels, and solar collectors. It is used as a fabric coating, protecting the fabric from the elements. PVF is resistant to UV-related degradation and unlike PVC, it is inherently flexible. While transparent, pigments can be added to give films and coatings with varying colors. Protective coatings are used on plywood, automotive parts, metal siding, lawn mower housings, house shutters, gutters, electrical insulation, and in packaging of corrosive chemicals. PVF has pizeoelectric properties generating a current when compressed.

(4) Polymerization techniques

Addition polymerizations can be carried out using several methods.

(ⅰ) **Bulk polymerization** In bulk polymerization, the system essentially consists of monomer (in liquid or gaseous form) and initiator. Polymerization of liquid monomers, such as phenylethene, raises all sorts of problems. Addition polymerizations are strongly exothermic and reaction may get out of control or even become explosive. Since polymers are usually soluble in their monomers, the viscosity of the system increases and difficulties in stirring can arise. Heat dissipation is a problem; in fact, localized overheating can lead to polymer degradation. Generally speaking, although bulk polymerization yields a relatively pure product, it is seldom used for preparing large batches of addition polymer.

(ⅱ) **Solution polymerization** Here, the monomer is dissolved in a suitable solvent prior to polymerization being carried out. In such a polymerizing system, the solvent can assist in the dissipation of (exothermic) heat of reaction. Disadvantages include the possibility of chain transfer to solvent occurring with the subsequent formation of lower relative molecular mass polymer, and the fact that the solvent has to be eventually removed from the resulting polymer. This latter problem can be overcome by using a solvent which will dissolve the monomer but not the polymer, hence, the polymer is obtained directly as a slurry.

(ⅲ) **Suspension polymerization** In this process the monomer (containing dissolved initiator) is dispersed as droplets in water. This is done by vigorous stirring during the reaction. Polymerization then takes place within the droplets (in other words, each isolated droplet is effectively undergoing a tiny bulk polymerization). The droplets are kept from sticking together by the addition of small amounts of stabilizers, such as talc or polyvinyl alcohol. Advantages of this system include the dissipation of heat of reaction into the aqueous phase, and the fact that the polymeric product is in the form of small granules which are easily handled and relatively uncontaminated.

(ⅳ) **Emulsion polymerization** Emulsion polymerization bears some resemblance to suspension polymerization but differs in that soap is added to stabilize the monomer droplets. The soap also forms aggregates of soap molecules or micelles. These micelles solubilize some of the monomer, i.e., take some of the monomer into the interior of the micelle. Initiator, which is dissolved in the aqueous phase, diffuses into the micelles, thereby initiating polymerization. The polymer molecule grows, taking further monomer from the aqueous phase. In this way, high relative molecular mass polymer can be formed. Again, the aqueous phase absorbs the heat of reaction.

习 题

1. 为什么单取代和1,1-双取代烯烃类单体容易聚合，而1,2-双取代烯烃类单体难聚合？
2. 为什么能够自由基聚合的烯烃类单体较多，而能离子聚合的烯烃类单体较少？
3. 为什么说传统自由基聚合的机理特征是慢引发、快增长、速终止？
4. 大致说明下列引发剂的使用温度范围，并写出分解反应式：

①异丙苯过氧化氢；②过氧化十二酰；③过氧化碳酸二环己酯；④过硫酸钾-亚铁盐；⑤过氧化二苯甲酰二甲基苯胺。

5. 等离子体对聚合和聚合物化学反应有何作用？传统聚合反应与等离子态聚合有何区别？

6. 推导自由基聚合动力学方程时，作了哪些基本假定？

7. 氯乙烯、苯乙烯、甲基丙烯酸甲酯聚合时，都存在自动加速现象，三者有何异同？这三种单体聚合的终止方式有何不同？氯乙烯聚合时，选用半衰期约为 2h 的引发剂，有望接近匀速反应，解释其原因。

8. 在自由基溶液聚合中，单体浓度增加 10 倍，聚合速率和数均聚合度会有什么变化？

9. 在自由基溶液聚合中，引发剂浓度减半，聚合速率和数均聚合度会有什么变化？

10. 动力学链长的定义？与平均聚合度的关系？链转移反应对动力学链长和聚合度有何影响？试举 2~3 例说明利用链转移反应来控制聚合度的工业应用，试用链转移常数数值来帮助说明。

11. 提高聚合温度和增加引发剂浓度，均可提高聚合速率，问哪一种措施更好？

12. 链转移反应对支链的形成有何影响？聚乙烯的长支链和短支链，以及聚氯乙烯的支链是如何形成的？

13. 苯乙烯和乙酸乙烯酯分别在苯、甲苯、乙苯、异丙苯中聚合，从链转移常数来比较不同自由基向相同溶剂链转移的难易程度和对聚合度的影响，并作出分子级的解释。

14. 为什么可以说丁二烯或苯乙烯是氯乙烯或乙酸乙烯酯聚合的终止剂或阻聚剂？比较乙酸乙烯酯和乙酸烯丙基酯的聚合速率和聚合产物的分子量，说明原因。

15. 可控和"活性"自由基聚合的基本原则是什么？简述原子转移自由基聚合法（ATRP）和可逆加成-断裂转移（RAFT）法可控自由基聚合的基本原理。

16. 本体法制备有机玻璃板和通用级聚苯乙烯，比较过程特征，如何解决传热问题以保证产品品质。

17. 溶液聚合多用于离子聚合和配位聚合，而较少用于自由基聚合，为什么？

18. 比较氯乙烯本体聚合和悬浮聚合的过程特征，产品品质有何异同？

19. 简述传统乳液聚合中单体、乳化剂和引发剂的所在场所，引发、增长和终止的场所和特征，胶束、乳胶粒、单体液滴和速率的变化规律。

20. 简述胶束成核、液滴成核、水相成核的机理和区别。

21. 60℃过氧化二碳酸二环己酯在某溶剂中分解，用碘量法测定不同时间的残留引发剂浓度，数据如下，试计算分解速率常数（s^{-1}）和半衰期（h）。

时间/h	0	0.2	0.7	1.2	1.7
DCPD 浓度/(mol·L^{-1})	0.0754	0.0660	0.0484	0.0334	0.0288

22. 在甲苯中不同温度下测定偶氮二异丁腈的分解速率常数，数据如下，求分解活化能。再求 40℃和 80℃下的半衰期，判断在这两个温度下聚合是否有效。

温度/℃	50	60.5	69.5
分解速率常数/s^{-1}	2.64×10^{-6}	1.16×10^{-5}	3.78×10^{-5}

23. 苯乙烯溶液浓度为 0.20mol/L，过氧类引发剂浓度为 4.0×10^{-3}mol/L，在 60℃下聚合，如引发剂半衰期为 44h，引发剂效率 $f=0.80$，$k_p=145$L/(mol·s)，$k_t=7.0\times10^7$L/(mol·s)，欲达到 50%转化率，需多长时间？

24. 计算苯乙烯乳液聚合速率和聚合度。已知 60℃时，$k_p=176$L/(mol·s)，[M] $=5.0$mol/L，$N=3.2\times10^{14}$/mL，$\rho=1.1\times10^{12}$个/(mL·s)。

25. 比较苯乙烯在 60℃下本体聚合和乳液聚合的速率和聚合度。乳胶粒数 $=1.0\times10^{15}$/mL，[M] $=5.0$mol/L，$\rho=5.0\times10^{12}$个/(mL·s)。两体系相同速率常数：$k_p=176$L/(mol·s)，$k_t=3.6\times10^7$L/(mol·s)。

26. 传统乳液聚合配方如下：苯乙烯 100g，水 200g，过硫酸钾 0.3g，硬脂酸钠 5g。试计算，(1) 溶于水中的苯乙烯分子数（个/mL），20℃溶解度为 0.02g/100g 水，阿伏伽德罗数 $N_A=6.023\times10^{23}$/mol；(2) 单体液滴数（个/mL），液滴直径 1000nm，苯乙烯溶解和增溶量共 2g，苯乙烯密度为 0.9g/mL；(3) 溶于水中的钠皂分子数（个/mL），硬脂酸钠 CMC 为 0.13g/L，分子量为 306.5；(4) 水中胶束数（个/mL），每胶束由 100 个肥皂分子组成；(5) 水中过硫酸钾分子数（个/mL），分子量为 270；(6) 初级自由基形成速率 ρ（个/mL·s），50℃时，$k_d=9.5\times10^{-7}$s^{-1}；(7) 乳胶粒数（个/mL），粒径 100nm，无单体液滴。苯乙烯密度为 0.9g/mL，聚苯乙烯密度为 1.05g/mL，转化率为 50%。

27. 传统乳液聚合经数十年发展，涌现出种子乳液聚合、核壳乳液聚合、无皂乳液聚合、反相乳液聚合、分散聚合等新技术。查阅文献，总结各种新兴乳液聚合技术的特点，聚合产物的结构特色和当前的应用状况。

英语习题

1. The decomposition of AIBN in xylene at 77℃ was studied by measuring the volume of N_2 evolved as a function of time. The volumes obtained at time t and $t=\infty$ are V_t and V_∞, respectively. Show that the manner of plotting used in following figure is consistent with the integrated first-order rate law and evaluate k_d.

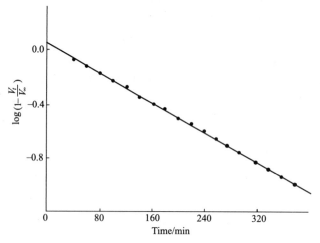

Volume of nitrogen evolved from the decomposition of AIBN at 77℃ plotted according to the first-order rate law as discussed above.

2. For an initiator concentration that is constant at $[I]_0$, the non-stationary-state radical concentration varies with time according to the following expression:

$$\frac{[P\cdot]}{[P\cdot]_S} = \frac{\exp[4fk_d k_t [I]_0^{1/2} t] - 1}{\exp[4fk_d k_t [I]_0^{1/2} t] + 1}$$

Calculate $[P\cdot]_S$ and the time required for the free-radical concentration to reach 99% of this value using the following as typical values for constants and concentrations: $k_d = 1.0 \times 10^{-4}\,s^{-1}$, $k_t = 3 \times 10^7\,L/mol$, $f = 1/2$, and $[I]_0 = 10^{-3}\,mol/L$. Comment on the assumption $[I] = [I]_0$ that was made in deriving this non-stationary-state equation.

3. The decomposition of benzoyl peroxide is characterized by a half-life of 7.3h at 70℃ and an activation energy of 29.7kcal/mol (1kcal/mol = 4.18kJ/mol). What concentration (mol/L) of this peroxide is needed to convert 50% of the original charge of a vinyl monomer to polymer in 6 hours at 60℃? [Data: $f = 0.4$; $k_p^2/k_t = 1.04 \times 10^{-2}\,L/(mol \cdot s)$ at 60℃]

4. For a new monomer 50% conversion is obtained in 500min when polymerized in homogeneous solution with a thermal initiator. How much time would be needed for 50% conversion in another run at the temperature but with four-fold initial initiator concentration?

5. If a 5% solution of a monomer A containing $10^{-4}\,mol/L$ of peroxide P is polymerized at 70℃, 40% of the original monomer charge is converted to polymer in 1h. How long will it take to polymerize 90% of the original monomer charge in a solution containing (initially) 10% A and peroxide P of $10^{-2}\,mol/L$?

6. There is evidence that thermal self-initiated polymerization of styrene may be of about five-halves order. Show that this is in agreement with the estabilished initiation mechanism involving a Diels-Alder dimer formation.

7. Estimate the chain transfer constants for styrene to isopropylbenzene, ethylbenzene, toluene, and benzene from the data presented in Figure 3-11. Comment on the relative magnitude of these constants in terms of the structure of the solvent molecules.

8. Many olefins can be readily polymerized by a free-radical route. On the other hand, isobutylene is usually polymerized by a cationic mechanis. Explain.

9. Draw the mechanisms for the following processes in the radical polymerization of styrene in toluene: (a) initiation by cumyl peroxide; (b) propagation; (c) termination by disproportionation; and (d) transfer to solvent.

10. Show the mechanisms of addition of a butadiene monomer to a poly(butadienyl) radical, to give each of the three possible geometric isomers.

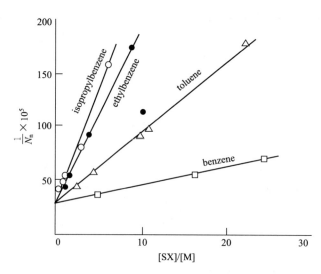

Figure 3-11　Effect of chain transfer to solvent for PS at 100℃

11. From a linear plot of $\ln([M]_0/[M])$ versus time, it has been reported that apparent propagation rate constant for the ATRP of styrene in bulk is on the order of 10^{-4} s^{-1}, where the apparent rate constant k_p is defined by $R_p = k_p[M]$. What is the order of magnitude of the concentration of active radicals at any time?

12. Show by chemical equations the polymerization of acrylonitrile initiated by the thermal decomposition of cumyl hydroperoxide.

13. Using ^{14}C-labeled AIBN as an initiator, a sample of styrene is polymerized to an number-average degree of polymerization of 1.52×10^4. The AIBN has an activity of 9.81×10^7 counts/(min·mol) in a scintillation counter. If 3.22g of the polystyrene has an activity of 203counts/min, what is the mode of termination?

14. Poly(vinyl acetate) of number-average molecular weight 100000 is hydrolyzed to poly(vinyl alcohol). Oxidation of the latter with periodic acid to cleave 1,2-diol linkages yields a poly(vinyl alcohol) with $X_n = 200$. Calculate the percentages of head-to-tail and head-to-head linkages in the poly (vinyl acetate).

15. For a radical polymerization with bimolecular termination, the polymer produced contains 1.30 initiator fragments per polymer molecule. Calculate the relative extents of termination by disproportionation and coupling, assuming that no chain-transfer reactions occur.

16. A solution 0.20mol/L in monomer and 4.0×10^{-3} mol/L in a peroxide initiator is heated at 60℃. How long will it take to achieve 50% conversion?

$k_p = 145$L/(mol·s), $k_t = 7.0 \times 10^7$ L/(mol·s), $f = 1$, and the initiator half-life is 44h.

17. Consider the polymerization of styrene initiated by di-t-butyl peroxide at 60℃. For a solution of 0.01mol/L peroxide and 1.0mol/L styrene in benzene, the initial rates of initiation and polymerization are 4.0×10^{-11} and 1.5×10^{-7} L/(mol·s), respectively. Calculate the values of (fk_d), the initial kinetic chain length, and the initial degree of polymerization. Indicate how often on the average chain transfer occurs per each initiating radical from the per-

oxide. What is the breadth of the molecular weight distribution that is expected, that is, what is the value of $X_w = X_n$? Use the following chain-transfer constants:

$C_M = 8.0 \times 10^{-5}$; $C_I = 3.2 \times 10^{-4}$; $C_P = 1.9 \times 10^{-4}$; $C_S = 2.3 \times 10^{-6}$.

18. For a particular application the molecular weight of the polystyrene obtained in Problem 17 is too high. What concentration of n-butyl mercaptan should be used to lower the molecular weight to 85000? What will be the polymerization rate for the reaction in the presence of the mercaptan?

19. Consider the polymerization reaction in Problem 17. Aside from increasing the monomer concentration, what means are available for increasing the polymerization rate? Compare the alternate possibilities with respect to any changes that are expected in the molecular weight of the product.

20. Describe how ATRP is used to graft styrene onto a vinyl chloride-vinylchloroacetate copolymer.

3.2 自由基共聚合

自由基共聚合

全地形轮胎（all terrain），简称 AT 轮胎，是越野爱好者使用最广的轮胎。全地形轮胎的花纹粗犷，胎牙间距比普通胎大，耐用性和在非铺装路面上的抓地力要强于公路胎，是越野和公路性能兼顾的轮胎。丁苯橡胶（SBR），由苯乙烯和丁二烯通过自由基共聚获得，是橡胶工业的骨干产品，也是最大的通用合成橡胶品种，其加工性能接近于天然橡胶，耐磨、耐热、耐老化则优于天然橡胶，是轮胎生产的主要原料。

本节目录

3.2.1　共聚物的类型和命名 / 123
3.2.2　研究共聚反应的意义 / 123
3.2.3　二元共聚物的组成 / 124
3.2.4　竞聚率 / 128
3.2.5　单体活性和自由基活性 / 129
3.2.6　$Q\text{-}e$ 概念 / 131
3.2.7　重要烯烃类共聚物 / 132

重点要点

无规共聚，交替共聚，嵌段共聚，接枝共聚，竞聚率，理想共聚，交替共聚，恒比共聚，单体和自由基的活性，$Q\text{-}e$ 概念，共聚合速率。

仅仅由一种单体进行的聚合反应，称作均聚（homopolymerization），聚合产物称作均聚物（homopolymer）；由两种及以上单体同时参与的聚合，称作共聚（copolymerization），聚合产物为成分复杂的共聚物（copolymer）。

3.2.1 共聚物的类型和命名

由两种单体聚合而成的最简单的二元共聚物有下列四种类型（图 3-12）。

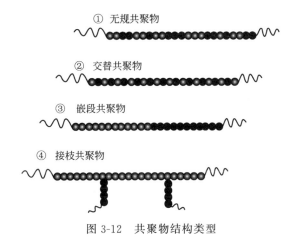

图 3-12 共聚物结构类型

① 无规共聚物（random copolymer） 两单体产生的结构单元 M_1、M_2 随机排列，无固定顺序。多数自由基共聚物属于这一类型，如氯乙烯-乙酸乙烯酯共聚物。

② 交替共聚物（alternating copolymer） 共聚物中 M_1、M_2 两单元严格交替相间，M_1 单元与 M_2 单元相邻。苯乙烯-马来酸酐共聚物即是这类共聚物的典型代表。

③ 嵌段共聚物（block copolymer） 由几百至几千个 M_1 结构单元构成的 A 链段和另一较长的由 M_2 结构单元构成的 B 链段通过共价键连接而成的大分子，称作 AB 型嵌段共聚物。也有 ABA 型（如苯乙烯-丁二烯-苯乙烯二嵌段共聚物 SBS）。

④ 接枝共聚物（graft copolymer） 主链由 M_1 单元组成，支链则由另一种 M_2 单元组成。例如，丁二烯-(苯乙烯-丙烯腈)接枝共聚物称作 ABS 树脂。

一般而言，无规和交替共聚物呈均相，聚合时遵循同一聚合原理；嵌段和接枝共聚物大多呈异相，可有多种聚合机理。

共聚物的命名原则是将两单体名称连以短横，前面冠以"聚"字，如聚(丁二烯-苯乙烯)，或称作丁二烯-苯乙烯共聚物。国际命名中常在两单体名之间插入 -ran-、-alt-、-b-、-g-，分别代表无规、交替、嵌段、接枝，无特定聚合方式的共聚物使用 -co- 表示。

3.2.2 研究共聚反应的意义

一种单体只能形成一种均聚物，种类有限，但是通过与第二种、第三种单体共聚，可以获得多种高分子结构。共聚是调整诸如柔韧性、玻璃化温度、熔点、溶解性能、染色性能、表面性能等材料性能的一种简单有效的方法。

在理论上，通过共聚研究，可以评价单体、自由基、碳阳离子、碳阴离子的活性，进一步了解单体活性与结构的关系。

研究均聚时，聚合速率、平均聚合度、聚合度分布等指标比较重要；研究共聚时，共聚物的组成和序列分布是首要问题。

3.2.3 二元共聚物的组成

20 世纪 40 年代，Mayo 等建立了共聚物组成方程及相关理论。建立共聚物组成方程时，设定了下列假定条件：

① 自由基等活性。即自由基活性仅仅与单电子所在的结构单元有关，与链长无关，同时与该结构单元以外的链结构无关。

② 聚合反应不可逆。即无解聚反应。

③ 长链假设。即共聚物聚合度很大，链增长是消耗单体和决定共聚物组成的主要过程，链引发和链终止对共聚物组成影响可以忽略。

④ 稳态。自由基总浓度和两种自由基的浓度都保持恒定不变。

M_1、M_2 两种单体二元共聚时有四种链增长反应：

$$\sim\!\!M_1^\bullet + M_1 \xrightarrow{k_{11}} \sim\!\!M_1 M_1^\bullet \qquad R_{11} = k_{11}[M_1^\bullet][M_1]$$

$$\sim\!\!M_1^\bullet + M_2 \xrightarrow{k_{12}} \sim\!\!M_1 M_2^\bullet \qquad R_{12} = k_{12}[M_1^\bullet][M_2]$$

$$\sim\!\!M_2^\bullet + M_2 \xrightarrow{k_{22}} \sim\!\!M_2 M_2^\bullet \qquad R_{22} = k_{22}[M_2^\bullet][M_2]$$

$$\sim\!\!M_2^\bullet + M_1 \xrightarrow{k_{21}} \sim\!\!M_2 M_1^\bullet \qquad R_{21} = k_{21}[M_2^\bullet][M_1]$$

式中，R 和 k 下标中第一个数字代表自由基，第二个代表单体，例如 R_{11} 和 k_{11} 分别代表自由基 $M_1^\bullet$ 和单体 M_1 反应的增长速率和增长速率常数。

研究自由基共聚时，将均聚自增长和共聚交叉增长速率常数之比称为竞聚率 r （reactivity ratio），以表征两单体的相对活性。

$$r_1 = k_{11}/k_{12}$$
$$r_2 = k_{22}/k_{21}$$

根据上述假设条件③，共聚物聚合度很大，引发和终止反应对共聚物组成的影响可忽略。所以，我们只考虑四个链增长反应动力学方程。于是，M_1 和 M_2 进入共聚物的速率仅取决于增长速率。

$$-\frac{d[M_1]}{dt} = R_{11} + R_{21} = k_{11}[M_1^\bullet][M_1] + k_{21}[M_2^\bullet][M_1]$$

$$-\frac{d[M_2]}{dt} = R_{12} + R_{22} = k_{12}[M_1^\bullet][M_2] + k_{22}[M_2^\bullet][M_2]$$

两式相除，两单体消耗速率比等于两单体进入共聚物的物质的量之比（n_1/n_2）。

$$\frac{n_1}{n_2} = \frac{d[M_1]}{d[M_2]} = \frac{R_{11} + R_{21}}{R_{12} + R_{22}} = \frac{k_{11}[M_1^\bullet][M_1] + K_{21}[M_2^\bullet][M_1]}{k_{12}[M_1^\bullet][M_2] + K_{22}[M_2^\bullet][M_2]}$$

依稳态假设，$M_1^\bullet$ 和 $M_2^\bullet$ 的浓度保持恒定需要两种自由基相互转化的速率相等，即：

$$R_{12} = R_{21} = k_{12}[M_1^\bullet][M_2] = k_{21}[M_2^\bullet][M_1]$$

代入上式，得：

$$\frac{d[M_1]}{d[M_2]} = \frac{R_{11}+R_{12}}{R_{21}+R_{22}} = \frac{k_{11}[M_1^\bullet][M_1]+k_{12}[M_1^\bullet][M_2]}{k_{21}[M_2^\bullet][M_1]+k_{22}[M_2^\bullet][M_2]}$$

$$= \frac{[M_1^\bullet]}{[M_2^\bullet]} \times \frac{k_{11}[M_1]+k_{12}[M_2]}{k_{21}[M_1]+k_{22}[M_2]}$$

利用 $R_{21}=R_{12}$，可推得两种自由基浓度之比为：

$$\frac{[M_1^\bullet]}{[M_2^\bullet]} = \frac{k_{21}[M_1]}{k_{12}[M_2]}$$

将 $r_1=k_{11}/k_{12}$，$r_2=k_{22}/k_{21}$ 两种单体的竞聚率代入上式，可简化为：

$$\frac{d[M_1]}{d[M_2]} = \frac{[M_1]}{[M_2]} \times \frac{k_{21}}{k_{12}} \times \frac{k_{11}[M_1]+k_{12}[M_2]}{k_{21}[M_1]+k_{22}[M_2]}$$

$$= \frac{[M_1]}{[M_2]} \times \frac{\dfrac{k_{11}}{k_{12}}[M_1]+[M_2]}{[M_1]+\dfrac{k_{22}}{k_{21}}[M_2]}$$

$$= \frac{[M_1]}{[M_2]} \times \frac{r_1[M_1]+[M_2]}{[M_1]+r_2[M_2]} \tag{3-4}$$

这就是以两种单体浓度表示的二元共聚物组成的微分方程。该方程描述的是二元共聚物瞬时组成与单体瞬时浓度之间的定量关系。在该方程中，竞聚率是非常重要的共聚反应参数，也是本章的重要概念。

如果进一步设定 f_1 和 f_2 分别代表单体 M_1 和 M_2 在某一瞬间占两单体总量的摩尔分数，即：

$$f_1 = \frac{[M_1]}{[M_1]+[M_2]} = 1-f_2$$

又设定 F_1 和 F_2 分别代表结构单元 M_1 和 M_2 在某一瞬间占共聚物中结构单元总量的摩尔分数，即：

$$F_1 = \frac{d[M_1]}{d[M_1]+d[M_2]} = 1-F_2$$

可推导出：

$$F_1 = \frac{r_1 f_1^2 + f_1 f_2}{r_1 f_1^2 + 2 f_1 f_2 + r_2 f_2^2} \tag{3-5}$$

在实际应用中，三元共聚和四元共聚也经常出现。许多多元共聚物一般以两种主要单体确定基本性能，再加少量第三或第四单体作特殊改性。例如，氯乙烯-乙酸乙烯酯共聚体系内添加 1%~2%马来酸酐可提高黏结性能；丙烯腈-丙烯酸甲酯共聚时添加 1%~2%衣康酸，可改善聚丙烯腈的染色性能；乙烯-丙烯共聚合成乙丙橡胶时加 2%~3%二烯烃，可提高乙丙橡胶中的双键含量，改善交联性能。

三元共聚物组成可以参照二元共聚方程进行推导，但是其共聚物组成微分方程非常复杂，此处不讨论三元共聚。

式(3-5)表示共聚物瞬时组成与瞬时单体浓度间的函数关系。竞聚率 r_1、r_2 是影响两者关系的主要参数，共聚组成曲线因竞聚率不同而不同，下面以六种典型特例加以分析。

（1）$r_1=r_2=1$（理想恒比共聚）

$$F_1 = \frac{f_1(f_1+f_2)}{(f_1+f_2)^2} = f_1$$

第 3 章 烯烃类单体的连锁聚合

表示两自由基的自增长和交叉增长的概率完全相同,不论单体物质的量比和转化率如何,共聚物组成与单体组成始终保持一致。共聚物组成曲线是一对角线,可称为理想恒比共聚。乙烯-乙酸乙烯酯、甲基丙烯酸甲酯-偏二氯乙烯和四氟乙烯-三氟氯乙烯的共聚具有恒比共聚特征。

(2) $r_1 r_2 = 1$ 或 $r_2 = 1/r_1$(一般理想共聚)

式(3-4)和式(3-5)可简化为:

$$\frac{d[M_1]}{d[M_2]} = r_1 \times \frac{[M_1]}{[M_2]}$$

$$F_1 = \frac{r_1 f_1}{r_1 f_1 + f_2}$$

上式表明,共聚物中两结构单元的组成比是原料两单体物质的量比的 r_1 倍。组成曲线处于恒比对角线的上方,与另一对角线(未画出)呈对称状态,如图 3-13 所示。60℃丁二烯($r_1 = 1.39$)-苯乙烯($r_2 = 0.78$)和偏氯乙烯($r_1 = 3.2$)-氯乙烯($r_2 = 0.3$)的共聚接近这种情况。结果表明,多数离子的共聚具有一般理想共聚(ideal copolymerization)的特征。

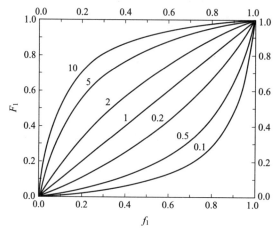

图 3-13 理想共聚曲线

$r_1 r_2 = 1$,图中数值为 r_1

(3) $r_1 = r_2 = 0$(交替共聚)

自由基不能与同种单体均聚,只能与异种单体共聚,因此共聚物中两单元严格交替相间。不论单体配比如何,共聚物组成均恒定,F_1-f_1 关系曲线是一条 $F_1 = 0.5$ 的水平线。原始物料中两种单体量不相等的情况下进行共聚时,含量少的单体消耗完毕,就停止聚合,留下的是多余的另一种单体。马来酸酐和乙酸-2-氯烯丙基酯的共聚属于典型的交替共聚(alternating copolymerization)。

60℃下苯乙烯($r_1 = 0.01$)和马来酸酐($r_2 = 0$)的共聚是一种衍生交替共聚的例子。因 $r_2 = 0$,式(3-4)可简化为:

$$\frac{d[M_1]}{d[M_2]} = 1 + r_1 \times \frac{[M_1]}{[M_2]}$$

当 $r_1 \times \dfrac{[M_1]}{[M_2]} \ll 1$ 时，才形成交替共聚物。单体 M_1 耗尽后，聚合也就停止。如 $[M_1]$ 和 $[M_2]$ 大致相同，则共聚物中 $F_1 > 50\%$。

(4) $r_1 r_2 < 1$，$r_1 < 1$，$r_2 < 1$（有恒比点共聚）

两种单体的共聚倾向高于均聚时，共聚曲线具有反"S"形态特征。曲线与对角线有一交点，该点的共聚物组成与单体组成相等，故称作恒比点，类似于汽液平衡中的恒沸点。

当 $r_1 = r_2 < 1$ 时，恒比点为 0.5，共聚物组成曲线相对于恒比点作点对称。丙烯腈（$r_1 = 0.83$）与丙烯酸甲酯（$r_2 = 0.84$）共聚属于这种情况。

$r_1 < 1$、$r_2 < 1$、$r_1 \neq r_2$ 时，共聚曲线对恒比点不再呈点对称，如图 3-14 所示。当 $r_1 < r_2$ 时，恒比点在对角线下部，反之则在上部。这类例子很多，如苯乙烯（$r_1 = 0.41$）与丙烯腈（$r_2 = 0.04$），丁二烯（$r_1 = 0.3$）与丙烯腈（$r_2 = 0.2$）等。

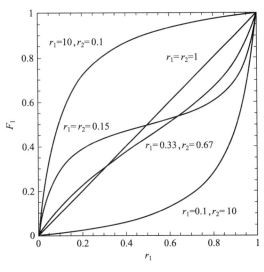

图 3-14 典型共聚曲线

（图中数值为 r_1）

$r_1 r_2$ 接近于零，则趋向于交替共聚；$r_1 r_2$ 接近于 1，则接近理想共聚。$0 < r_1 = r_2 < 1$ 的共聚曲线介于交替共聚（$F_1 = 0.5$）和恒比对角线（$F_1 = f_1$）之间。

(5) $r_1 > 1$，$r_2 < 1$（无恒比点共聚）

共聚曲线不与对角线相交，处于另一对角线的上方或下方。从竞聚率一个大于 1，一个小于 1 可以看出，这类共聚物只是一种"镶嵌"着少许第二结构单元的"均聚物"，因此也称作"嵌均共聚"。这类共聚的实例很多，如氯乙烯（$r_1 = 1.68$）-乙酸乙烯酯（$r_2 = 0.23$），甲基丙烯酸甲酯（$r_1 = 1.91$）-丙烯酸甲酯（$r_2 = 0.5$），苯乙烯（$r_1 = 55$）-乙酸乙烯酯（$r_2 = 0.01$），等等。

(6) $r_1 > 1$，$r_2 > 1$（"嵌段"共聚）

两种链自由基都倾向于加上同种单体，形成"嵌段"共聚物，链段长短决定于 r_1 和 r_2 的大小。但 M_1 和 M_2 的链段都不长，很难用这种方法来制备真正嵌段共聚物。$r_1 > 1$ 且 $r_2 > 1$ 的实例很少，苯乙烯（$r_1 = 1.38$）与异戊二烯（$r_2 = 2.05$）的共聚属于这种情况。

3.2.4 竞聚率

竞聚率是共聚物组成方程中的重要参数，利用竞聚率可从单体组成来计算共聚物组成。文献手册已经有大量竞聚率数据，部分数据见表 3-3。

表 3-3 常用单体竞聚率

M_1	M_2	温度/℃	r_1	r_2
丁二烯	异戊二烯	5	0.75	0.85
	苯乙烯	50	1.35	0.58
	丙烯腈	40	0.3	0.02
	甲基丙烯酸甲酯	90	0.75	0.25
	氯乙烯	50	8.8	0.035
苯乙烯	异戊二烯	50	0.8	1.68
	丙烯腈	60	0.4	0.04
	甲基丙烯酸甲酯	60	0.52	0.46
	氯乙烯	60	17	0.02
	乙酸乙烯酯	60	55	0.01
丙烯腈	甲基丙烯酸甲酯	80	0.15	1.22
	氯乙烯	60	2.7	0.04
	乙酸乙烯酯	30	4.2	0.05
甲基丙烯酸甲酯	丙烯酸甲酯	130	1.91	0.5
	氯乙烯	68	2.35	0.24
	乙酸乙烯酯	60	20	0.015
氯乙烯	乙酸乙烯酯	60	9	0.1
马来酸酐	苯乙烯	50	0.04	0.015
	丙烯腈	60	0	6
	甲基丙烯酸甲酯	75	0.02	6.7
	乙酸乙烯酯	75	0.055	0.003

实验室测定竞聚率，需在低转化率（<5%）下测定若干单体配比条件下的共聚物的组成或残留单体组成，有时两者需同时分析，然后求解二元方程组，可计算出两个单体的竞聚率。共聚物组成可以选用元素分析、红外、紫外或浊度滴定来分析，残留单体组成则多用气相色谱法测定。

竞聚率是自由基均聚和共聚反应的增长速率常数之比，如温度、压力、溶剂等影响增长速率常数的因素都会影响竞聚率。

(1) 温度的影响

$$竞聚率 \quad r_1 = \frac{k_{11}}{k_{12}}$$

则：

$$\frac{d(\ln r_1)}{dT} = \frac{E_{11} - E_{12}}{RT^2}$$

式中，E_{11}、E_{12} 分别是自增长和交叉增长的活化能，链增长活化能本身就小（21～34kJ·mol^{-1}），$E_{11}-E_{12}$ 更小，因此温度对竞聚率的影响较小。

(2) 压力的影响

压力对竞聚率影响的报道不多。从现有数据看来，竞聚率随压力的变化有点与温度的影响类似。升高压力使共聚向理想共聚方向移动。例如在 0.001MPa、0.1MPa、10MPa 下，甲基丙烯酸甲酯-丙烯腈进行共聚，其 r_1 均分别为 0.16，0.54，0.91。

(3) 溶剂的影响

溶剂的极性对竞聚率有影响，如表 3-4 所示。

表 3-4 苯乙烯（M_1）和甲基丙烯酸甲酯（M_2）在不同溶剂中的竞聚率

溶剂	r_1	r_2	溶剂	r_1	r_2
苯	0.57	0.46	苯甲醇	0.44	0.39
苯甲腈	0.48	0.49	苯酚	0.35	0.35

对于离子聚合，溶剂极性将影响离子对的性质，对增长速率和竞聚率的影响都较大。

3.2.5 单体活性和自由基活性

在讨论自由基均聚反应时，常常使用链增长速率常数的大小来比较单体的反应活性。这个方法并不准确。例如，苯乙烯是典型的活泼单体，乙酸乙酯是典型的不活泼单体，可是后者的均聚链增长速率常数约为前者的 16 倍。自由基聚合反应的链增长是自由基与单体间的反应，链增长速率常数不仅与单体反应活性有关，还与自由基的反应活性相关。因此，比较两单体活性时，需比较它们与同种自由基反应的速率常数；比较两自由基活性时，需考虑与同种单体反应的速率常数。在研究自由基聚合反应时，共聚反应体系中的重要参数——竞聚率，常常用来对比各种单体、自由基的活性。

• 单体活性。竞聚率的倒数 $1/r_1=k_{12}/k_{11}$，表示同一自由基和异种单体共聚的链增长速率常数与均聚链增长速率常数之比，可用来衡量两种单体的相对活性。从表 3-5 可看出，对于同一自由基而言，单体活性有较大差异。

表 3-5 乙烯基单体对自由基的活性（$1/r$）

单体	链自由基						
	B·	S·	VAc·	VC·	MMA·	MA·	AN·
B		1.7	29	4	20		50
S	0.4		100	50	2.2	6.7	25
MMA	1.3	1.9	67	10		2	6.7
AN	3.3	2.5	20	25	0.82		
MA		1.4	10	17	0.52		0.67
VDC		0.54	10		0.39		1.1
VC	0.11	0.059	4.4		0.10	0.25	0.37
VAc		0.019		0.59	0.050	0.11	0.24

- 自由基活性。表 3-6 中的 k_{12} 值可以用来比较自由基的相对活性，从左到右依次增加。纵列数据则可比较单体活性，从上而下依次减弱。从取代基的影响来看，单体活性次序与自由基活性次序恰好相反，但变化的倍数并不相同。例如苯乙烯单体的活性是乙酸乙烯酯单体的 50~100 倍，但乙酸乙烯酯自由基的活性却是苯乙烯自由基的 100~1000 倍。可见，取代基对自由基活性的影响比对单体活性的影响要大得多。

表 3-6　同一自由基与不同单体反应的 k_{12}　　　单位：L·mol^{-1}·s^{-1}

单体	链自由基						
	B·	S·	MMA·	AN·	MA·	VC·	VAc·
B	100	246	2820	98000	41800	20	357000
S	40	145	1550	49000	14000	230000	615000
MMA	130	276	705	13100	4180	154000	123000
AN	330	435	578	1960	2510	46000	178000
MA	130	203	367	1310	2090	23000	209000
VC	0.11	0.059	4.4		0.10	0.25	0.37
VAc		3.9	35	230	230	2300	7760

取代基的共轭效应、极性效应和位阻效应对单体活性和自由基活性均有影响。

① 共轭效应。按表 3-6 所列自由基活性的次序，可见共轭效应对自由基活性的影响很大。苯乙烯自由基中的苯环与单电子共轭稳定，使活性降低，几乎成为烯类自由基中活性最低的一员。—CN、—COOH、—COOR 等基团对自由基均有共轭效应，这类自由基的活性也不是很高。相反，卤素、乙酰基、醚等基团只有卤素、氧原子上的未键合电子对自由基稍有作用，因此氯乙烯、乙酸乙烯酯、乙烯基醚等自由基就很活泼。单体活性与自由基活性次序正好相反，即苯乙烯单体活泼，而乙酸乙烯酯单体并不活泼。

② 极性效应。有些极性单体，如丙烯腈，在单体和自由基活性次序中出现反常现象。供电子基使烯类单体双键带负电，吸电子基团则使其带正电，这两类单体易进行共聚，并有交替倾向，称作极性效应。

一些难均聚的单体，如顺丁烯二酸酐（马来酸酐）、反丁烯二酸二乙酯，却能与极性相反的单体，如苯乙烯、乙烯基醚等共聚。反二苯基乙烯（电子给体）和马来酸酐（电子受体）两单体虽然不能均聚，却往往形成电荷转移络合物而交替共聚。

③ 位阻效应。自由基与单体的共聚速率还与位阻效应有关。如果烯类单体的两个取代基处于同一碳原子上，位阻效应并不显著，两个取代基电子效应的叠加反而使单体活性增加。如果两个取代基处在不同的碳原子上，则因位阻效应使活性减弱。例如，与氯乙烯相比，偏二氯乙烯和多种自由基反应的活性要增加 2~10 倍，而 1,2-二氯乙烯的活性则降低 2~20 倍。

因位阻关系，1,2-双取代乙烯不能均聚，却能与苯乙烯、丙烯腈、乙酸乙烯酯等单取代乙烯共聚，共聚速率比 1,1-双取代乙烯要低。

比较顺式和反式1,2-二氯乙烯,可以看出反式异构体的活性要高大约6倍。主要原因是顺式异构体不易成平面型,因而活性较低。氟原子体积小,位阻效应小,因此四氟乙烯和三氟氯乙烯不仅容易均聚,还容易共聚。

3.2.6 Q-e 概念

竞聚率是共聚物组成方程中的重要参数。每一对单体都可由实验测得一对竞聚率,100种单体将构成4950对竞聚率,全面测定r_1、r_2值,将带来很多工作量。因此希望建立单体结构与活性间的定量关系式来估算竞聚率。最通用的关系式是Alfrey-Frice的 $Q\text{-}e$ 式($Q\text{-}e$ scheme)。该式将自由基-单体间的反应速率常数与共轭效应、极性效应关联起来。其中Q值是自由基和单体活性的共轭效应度量,e值是自由基和单体活性的极性度量。

$$k_{12}=P_1Q_2\exp[-e_1(e_1-e_2)]$$

因此规定苯乙烯的$Q=1.0$、$e=-0.8$作基准,利用$Q\text{-}e$式就可求出其他单体的Q、e值。以Q值为横坐标,e为纵坐标,将各单体的Q、e值标绘在$Q\text{-}e$图上(图3-15)。图中右边和左边距离较远,Q值相差较大,难以共聚。Q、e相近的一对单体,往往接近理想共聚,如苯乙烯-丁二烯,氯乙烯-乙酸乙烯酯。e值相差较大的一对单体,如苯乙烯-马来酸酐,苯乙烯-丙烯腈,则有较大的交替共聚倾向。

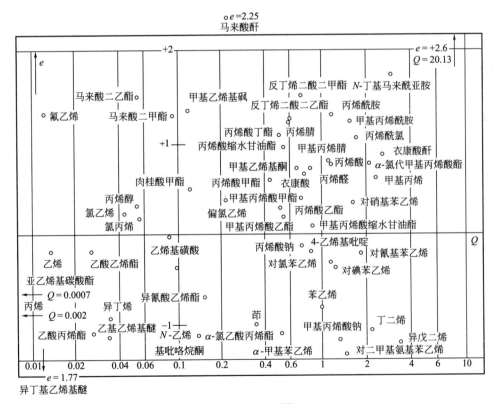

图 3-15 $Q\text{-}e$ 图

3.2.7 重要烯烃类共聚物

(1) 丁苯橡胶

丁苯橡胶（SBR）在工业生产中采用乳液聚合的方法合成。50℃下丁二烯（M_1）和苯乙烯（M_2）的竞聚率分别是 1.59 和 0.44，共聚曲线类似一般理想共聚（图 3-13）。共聚物的丁二烯结构单元中含顺-1,4 结构 72%，反-1,4 结构 12%，1,2 结构 16%。数均分子量约 15 万~40 万，T_g 约为 -57~$-52℃$。聚合物分子链中苯乙烯结构增多，会导致 T_g 上升，柔性下降；如丁二烯结构增多，T_g 下降，柔性提高。按照通用级乳液丁苯橡胶中有 23.5%苯乙烯考虑，丁二烯/苯乙烯单体配比常取(70~71)/(30~29)。SBR 主要用于轮胎制造业，因耐油性不佳，很少用作密封材料或减震材料。

按聚合温度，丁苯橡胶合成有热法（50℃）和冷法（5℃）两类，热法采用单一过硫酸盐作引发剂，冷法则选用氧化还原引发体系。多种阴离子表面活性剂组成乳化体系，十二硫醇作分子量调节剂，还有其他助剂。乳液丁苯橡胶的生产过程多采用 8~12 釜串联连续操作，单体在首釜进入，调节剂和乳化剂可以在不同釜次分点补加，使分子量分布更均匀一些，使乳液更稳定一些。最终转化率约为 60%，以减少交联、凝胶的产生。聚合时间为 8~12h。

(2) 丁腈橡胶

丁腈橡胶（NBR）是丁二烯和丙烯腈的无规共聚物。其中丁二烯单元以反-1,4 结构为主，随着聚合温度的提高，反-1,4 结构减少，而顺-1,4 和 1,2 结构增加。丁腈橡胶耐油性、耐老化性、耐磨性和耐热性较丁苯橡胶优良，T_g 也较丁苯橡胶高。丙烯腈含量为 26%的丁腈橡胶的 T_g 约为 $-52℃$，丙烯腈含量 40%的丁腈橡胶的 T_g 约为 26℃。丁腈橡胶比天然橡胶、丁苯橡胶耐热，但是耐寒性却降低。丁腈橡胶采用乳液聚合法，也分高温聚合（30~50℃）和低温聚合（5~10℃）两种技术，都采用氧化还原体系。丁二烯和丙烯腈的竞聚率都小于 1，有交替聚合的倾向，有恒比点，例如 5℃，$r_1 = 0.18$~0.08，$r_2 = 0.02$~0.03，恒比组成=56/44。在恒比点以上和以下，单体组成或共聚物组成与转化率的变化走向相反，应该采取不同的控制措施。丁腈橡胶主要用于耐油管材和密封零部件等用途。

(3) 苯乙烯-丙烯腈共聚物和 ABS 树脂

苯乙烯-丙烯腈共聚物（SAN），可以克服聚苯乙烯的脆性。与均聚苯乙烯相比，SAN 的耐溶剂性、耐油性、软化点、耐应力、抗冲击性能等均有很大的提高，可以用作结构材料。进一步发展的共聚物是丙烯腈-丁二烯-苯乙烯三元接枝共聚物，简称 ABS 树脂。

苯乙烯和丙烯腈的竞聚率分别为 $r_1 = 0.41$，$r_2 = 0.04$，有交替倾向，恒比点处于 S/AN=75/25（质量比）。SAN 和 ABS 树脂中苯乙烯和丙烯腈含量比常取恒比点附近的组成，单体组成就很容易配制，随转化率的变化较小，容易控制。

ABS 树脂有多种制备方法，最常用的是先将丁二烯经乳液聚合，制成聚丁二烯胶乳。然后加入苯乙烯和丙烯腈，进行接枝共聚，部分苯乙烯-丙烯腈接枝在聚丁二烯胶粒表面，更多的苯乙烯-丙烯腈在接枝层上无规共聚，形成了所谓的核壳结构。共聚结束后，经凝聚分离干燥，就成为 ABS 母料。再与悬浮法或本体法苯乙烯-丙烯腈共聚物共混成 ABS 树脂商品。ABS 抗冲击性能优于 HIPS，属工程塑料的范畴，应用范围更广。

（4）聚偏氯乙烯共聚物

偏氯乙烯均聚物是结晶性聚合物，熔点约为220℃，玻璃化转变温度为23℃，熔点附近开始热分解，也不溶于有机溶剂，无法成型加工。若偏氯乙烯与少量氯乙烯（10%～20%）或丙烯酸甲酯（8%～10%）等第二单体共聚，适当降低结晶度，熔点和玻璃化转变温度（-5℃）则显著降低，改善了熔体的流动性能和加工性能，就可以用常规的塑料成型设备加工成薄膜和纤维，广泛用作食品、医药、军需品的包装材料。利用其优异的耐溶剂性能、高耐磨和阻燃性能，共聚物还可以用作管材、滤布以及特殊的军用场合。

45℃下偏氯乙烯和氯乙烯的竞聚率分别为2.98和0.175，接近理想共聚行为。偏氯乙烯/氯乙烯质量配比约为(80～85)/(20～15)，一次投料，最终转化率小于85%，共聚物组成分布还不致太宽，符合工业产品要求。偏氯乙烯和氯乙烯共聚多选悬浮聚合，工艺过程与氯乙烯悬浮聚合相似，具有沉淀聚合的特征。

（5）含氟共聚物

四氟乙烯、全氟丙烯、偏氟乙烯是用于合成含氟聚合物的主要单体。四氟乙烯共聚物可保留聚四氟乙烯热稳定性和化学稳定性的优点，适当破坏结构的规整性，改善溶解、熔融流动和加工性能。四氟乙烯-全氟丙烯共聚物是含氟共聚物中的第一大品种，可在285℃塑化，在-260～205℃较广的温度范围内长期使用。四氟乙烯和全氟丙烯的竞聚率差异较大，共聚物组成不容易控制。以过硫酸钾作引发剂，可在90～95℃下进行水溶液聚合。

偏氟乙烯-全氟丙烯共聚物是重要的含氟弹性体。随着组成的不同，共聚物可表现出塑料和弹性体的特性。50%～70%（摩尔分数）偏氟乙烯的共聚物是弹性体，玻璃化转变温度为-15～0℃。偏氟乙烯与全氟丙烯共聚物（70:30），可用 BPO/ZnO、胺类或二异氰酸酯交联成橡胶，耐油，耐化学品，可在-50～200℃之间长期使用。

英语读译资料

（1）Copolymerization

While the mechanism of copolymerization is similar to that discussed for the polymerization of one reactant, called as homo-polymerization, the reactivities of monomers may differ when more than one is present in the feed, that is, reaction mixture. Copolymers may be produced by step reaction or by chain-reaction polymerization. It is important to note that if the reactant species are M_1 and M_2 then the composition of the copolymer is not a physical mixture or blend.

Many naturally occurring polymers are largely homo-polymers, but proteins and nucleic acids are copolymers composed of a number of different units. While many synthetic polymers are homo-polymers, the most widely used synthetic rubber (SBR) is a copolymer of styrene (S) and butadiene (B), with the "R" representing "rubber." There are many other important copolymers.

Copolymers may be random in the placement of units, alternating in the placement of units, block in either or both of the different units in which there are long sequences of the same repeating unit in the chain, or graft copolymers in which the chain extension of the second monomer is present as branches.

Each of these types of copolymers offers different physical properties for a particular copolymer combination. It is interesting to note that block copolymers may be produced from one monomer only if the arrangement around the chiral carbon atom changes sequentially. These copolymers are called stereo-block copolymers.

(2) Commercial copolymers

One of the first commercial copolymers, introduced in 1928, was made of VC (87%) and VAc (13%) (vinylite). As the presence of the VAc mers disrupted the regular structure of PVC, the copolymer was more flexible and more soluble than PVC itself.

Copolymers of VC and vinylidene chloride were introduced in the 1930s. The copolymer with very high VC content is used as a plastic film (Pliovic), and the copolymer with high vinylidene chloride content is used as a film and filament (Saran).

Polybutadiene, produced in emulsion polymerization, is not useful as an elastomer. However, the copolymers with styrene (SBR) and acrylonitrile (Buna-N) are widely used as elastomers.

Ethylene-propylene monomers (EPM), polyethylene-co-propylene, show good resistance to ozone, heat, and oxygen and are used in blends to make today's external automotive panels. Two general types of EPMs are commercially available. EPMs are saturated and require vulcanization if used as a rubber. They are used in a variety of automotive applications including as body and chassis parts, bumpers, radiator and heater hoses, seals, mats, and weather strips. EPMs are produced using Ziegler-Natta catalysts (ZNC).

The second type of EPM is the EPDM, which is made by polymerizing ethylene, propylene, and a small amount (3%~10%, mole fraction) of nonconjugated diolefine employing ZNCs. The side chains allow vulcanization with sulfur. They are employed in the production of appliance parts, wire and cable insulation, coated fabrics, gaskets, hoses, seals, and high-impact PP.

(3) Thermoplastic elastomers

A number of thermoplastic elastomers (TPE) have been developed since the mid-1960s. The initial TPEs were derived from plasticized PVC and are called plastisols. Plastisols are formed from the fusing together of PVC with a compatible plasticizer through heating. The plasticizer acts to lower the T_g to below room temperature (RT). Conceptually, this can be thought of as the plasticizer acting to put additional distance between the PVC chains thus lowering the inter- and intrachain forces, as well as helping solubilize chain segments. The resulting materials are used in a number of areas including construction of boot soles.

The hard-soft segment scenario is utilized in the formation of a number of industrially important TPEs. TPEs contain two or more distinct phases and their properties depend on these phases being intimately mixed and small. These phases may be chemically or physically connected. For the material to be a TPE at least one phase must be soft or flexible under the operating conditions and at least one phase must be hard in the hard phase(s) becoming soft (or fluid) at higher temperatures. Often the hard segments or phases are crystalline thermoplastics, while the soft segments or phases are amorphous. In continuous chains containing blocks of hard and soft segments the molecular arrangement normally contains crystalline regions where there is sufficient length in the hard segment to form the crystalline regions or phases where the soft segments form amorphous regions. Such hard-soft scenarios can also be achieved through employing grafts where the pendant group typically acts as the hard segment with the backbone acting as the soft segment. In order for an effective network to be formed each "A" chain needs to have at least two "B" grafts to allow for formation of a continuous interlinked network. Although much research has been done with such graft materials they have not yet become very important commercially.

(4) Advantages of copolymerisation

The simplest types of polymers are homopolymers where the repeating units are all of the same type or structure, in other words, homopolymers can be given a general formula of the type X(A) nY, where X and Y are the end or terminal groups (which we have already discussed) and A represents the repeating unit. Hence, polymers such as polyethene and polyvinyl chloride are both examples of homopolymers. If, however, two or more suitable monomers are polymerized together to give polymers containing more than one type of structural unit, a copolymer can be formed. In this case, we can write the general formula X(A) n(B) m(C) ···Y, where A, B, C, etc. represent the various structural units depending on the different monomers used. If monomers A and B are reacted together to form a copolymer, then the copolymer often shows very different properties from those of a physical mixture of the separate homopolymers of A and B. Sometimes, the good qualities of each homopolymer can be combined or retained in the copolymer, and this is one of the obvious advantages which copolymerisation offers.

We will now discuss some of the ways in which copolymers can be produced. Random copolymers are prepared by polymerizing the appropriate mixture of monomers, examples include chloroethene-ethenyl ethanoate (vinyl chloride-vinyl acetate) and phenylethene-buta-1,3-diene copolymers.

In the case of the chloroethene-ethenyl ethanoate copolymer, the presence of the ethenyl ethanoate increases solubility and improves moulding characteristics (by improving flow properties) as compared to the homopolymer of chloroethene. The properties of a copolymer produced from the monomers, A and B, will clearly depend on the distribution of A and B units in the chains of our copolymer. This distribution is not necessarily the same as the ratio of the concentrations of A to B in the original monomer mixture. Generally speaking, if two monomers A and B react to form a copolymer, and A is the more reactive monomer, then the copolymer formed in the early stages of polymerization will be richer in A than in B.

In the later stages of reaction, as the concentration of monomer A becomes low, the copolymer formed becomes richer in B. This problem of the copolymer composition changing during polymerization may be reduced by adding the more reactive monomer gradually. The question of monomer reactivity can be treated quantitatively by studying the kinetics of copolymerisation.

习 题

1. 简述提高温度对竞聚率 r_1 和 r_2 的影响。

2. 说明竞聚率 r_1、r_2 的定义，指明理想共聚、交替共聚、恒比共聚时竞聚率数值的特征。

3. 考虑 $r_1=r_2=1$、$r_1=r_2=0$、$r_2=0$、$r_1r_2=1$ 等情况，说明共聚物组成曲线的特征。

4. 乙酸烯丙基酯（$e=-1.13$，$Q=0.028$）和甲基丙烯酸甲酯（$e=0.41$，$Q=0.74$）等物质的量共聚，是否合理？

5. 氯乙烯和乙酸乙烯酯，甲基丙烯酸甲酯和苯乙烯两对单体共聚，若两体系中乙酸乙烯酯和苯乙烯的浓度均为 15%（质量分数），根据文献报道的竞聚率，试求共聚物起始组成。

6. 氯乙烯（$r_1=0.167$）与乙酸乙烯酯（$r_2=0.23$）共聚，希望获得初始共聚物瞬时组成和 85% 转化率时共聚物平均组成为 5%（摩尔分数）的乙酸乙烯酯，分别求两单体的初始配比。

7. 0.3mol 甲基丙烯腈和 0.7mol 苯乙烯进行自由基共聚，求共聚物中每种单元的链段长。

8. 不饱和树脂使用马来酸酐和二元醇等进行缩聚制备聚酯低聚物，溶于苯乙烯后再进行自由基聚合。聚酯低聚物中马来酸酐中的双键结构（M_2）和苯乙烯（M_1）进行自由基共聚时的活性类似于富马酸二甲酯，$r_1=0.39$，$r_2=0.03$，分析苯乙烯和聚酯交联网络的结构特征。

英语习题

1. Draw representative structures for ① homopolymers, ② alternation copolymers, ③ random copolymers, ④ AB block copolymers, and ⑤ graft copolymers of styrene and acrylonitrile.

2. If equimolar quantities of M_1 and M_2 are used in an azeotropic copolymerization, what is the composition of the feed after 50% of the copolymer has formed?

3. Do the r_1 and r_2 values increase or decrease during copolymerization?

4. What will be the composition of copolymers produced in the first part of the polymerization of equimolar quantities of vinylidene chloride and vinyl chloride?

5. What is the value of r_1 and r_2 for an ideal random copolymer?

6. What is the composition of the first copolymer chains produced by the copolymerization of equimolar quantities of styrene and methyl methacrylate in free radical copolymerization?

7. What is the composition of the first copolymer butyl rubber chains produced from

equimolar quantities of the two monomers?

8. What is the composition of the first copolymer butyl rubber chains produced from a feed containing 9mol of isobutylene and 1mol of isoprene?

9. What is the composition of the first polymer chains produced by the copolymerization of equimolar quantities of VC and VAc?

10. What are the advantages, if any, of the VC/VAc copolymer over PVC itself?

11. What product is obtained if 1.5mol of styrene is copolymerized with 1mol of maleic anhydride in benzene?

12. How would the properties of two polymers containing the same amounts of monomer A and B differ if one polymer is a simple random copolymer and the other polymer is a block copolymer?

3.3 离子聚合

阳离子聚合

阴离子聚合

双层真空玻璃在除掉两层玻璃间的空气后可长时间保持密封低压状态，可有效阻隔室内外冷热传导，减少冷热辐射，节能环保。由于真空层的存在，还可以阻隔声音传递，起到隔声降噪的效果。高分子量聚异丁烯（polyisobutylene，PIB）耐热、耐光、耐臭氧老化性好，具有理想的化学稳定性，可作为玻璃密封胶。在室温下，对稀碱和浓酸、浓碱、盐的作用稳定，在较高温度下仍然具有优良的防水性和气密性。高分子量聚异丁烯一般按阳离子聚合机理合成。

本节目录

3.3.1 阳离子聚合 / 139
3.3.2 阴离子聚合 / 143

重点要点

阳离子聚合，阳离子聚合引发体系，阳离子聚合机理，聚异丁烯，丁基橡胶，阴离子聚合，活性聚合，阴离子聚合引发体系，阴离子聚合引发体系与单体的匹配，阴离子聚合机理。

3.3.1 阳离子聚合

连锁聚合中活性物种为碳阳离子（carbocation）进行的聚合反应为阳离子聚合（cationic polymerization）。阳离子聚合的基本机理就是亲电试剂（electrophile）对烯烃类单体的亲电加成反应（electrophilic addition reaction）。和自由基聚合反应不同，阳离子聚合反应的活性种碳阳离子总伴随着反离子。如果碳阳离子和反离子比较稳定，就可能发生碳阳离子对烯类单体的加成反应，形成聚合物。所以，碳阳离子及其反离子的结构稳定性是阳离子聚合能否成功的关键因素。

阳离子聚合的烯类单体只限于带有供电子基团的异丁烯、烷基乙烯基醚，以及有共轭结构的苯乙烯类、二烯烃等少数几种。供电子基团一方面使碳碳双键电子云密度增加，有利于阳离子活性种的进攻，另一方面又使生成的碳阳离子电子云分散而稳定，减少副反应。

自由基聚合反应的引发剂一般需要加热分解才能产生自由基活性种。阳离子引发剂产生阳离子活性种所需要的活化能较低，一般可以在较低温度下进行聚合。

因碳阳离子的反应活性较高，阳离子聚合反应过程中容易发生链转移等副反应，为了得到较高的分子量，抑制各种副反应，阳离子聚合一般在低于 $-50℃$ 的低温条件下进行。

由于自由基活性种呈电中性，溶剂的极性对聚合反应性能影响较小。与其相比，溶剂的极性能够左右离子对状态，对阳离子聚合反应的影响较大。如下所示，接触离子对（contact ion pair）的反应活性较低，而溶剂分离离子对（solvent-separated ion pair）和自由离子（free ion）的聚合反应活性较高。非极性溶剂中容易产生接触离子对；极性溶剂导致溶剂分离离子对和自由离子的产生。阳离子聚合在极性溶剂中进行更倾向于增大聚合速率和分子量。

接触离子对　　溶剂分离离子对　　自由离子

自由基聚合反应的链终止方式为双基终止，阳离子聚合反应中由于电荷排斥作用，不会发生双基终止。大多数阳离子聚合以单基终止为主，聚合速率与引发剂浓度成正比。

(1) 阳离子聚合单体

一般而言，具有供电子取代基团，双键结构中电子云密度较大（e 值为负）的单体容易进行阳离子聚合。共轭结构对单体的阳离子聚合活性影响不大。

异丁烯有两个供电子甲基，碳碳双键电子云密度较大，易受阳离子进攻进行加成反应。链中—CH_2—上的氢受两边四个甲基的保护，不易被夺取，减少了转移、重排、支化等副反应。实际上，异丁烯几乎成为 α-烯烃中唯一能阳离子聚合的单体，而且异丁烯也只能阳离子聚合。异丁烯这一特性可用来判断聚合是否属于阳离子机理。

乙烯基烷基醚是容易阳离子聚合的另一类单体。烷氧基的诱导效应使双键的电子云密度降低，但氧原子上未共用电子对与双键形成的 p-π 共轭效应，使双键电子云密度增加，相比之下，共轭效应占主导地位。因此，乙烯基烷基醚更容易进行阳离子聚合。

苯乙烯、α-甲基苯乙烯、丁二烯、异戊二烯等共轭烯类虽然能够阳离子聚合，但其反应活性远不及异丁烯和乙烯基烷基醚。共轭烯类单体多用于共聚，如异丁烯与少量异戊二烯共聚，制备丁基橡胶。

N-乙烯基咔唑、乙烯基吡咯烷酮、茚和古马隆等都是可进行阳离子聚合的活泼单体。

（2）阳离子聚合的引发体系

阳离子聚合的引发剂主要有质子酸和 Lewis 酸两大类，都属于亲电试剂。

① 质子酸。质子酸使烯烃质子化，能引发阳离子聚合，但是酸中阴离子的亲核性不应太强（如卤氢酸），以免与质子或阳离子共价结合而终止。

浓硫酸、磷酸、高氯酸、氯磺酸、氟磺酸、三氯代乙酸、三氟代乙酸、二氟甲基磺酸等强质子酸在非水介质中部分电离，产生质子 H^+，能引发一些烯类聚合。

② Lewis 酸。Lewis 酸是最常用的阳离子引发剂，种类很多，主要有 $AlCl_3$、$AlBr_3$、$FeCl_3$、$TiCl_4$、$SnCl_4$ 等。Lewis 酸常需添加微量共引发剂才能以较快的速度引发阳离子聚合。共引发剂有质子供体和碳阳离子供体两类。

H_2O、$RCOH$、$RCOOH$、HX 等质子供体，与 Lewis 酸先形成络合物和离子对，如三氟化硼-水体系，然后引发异丁烯聚合。

$$BF_3 + H_2O \rightleftharpoons [H_2O \cdot BF_3] \rightleftharpoons H^+(BF_3OH)^-$$

异丁烯 $\xrightarrow{H^+(BF_3OH)^-}$ $H_3C-\overset{CH_3}{\underset{CH_3}{C^+}}$ $(BF_3OH)^-$

RX、RCOX、$(RCO)_2O$ 等（R 为烷基）碳阳离子供体，引发反应在机理上相似，如 $SnCl_4$-RCl 体系。

$$SnCl_4 + RCl \rightleftharpoons R^+(SnCl_5)^-$$

异丁烯 $\xrightarrow{R^+(SnCl_5)^-}$ $H_3C-\overset{CH_3}{\underset{CH_3}{C^+}}$ $(SnCl_5)^-$

③ 其他。其他阳离子引发剂还有碘、硫鎓离子等比较稳定的阳离子盐。其中二芳基碘鎓盐和三芳基硫鎓盐都是较常用的阳离子引发剂，它们的反离子是一些非亲核性的阴离子，一般在光照下活化使用。

$(C_6H_5)_3S^+$ $(C_6H_5)_2I^+$ BF_4^- SbF_6^- PF_6^-

（3）阳离子聚合的溶剂

适合阳离子聚合的有机溶剂应该对引发剂、单体、聚合物都有较好的溶解性，一般选择极性较大的溶剂。常用的溶剂有正己烷、四氯化碳、氯仿、甲苯、硝基苯等。水、乙酸、醇类以及一些碱性的有机溶剂不适合阳离子聚合。

(4) 阳离子聚合机理

阳离子聚合机理可以概括为快引发、快增长、易转移、难终止，转移是终止的主要方式，是影响聚合度的主要因素。

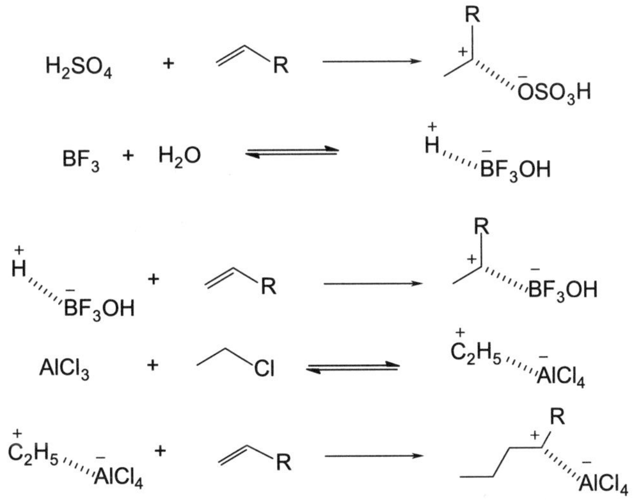

① 链引发。一般情况下，Lewis 酸（C）先与质子供体（RH）或碳阳离子供体（RX）形成络合物离子对，小部分解离成质子（自由离子），两者构成平衡，然后引发单体 M。

阳离子引发极快，几乎瞬间完成，引发活化能 $E_i=8.4\sim21kJ/mol$，与自由基聚合中的慢引发截然不同（$E_d=105\sim125kJ/mol$）。

② 链增长。引发生成的碳阳离子活性种与反离子形成离子对，单体分子不断插入其中而增长。

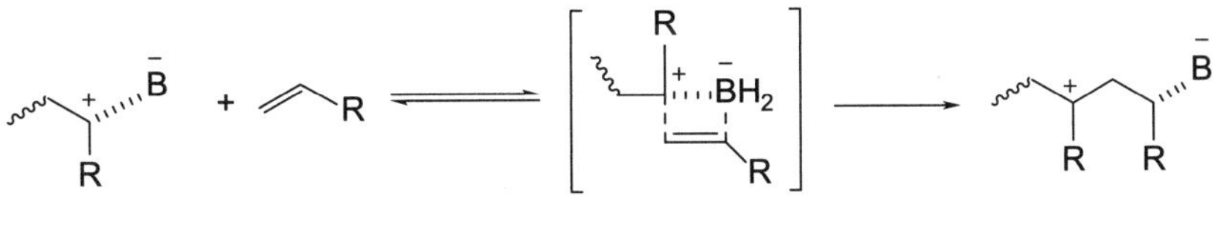

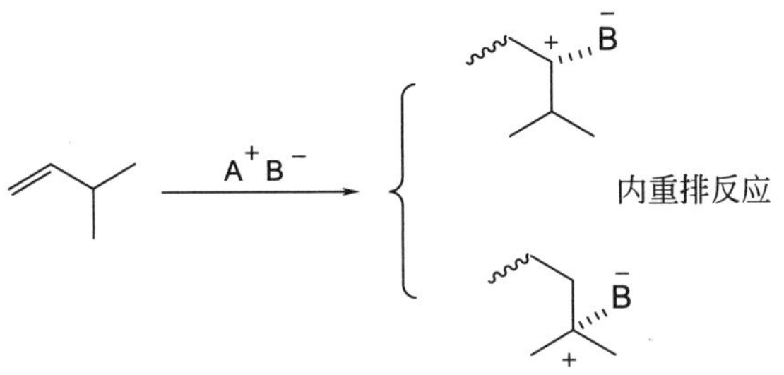

内重排反应

阳离子聚合的增长反应有下列特征：增长速率快，活化能低（$E_p=8.4\sim21kJ/mol$），"低温高速"；单体按头-尾结构插入离子对，对聚合物结构有一定影响；存在较多副反应，例如分子内重排反应（rearrangement reaction）等，所以一般在较低温度下进行。

③ 链转移。碳阳离子活性种很活泼，容易向单体或溶剂链转移，形成带不饱和端基的大分子和仍有引发能力的离子对，动力学链不终止，但聚合度降低。阳离子聚合往往在低温（例如$-100℃$）下进行，以减弱链转移，提高分子量。

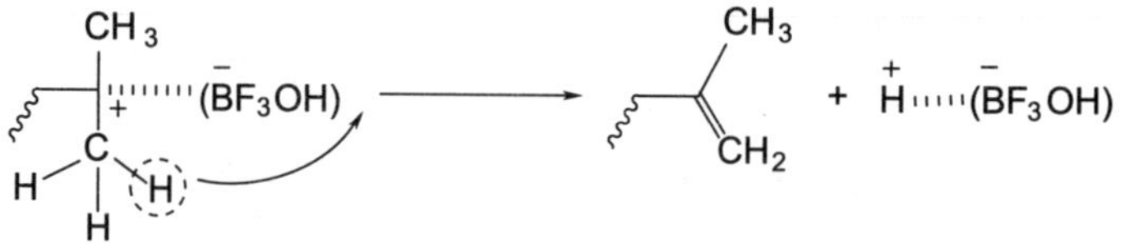

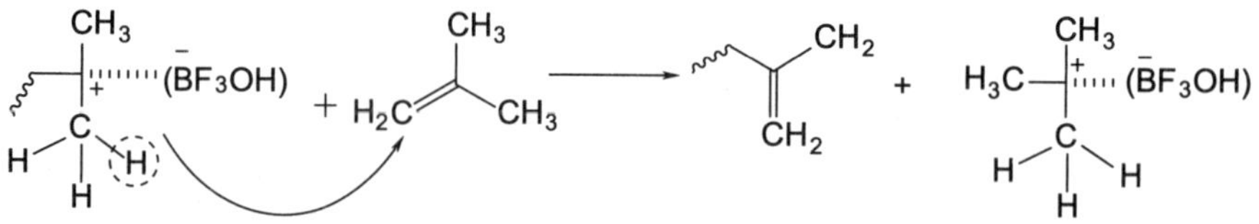

④ 链终止。碳阳离子活性种带有电荷，因电荷相斥作用，不能双基终止，因此也无凝胶效应。这是与自由基聚合显著不同之处。阳离子聚合的终止方式有如下几种。

a. 碳阳离子稳定化。一些反离子能与碳阳离子活性种形成稳定络合物，使增长离子对失去活性。

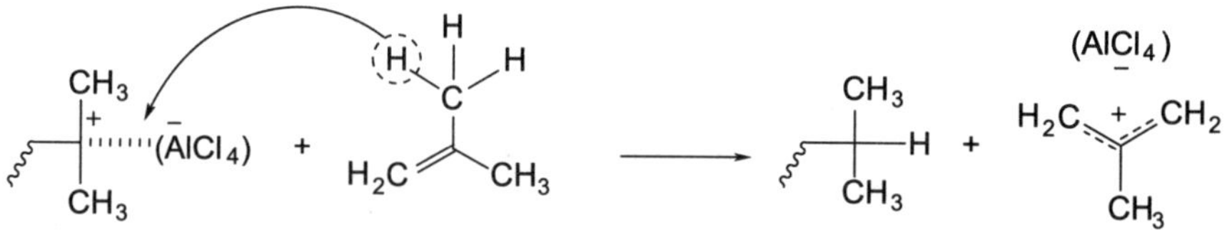

b. 反离子加成。当反离子的亲核性足够强时，将与增长碳阳离子共价结合而终止。

c. 活性中心与反离子中的一部分结合而终止。

以上众多阳离子聚合终止方式往往都难以顺利进行，因此有"难终止"之称，但未达到完全无终止的程度。实际上，经常会添加水、醇、酸等终止聚合反应。

阳离子聚合机理特征为快引发、快增长、易转移、难终止；动力学特征是低温高速，高分子量。

（5）阳离子聚合动力学

以 M 代表单体，H^+X^- 为引发剂，P 为聚合产物，阳离子聚合反应可简单表示如下。

$$M + H^+X^- \xrightarrow{k_i} HM^+\cdots X^-$$

$$M_n^+\cdots X^- + M \xrightarrow{k_p} HM_{n+1}^+\cdots X^-$$

$$M_n^+\cdots X^- \xrightarrow{k_{tr}} PH^+X^-$$

$$M_n^+\cdots X^- \xrightarrow{k_t} P$$

$$R_i = k_i[M][H^+X^-]$$

$$R_p = k_p[M][M^+]$$

$$R_t = k_t[M^+]$$

$$R_{tr} = k_{tr}[M^+][M]$$

阳离子聚合动力学研究因引发反应较自由基聚合复杂，并且聚合速率极快，很难确定真正的终止反应，大多数不适用于稳态假定。当选用低活性引发剂，如 $SnCl_4$，可按自由基聚合类似的假定方法，作稳态假定，$R_i = R_t$，可以解得离子对 [M^+] 浓度。推导动力学公式为：

$$-\frac{d[M]}{dt} = R_p = \frac{k_i k_p}{k_t}[M]^2[H^+X^-]$$

该式表明在自终止的条件下，速率对引发剂和共引发剂浓度呈一级反应，对单体浓度则呈二级反应。

如果是单体转移为主要终止方式，由 $R_i = R_{tr}$，也可导出类似速率方程式，只是与单体浓度一次方成正比。

$$-\frac{d[M]}{dt} = R_p = \frac{k_i k_p}{k_{tr}}[M][H^+X^-]$$

阳离子聚合反应过程中，除活性离子自发终止反应外，向单体转移（k_{trM}）和向溶剂转移（k_{trS}）也是主要的链终止方式，因此，聚合度可表示为：

$$X_n = \frac{R_p}{R_t + R_{trM} + R_{trS}}$$

其中：

$$R_{trM} = k_{trM}[M^+][M]$$
$$R_{trS} = k_{trS}[M^+][S]$$

得：

$$\frac{1}{X_n} = \frac{k_t}{k_p[M]} + \frac{k_{trM}}{k_p} + \frac{k_{trS}[S]}{k_p[M]} = \frac{k_t}{k_p[M]} + C_M + C_S\frac{[S]}{[M]}$$

上式右边各项分别代表单基终止、向单体转移和向溶剂转移终止对聚合度的贡献。

(6) 聚异丁烯和丁基橡胶

由异丁烯合成聚异丁烯和丁基橡胶是阳离子聚合的重要工业应用。

以 $AlCl_3$ 为引发剂，在 $0 \sim -40℃$ 下，异丁烯经阳离子聚合，可合成低分子量聚异丁烯（$M_n < 5×10^4$），该产物是黏滞液体或半固体，主要用作黏结剂、嵌缝材料、密封材料、动力油料的添加剂，以改进黏度。异丁烯在 $-100℃$ 下低温聚合，则得橡胶状固态的高分子量聚异丁烯 [$M_n = 5×(10^4 \sim 10^6)$]。

以氯甲烷为溶剂，$-100℃$，$AlCl_3$ 为引发剂，异丁烯和异戊二烯（1%～6%）进行共聚，可合成丁基橡胶，反应几乎瞬间完成。丁基橡胶不溶于氯甲烷，以细粉状沉析出来，属于淤浆聚合，俗称悬浮聚合。保证传热和悬浮分散是技术关键。丁基橡胶分子量在 $20×10^4$ 以上才不发黏，低温下并不结晶，$-50℃$ 下仍能保持柔软，具有耐臭氧、气密性好等优点，主要用来制作内胎。

3.3.2 阴离子聚合

连锁聚合中活性种为碳阴离子（carbanion）进行的聚合反应为阴离子聚合（anionic polymerization）。阴离子聚合的基本机理就是亲核试剂（nucleophile）对烯烃类单体的亲核加成反应（nucleophilic addition reaction）。

阴离子活性种末端近旁一定伴有阳离子作为反离子（counter cation），形成离子对。由于阴离子活性种的电荷排斥作用，阴离子聚合反应中不存在类似自由基的双基终止反应。此外，由于碳阴离子活性种末端不易脱除氢负离子（H⁻，hydride），所以阴离子聚合较容易获得高分子量聚合物。这一点和阳离子聚合易转移，较难得到高分子量的特点形成鲜明对比。和阳离子聚合类似，溶剂种类和反离子的结构对阴离子聚合的聚合速率以及聚合产物的立体规整性（stereoregularity, tacticity）等产生较大影响。一般而言，阴离子聚合需要使用惰性气体保护，还要充分排除酸性以及卤代烷烃类化合物。

（1）阴离子聚合的烯类单体

烯类单体带有吸电子取代基时较容易阴离子聚合。吸电子取代基能减弱烯烃单体双键的电子云密度，有利于阴离子活性种的亲核进攻，并能使碳阴离子的电子云密度分散而稳定。

按阴离子聚合反应活性次序，可将烯类单体分成三组。图 3-16 中的单体活性从上而下递增，Ⅰ组为共轭烯类，如苯乙烯、丁二烯类，活性较弱；Ⅱ组为（甲基）丙烯酸酯类，活性较强；Ⅲ组为具有硝基或双取代吸电子基的单体，活性最强。

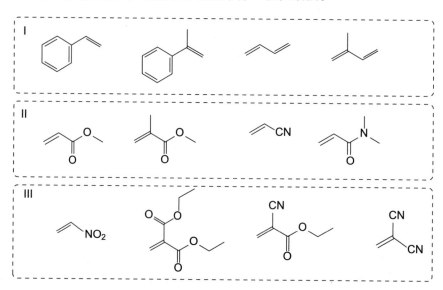

图 3-16 阴离子聚合单体结构及反应活性

（2）阴离子聚合引发剂

阴离子聚合引发剂有碱金属、碱金属和碱土金属的有机化合物、三级胺等碱类或亲核试剂，其活性可参见图 3-17，其反应活性从上向下递减。

(a) 组是还原性极强的碱金属或碱土金属、亲核性非常强的金属烷基化合物和氨基金属化合物等反应活性极高的阴离子引发剂。(b) 组包括格氏试剂、金属烷氧化合物、芳香族酮阴离子自由基等活性较高的引发剂。(c) 组是具有与叔胺类似反应活性的弱路易斯碱类化合物。

阴离子聚合引发剂的反应活性，即引发单体的能力，与碱性强度有关。利用有机化学中的 pK_a 值可以判断阴离子聚合引发剂的活性。pK_a 值愈大，表示碱性愈大，或亲电性愈小。pK_a 值大的烷基金属化合物可以引发 pK_a 值较小的单体，反之则不能。

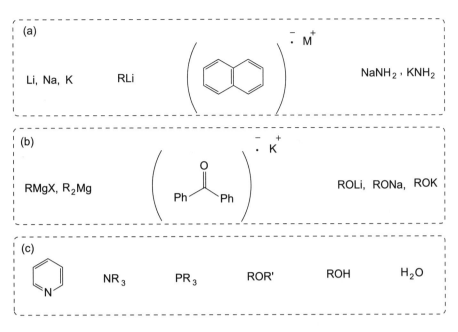

图 3-17 阴离子聚合引发剂结构及反应活性

自由基聚合中曾有单体活性次序与自由基活性次序相反的规律，阴离子聚合中也类似，即单体活性愈低，其阴离子活性愈高。

阴离子聚合的引发剂和单体的活性差别很大，需要适当的匹配才能聚合。一般而言，活性最高的（a）类引发剂可以引发Ⅰ、Ⅱ、Ⅲ组中所有单体的阴离子聚合。（b）类引发剂可以对应Ⅱ和Ⅲ组，而（c）类引发剂仅能引发反应活性最强的Ⅲ组单体。

（3）阴离子聚合的溶剂

阴离子聚合使用的有机溶剂需要对高活性引发剂或活性种端基有较好的稳定保护作用。比较典型的溶剂有，正己烷、环己烷、苯、甲苯等非极性烷烃类或芳香族化合物溶剂，乙醚、四氢呋喃（THF）等醚类溶剂，以及吡啶、二甲基甲酰胺（DMF）、N-甲基吡咯烷酮（NMP）等碱性的极性溶剂，等等（图 3-18）。四氯化碳、氯仿、水、醋酸等比较容易和阴离子活性种反应的有机溶剂则不适用于阴离子聚合反应。

阴离子聚合过程中使用的有机溶剂对阴离子活性端的解离状态及聚合速率有较大影响。一般而言，极性溶剂可以增加阴离子聚合速率。极性较低的有机溶剂中如果混入少量的极性有机溶剂也会大幅改变聚合产物的立体规整性。

例如，以丁基锂为引发剂聚合苯乙烯，在极性的 THF 溶剂中，在 −78℃ 低温条件下大约需要 1 分钟完成聚合，而在非极性的烷烃类溶剂中，30℃ 条件下需要几个小时才能完成。

（4）活性阴离子聚合机理

阴离子聚合和已经学习过的自由基聚合和阳离子聚合一样，具有链引发、链增长、链终止、链转移等基元反应。值得注意的是，阴离子活性端反应活性较强，容易和氧气、二氧化碳、水等微量杂质反应失去活性，因此，大多数阴离子聚合需要将溶剂和单体高度提纯，并在氮气或惰性气体保护下，或在高度减压的封闭容器中进行。其实验室操作可以参考格氏试剂，但是由于聚合引发剂的使用浓度较低，所以要求会更加严格。

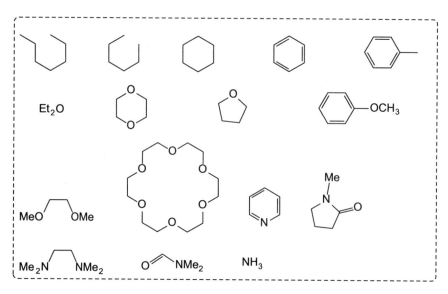

图 3-18 阴离子聚合常用有机溶剂

阴离子聚合单体一经引发成阴离子活性种，就以相同的模式进行链增长，一般无终止和无链转移，直至单体耗尽，因此称作活性聚合。难终止的原因有：活性链末端都是阴离子，无法双基终止；反离子为金属离子，无 H^+ 可供夺取而终止；夺取活性链中的 H^- 需要很高的能量，难以进行。

因此，活性阴离子聚合的机理特征是快引发、慢增长，无终止、无链转移，成为最简单的聚合机理。一般情况下，可按此机理来处理动力学问题。

① 链引发

阴离子聚合的引发反应可以分为电子直接转移引发和电子间接转移引发两大类。

如图 3-19 所示，丁基锂的碳阴离子可直接与苯乙烯发生亲核加成反应，形成单体阴离子活性种并引发聚合，在聚合链的一侧持续进行链增长反应。

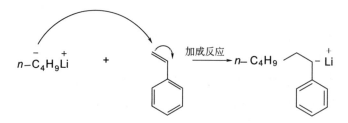

图 3-19 丁基锂引发苯乙烯阴离子聚合
电子直接转移，聚合链单侧链增长

又如图 3-20 所示，碱金属先向萘分子供给一个电子，形成萘自由基阴离子。加入苯乙烯后，萘自由基阴离子将多余的电子转移给苯乙烯单体，形成苯乙烯自由基阴离子。两个苯乙烯阴离子自由基偶合，形成苯乙烯双阴离子，后续引发苯乙烯双向链增长。

② 链增长

和阳离子聚合相似，溶剂的极性对阴离子聚合反应的影响较大。在极性较强的有机溶剂中，溶剂分子与反离子容易形成络合结构，形成溶剂分离离子对或自由离子，它们的聚合反

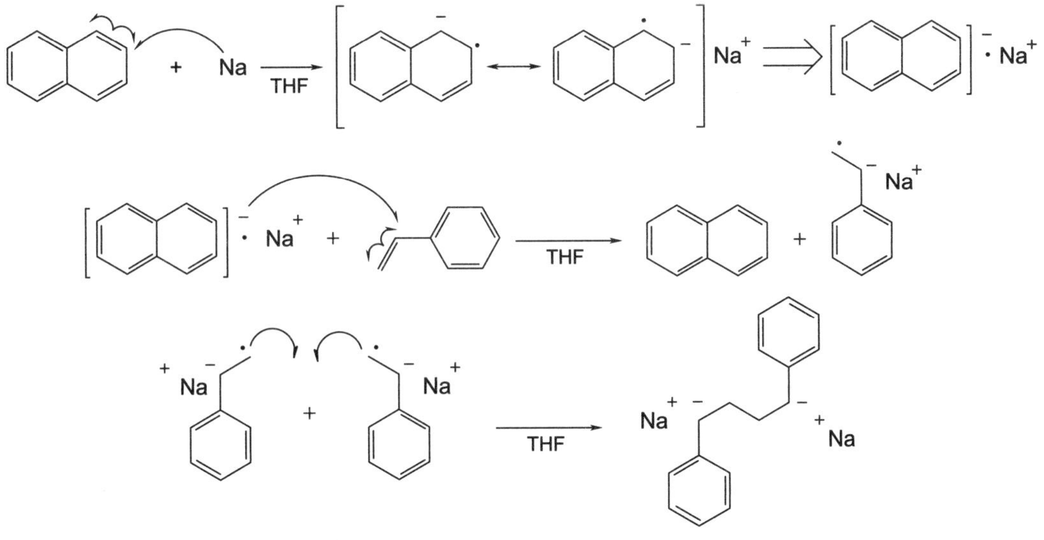

图 3-20　萘钠体系引发苯乙烯阴离子聚合
电子间接转移，聚合链双侧链增长

应活性较高。相反，非极性溶剂中容易产生接触离子对，聚合反应活性较低。

阴离子聚合反应一般具有快引发、慢增长、无终止、无转移的机理特征。阴离子聚合引发剂，在聚合前预先全部转变成阴离子活性种，然后以同一速率引发单体增长。在增长过程中，活性种数量保持不变，每一活性种所连接的单体数基本相等，聚合度就等于单体物质的量除以引发剂物质的量，而且比较均一，分布窄。如无杂质，则不终止，聚合将一直进行到单体耗尽。

$$B^- A^+ + M \longrightarrow BM^- A^+$$

$$BM^- A^+ + nM \longrightarrow BM_n^- A^+$$

$$-\frac{d[M]}{dt} = R_p = k_p[B^-][M]$$

式中，k_p 为表观链增长速率常数；$[B^-]$ 为阴离子活性种浓度；$[M]$ 为单体浓度。因 $[B^-]$ 等于引发剂浓度 $[C]$，且在聚合过程中始终保持恒定，聚合速率对单体呈一级反应。如将上式积分，就可导得单体浓度（或转化率）随时间作线性变化的关系式：

$$n\frac{[M_0]}{[M]} = k_p[C]t$$

式中，引发剂浓度 $[C]$ 和起始单体浓度 $[M_0]$ 已知，只要测得 t 时的残留单体浓度 $[M]$，就可求出增长速率常数 k_p。在适当溶剂中，苯乙烯阴离子聚合的 k_p 值与自由基聚合相近，但阴离子聚合无终止，活性种浓度（$10^{-3} \sim 10^{-2}$ mol/L）比自由基聚合（$10^{-9} \sim 10^{-7}$ mol/L）高得多，因此阴离子聚合速率比自由基聚合快。前文所述阴离子聚合反应的快引发、慢增长的机理特征中，慢增长是相对阴离子聚合的引发反应速率而言的。

根据阴离子聚合机理，所消耗的单体平均分配在每个活性端上，活性聚合物的平均聚合度就等于消耗单体数：

$$\overline{X}_n = \frac{[M_0] - [M]}{[B^-]/n} = \frac{n([M_0] - [M])}{[C]}$$

式中，[C] 为引发剂浓度；n 为每一大分子所带有的活性端基数。双侧阴离子聚合，$n=2$；单侧阴离子聚合，$n=1$。如果聚合至结束，单体全部耗尽，则 [M]=0。

聚合度分布服从 Flory 分布或 Poissen 分布，重均和数均聚合度之比为：

$$\frac{\overline{X}_w}{\overline{X}_n} = 1 + \frac{\overline{X}_n}{(\overline{X}_n+1)^2} \approx 1 + \frac{1}{\overline{X}_n}$$

当 $\overline{X}_n$ 很大时，$\overline{X}_w/\overline{X}_n$ 接近于 1，表示分布很窄。例如以萘钠-四氢呋喃引发经阴离子聚合制得聚苯乙烯时，$\overline{X}_w/\overline{X}_n \approx 1.09$，可用来制备分子量测定中的标样。

③ 链终止和链转移

阴离子聚合中如果发生链终止和链转移反应，需要产生极不稳定的氢负离子（H^-），这在热力学上也很难发生。实际上，阴离子聚合反应过程中也存在一些特殊的链终止和链转移反应。

例如，甲基丙烯酸甲酯等极性单体中的羰基能与碳阴离子反应，成为链转移反应。

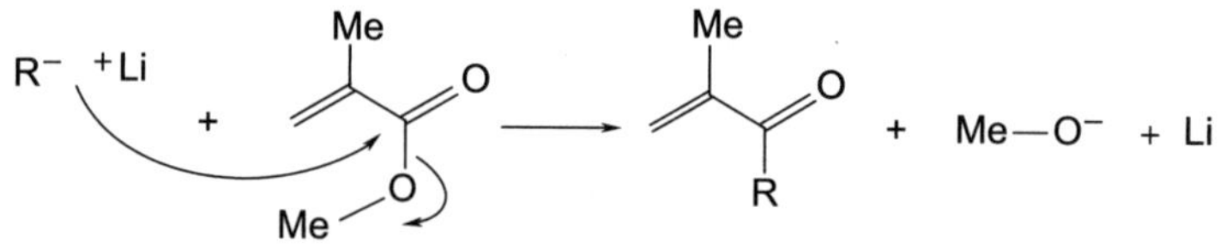

增长的碳阴离子分子内的"回咬"反应也是较典型的阴离子聚合中的副反应。

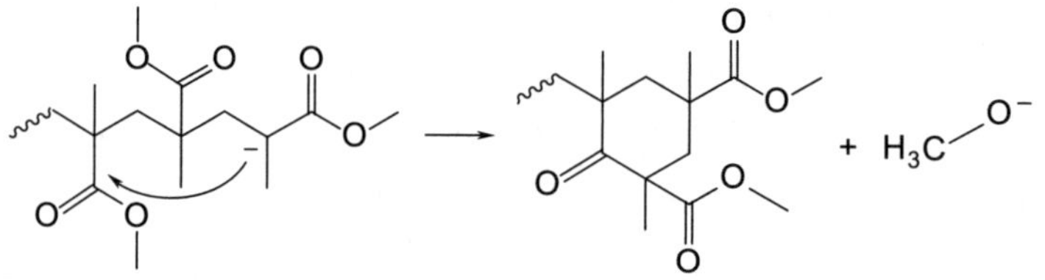

这些反应使速率和聚合度降低，并使分子量分布变宽。降低聚合温度（小于 $-50 \sim -70$℃）或以极性溶剂乙醚代替烃类，均可抑制上述副反应，获得活性聚合物。

氧、水、二氧化碳等含氧杂质均可使阴离子终止。对于苯乙烯-萘钠体系，微量水就可使聚合速率和聚合度显著降低，甚至终止。因此，阴离子聚合需要除尽聚合液体内痕量级的杂质。单体、溶剂要严格纯化，反应容器要反复真空干燥，甚至用少量活性聚合物溶液来洗涤。

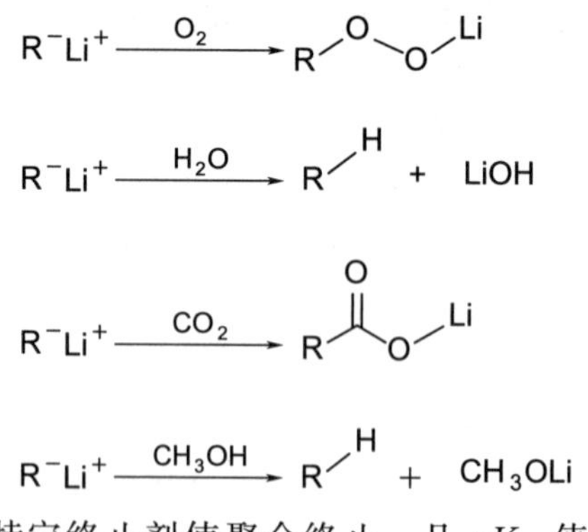

活性聚合结束时，需加特定终止剂使聚合终止。凡 pK_a 值比单体小的化合物都能终止阴离子聚合。例如甲醇（$pK_a=16$）能与活性阴离子反应生成甲醇锂，其反应活性低，不再引发烯类单体聚合。

英语读译资料

(1) Cationic polymerization of isobutylene

Butyl rubber is widely used for inner tubes and as a sealant. It is produced using the cationic polymerization with the copolymerization of isobutylene in the presence of a small amount (10%) of isoprene. Thus, the random copolymer chain contains a low concentration of widely spaced isolated double bonds, from the isoprene, that are later cross-linked when the butyl rubber is cured. The number of units derived from isobutylene units greatly outnumbers the number the units derived from the isoprene monomer. The steric requirements of the isobutylene-derived units cause the chains to remain apart giving it a low stress/strain value and a low T_g.

The cationic polymerization of vinyl isobutyl ether at $-40℃$ produces stereoregular polymers. The carbocations of vinyl alkyl ethers are stabilized by the delocalization of "p" valence electrons in the oxygen atom, and thus these monomers are readily polymerized by cationic initiators. Poly(vinyl isobutyl ether) has a low T_g because of the steric hindrance offered by the isobutyl group. It is used as an adhesive and as an impregnating resin.

This production of stereoregular structures has been known for some time and is especially strong for vinyl ethers. Several general observations have been noted. First, the amount of stereoregularity is dependent on the nature of the initiator. Second, stereoregularity increases with a decrease in temperature. Third, the amount and type (isotactic or syndiotactic) is dependent on the polarity of the solvent. For instance, the isotactic form is preferred in nonpolar solvents, but the syndiotactic form is preferred in polar solvents.

(2) Living Anionic Polymerization

An important feature of anionic polymerization is the absence of inherent termination processes. As for cationic polymerization, bimolecular termination between two propagating chains is not possible due to charge repulsion, but unlike cationic polymerization, termination by ion-pair rearrangement does not occur because it requires the highly unfavourable elimination of a hydride ion. Furthermore, the alkali metal (or alkaline earth metal) counter-ions used have no tendency to combine with the carbanionic active centers to form unreactive covalent bonds. Thus, in the absence of chain transfer reactions, the propagating polymer chains retain their active carbanionic end groups and are truly living, unlike reversible-deactivation radical and cationic polymerizations in which termination is still possible. Interest in anionic polymerization grew enormously following the work of Szwarc in the mid-1950s. He demonstrated that, under carefully controlled conditions, carbanionic living polymers could be formed using electron transfer initiation. If more monomer is added after complete conversion of the initial quantity, the chains grow further by polymerization of the additional monomer and will again remain active. Such polymer molecules, which permanently retain their active centers in chain polymerization and continue to grow so long as monomer is available, are termed living polymers.

Organolithium compounds (e.g. butyllithium) are the most widely used organometallic initiators. They are soluble in non-polar hydrocarbons but tend to aggregate in such media. Electron transfer initiation involves donation of single electrons to molecules of monomer to form monomeric radical-anion species which then couple together to give dicarbanion species that initiate polymerization of the remaining monomer. Homogeneous initiation by electron transfer can be achieved in ether solvents, such as THF, using soluble electron-transfer complexes formed by reaction of alkali metals with aromatic compounds (e. g. naphthalene, biphenyl). Sodium naphthalide was one of the first to be used and is formed by the addition of sodium to an ethereal solution of naphthalene. An atom of sodium donates an electron to naphthalene producing the naphthalide radical-anion, which is green in colour and stabilized by resonance.

(3) Comparison of radical and ionic polymerization

Almost all monomers containing the carbon-carbon double bond undergo radical polymerization, while ionic polymerizations are highly selective. Cationic polymerization is essentially limited to those monomers with electron-releasing substituents such as alkoxy, phenyl, vinyl, and 1,1-dialkyl. Anionic polymerization takes place with monomers possessing electron-withdrawing groups such as nitrile, carbonyl, phenyl, and vinyl. The selectivity of ionic polymerization is due to the very strict requirements for stabilization of anionic and cationic propagating species. The commercial utilization of cationic and anionic polymerizations is rather limited because of this high selectivity of ionic polymerizations compared to radical polymerization. Ionic polymerizations, especially cationic polymerizations, are not as well understood as radical polymerizations because of experimental difficulties involved in their study. The nature of the reaction media in ionic polymerizations is often not clear since heterogeneous inorganic initiators are often involved. Further, it is extremely difficult in most instances to obtain reproducible kinetic data because ionic polymerizations proceed at very rapid rates and are extremely sensitive to the presence of small concentrations of impurities and other adventitious materials. The rates of ionic polymerizations are usually greater than those of radical polymerizations. These comments generally apply more to cationic than anionic polymerizations. Anionic systems are more reproducible because the reaction components are better defined and more easily purified.

Cationic and anionic polymerizations have many similar characteristics. Both depend on the formation and propagation of ionic species, a positive one in one case and a negative one in the other. The formation of ions with sufficiently long lifetimes for propagation to yield high-molecular-weight products generally requires stabilization of the propagating centers by solvation. Relatively low or moderate temperatures are also needed to suppress termination, transfer, and other chain-breaking reactions which destroy propagating centers.

Although solvents of high polarity are desirable to solvate the ions, they cannot be employed for several reasons. The highly polar hydroxylic solvents react with and destroy most ionic initiators. Other polar solvents such as ketones prevent initiation of polymerization by forming highly stable complexes with the initiators. Ionic polymerizations are, therefore, usually carried out in solvents of low or moderate polarity such as tetrahydrofuran, ethylene dichloride, and pentane, although moderately high polarity solvents such as nitrobenzene are also used. In such solvents one usually does not have only a single type of propagating species. For any propagating species such as ~BA in cationic polymerization, one can visualize the range of behaviors from one extreme of a completely covalent species (Ⅰ) to the other of a completely free (and highly solvated) ion (Ⅳ).

$$\sim BA\ (Ⅰ),\ \sim B^+A^-\ (Ⅱ),\ \sim B^+ \parallel A^-\ (Ⅲ),\ \sim B^+ + A^-\ (Ⅳ)$$

The intermediate species include the tight or contact ion pair (Ⅱ) (also referred to as the intimate ion pair) and the solvent-separated or loose ion pair (Ⅲ). The intimate ion pair has a counter or gegenion of opposite charge close to the propagating center (unseparated by solvent). The solvent-separated ion pair involves ions that are partially separated by solvent molecules. The propagating cationic chain end has a negative counterion. For an anionic polymerization the charges in species Ⅱ-Ⅳ are reversed; that is, B carries the negative charge and A the positive charge. There is a propagating anionic chain end with a positive counterion. Alternate terms used for free ion and ion pair are unpaired ion and paired ion, respectively.

Most ionic polymerizations involve two types of propagating species, an ion pair and a free ion Ⅳ, coexisting in equilibrium with each other. The identity of the ion pair (i.e., whether the ion pair is best described as species Ⅱ or Ⅲ) depends on the particular reaction conditions, especially the solvent employed. Increased solvent polarity favors the loose ion pair while the tight ion pair predominates in solvents of low polarity. The ion pairs in cationic polymerization tend to be loose ion pairs even in solvent of low or moderate polarity since the counterions (e.g., bisulfate, $SbCl_6^-$, perchlorate) are typically large ions. The lower charge density of a large counterion results in smaller electrostatic attractive forces between the propagating center and counterion. The nature of the ion pairs is much more solvent-dependent in anionic polymerizations where the typical counterion (e.g., Li^+, Na^+) is small. The covalent species I is generally ignored since it is usually unreactive (or much lower in reactivity) compared to the other species. Free-ion concentrations are generally much smaller than ion pair concentrations but the relative concentrations are greatly affected by the reaction conditions. Increased solvent polarity results in a shift from ion pairs to free ions. The nature of the solvent has a large effect in ionic polymerization since the different types of propagating species have different reactivities. Loose ion pairs are more reactive than tight ion pairs. Free ions are orders of magnitude higher in reactivity than ion pairs in anionic polymerization. Ion pairs are generally no more than an order of magnitude lower in reactivity compared to free ions in cationic polymerization.

习 题

1. 试从单体结构来解释丙烯腈和异丁烯离子聚合行为的差异，选用何种引发剂？丙烯酸、烯丙醇、丙烯酰胺、氯乙烯能否进行离子聚合，为什么？
2. 在离子聚合中，活性种离子和反离子之间的结合可能有几种形式？其存在形式受哪些因素影响？不同形式对单体的聚合机理、活性和定向能力有何影响？

3. 进行阴、阳离子聚合时，分别叙述控制聚合速率和聚合物分子量的主要方法。离子聚合中有无自动加速现象？离子聚合物的主要微观构型是头-尾还是头-头连接？聚合温度对立构规整性有何影响？

4. 甲基丙烯酸甲酯分别在苯、四氢呋喃、硝基苯中用萘钠引发聚合，试问在哪种溶剂中的聚合速率最大？

5. 由阳离子聚合来合成丁基橡胶，如何选择共单体、引发剂、溶剂和温度，为什么？

6. 用 BF_3 引发异丁烯聚合，如果将氯甲烷溶剂改成苯？预计会有什么影响？

7. 比较阴离子聚合、阳离子聚合、自由基聚合的主要差别，哪一种聚合的副反应最少？说明溶剂种类的影响，讨论原因和本质。

8. 将 $1.0×10^{-3}$ mol 萘钠溶于四氢呋喃中，然后迅速加入 2.0mol 苯乙烯，溶液的总体积为 1L。假如单体立即混合均匀，发现 2000s 内已有一半单体聚合，计算聚合 2000s 和 4000s 时的聚合度。

9. 苯乙烯单体加到萘钠的四氢呋喃溶液中，苯乙烯和萘钠的浓度分别为 0.2mol/L 和 $1×10^{-3}$ mol/L。在 25℃下聚合 5s，测得苯乙烯的浓度为 $1.73×10^{-3}$ mol/L，试计算：①链增长速率常数；②引发速率；③聚合反应 10s 的聚合速率；④10s 的数均聚合度。

10. 异丁烯阳离子聚合时，以向单体链转移为主要终止方式，聚合物末端为不饱和端基。现在 4.0g 聚异丁烯恰好使 6.0mL 0.01mol/L 溴-四氯化碳溶液褪色，试计算聚合物的数均分子量。

11. 异丁烯阳离子聚合时的单体浓度为 2.0mol/L，链转移剂浓度分别为 0.2mol/L、0.4mol/L、0.6mol/L、0.8mol/L，所得聚合物的聚合度依次是 25.34、16.01、11.70、9.20，向单体和向链的转移是主要终止方式，试用作图法求转移常数 C_M 和 C_S。

12. 某单体 M 在 Y 引发体系下聚合，观察到如下实验现象，论述它们分别是逐步聚合、自由基聚合、阳离子聚合、阴离子聚合的哪一种。

①聚合度随聚合温度增加而增加；②聚合度受溶剂影响较大；③聚合度和单体浓度呈线性关系；④聚合速率随反应温度增加而加快。

13. 写出合成如下嵌段聚合物的化学反应式。

① ABA；②CABAC。

其中 A、B、C 分别为苯乙烯、丁二烯和异戊二烯。

14. 聚异丁烯的合成过程中，分析应该利用什么方法提高聚合物的分子量？

英语习题

1. Which of the following could be used to initiate the polymerization of isobutylene：(a) sulfuric acid；(b) boron trifluoride etherate；(c) water；(d) butyllithium.

2. When termination is by chain transfer during a cationic chain polymerization, what is the relationship of average DP and the kinetic chain length?

3. What would be the composition of the product obtained by the cationic low-temperature polymerization of a solution of isobutylene in ethylene?

4. What is the relationship between the rate of initiation to the monomer concentration

in ionic chain polymerization?

5. What effect will the use of a solvent with a higher dielectric constant have on the rate of propagation in ionic chain polymerization?

6. How does the rate constant k_p change as the yield of polymer increases?

7. Which will have the higher T_g value: PS or PIB?

8. Which of the following could serve as an initiator for an anionic chain polymerization? (a) $AlCl_3/H_2O$, (b) BF_3/H_2O, (c) butyllithium, or (d) sodium metal?

9. What species, in addition to a dead polymer, is produced in a chain transfer reaction with a macrocarbocation in cationic chain polymerization?

10. For a cationic polymerization, what is the relationship between average DP and R_p and R_t?

11. What is the relationship between the average DP and initiator concentration in cationic chain polymerization?

3.4 配位聚合

配位聚合

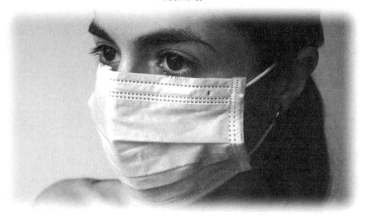

对抗新冠疫情中，个人防护口罩成为居民防护的必需用品。一次性防护口罩的关键材料是聚丙烯熔喷无纺布，纤维直径大约 5μm，经静电处理后，可吸附 90%以上的气溶胶颗粒。聚丙烯是利用配位聚合生产的重要聚合物，用于制纤维、薄膜、注塑件、管材等。

本节目录

3.4.1 配位聚合概述 / 155

3.4.2 配位聚合物的立体规整性 / 155

3.4.3 配位聚合催化剂 / 156

3.4.4 配位聚合的定向机理 / 159

3.4.5 配位聚合的反应历程及反应动力学 / 162

3.4.6 重要配位聚合产物 / 163

重点要点

配位聚合，立体结构，构型，构象，光学异构体，对映异构体，几何异构体，手性中心，全同立构聚合物，间同立构聚合物，无规立构聚合物，顺式构型，反式构型，立构规整度，全同指数，配位聚合引发体系，Ziegler-Natta 催化剂。

3.4.1 配位聚合概述

20 世纪 50 年代，Ziegler 和 Natta 分别报道了聚乙烯和聚丙烯的配位聚合（coordination polymerization）方法，开创了配位聚合合成高分子的新时代。20 世纪 60 年代，用配位聚合法合成的顺-1,4-聚丁二烯橡胶和乙丙橡胶出现，丰富了合成橡胶产品的种类。70 年代初期，高效的配位聚合催化剂用于低压聚乙烯的工业生产，此后，高效载体催化剂用于丙烯本体聚合，为聚丙烯合成工业带来了活力。如今，聚乙烯、聚丙烯等利用配位聚合合成的各种聚烯烃超过 1 亿吨，占全部高分子合成材料总量的三分之一，成为产量最大的高分子产品。

配位聚合的反应活性中心不是自由基，也不是阴阳离子，而是由烷基有机物活化的过渡金属元素的 d 轨道。烯烃单体的 π 电子在空的 d 轨道上配位活化，并和配位中间体中已有的烷基发生移位而进行链增长，所以称作配位聚合。配位聚合链增长过程中，单体分子首先在活性中心的空位配位形成配合物，烯烃单体和活性中心的配位作用决定单体在配合物上的位置和立体构型，而后插入到活性中心和聚合物分子链之间，因此，配位聚合能够合成高度立体规整的聚合物，具有立体定向性。配位聚合也因此被称作定向聚合。配位聚合的引发剂中含有过渡金属，有许多学者把配位聚合引发剂称为配位聚合催化剂，一般而言，配位聚合具有以下特点。

① 催化剂为过渡金属和有机化合物组成的配合催化剂。以 Ziegler-Natta 催化剂为主，其主要成分为 Ti、V、Cr、Mn、Fe、Ni、Zr、Mo、W 等 Ⅳ 族到 Ⅷ 族的卤盐化合物。助剂成分为 Ⅰ 族到 Ⅲ 族的烷基金属化合物。助催化剂还原过渡金属离子，并提供烷基形成活性中心。

② 具有定向性。α-烯烃或二烯烃经配位聚合可形成立体规整性极高的聚合物。例如，丙烯配位聚合可合成等规聚丙烯，丁二烯配位聚合可合成高顺-1,4-聚丁二烯等。

③ 适用单体较少。目前仅有 α-烯烃或二烯烃适用配位聚合。大多数含氧、氮的单体和极性大的含卤素单体因极性基团参与催化剂配位，干扰链增长反应，不适合配位聚合。聚乙烯和聚丙烯是利用配位聚合生产的重要聚合物。其他烯烃类单体、炔类单体、环状单体和一些极性单体在催化剂、溶剂、温度等条件适当情况下也可进行配位聚合。丙烯酸酯类极性单体有很强的配位能力，利用均相催化剂可形成全同聚合物。极性较强单体中的 O 或 N 等给电子原子较烯烃单体的 π 键更易与 Ziegler-Natta 催化剂形成稳定络合物，使催化剂失效。苯乙烯具有弱极性，可通过配位聚合合成全同聚合物，也可以合成间规聚苯乙烯。

④ 原料要求苛刻。配位聚合只能选用脂肪烃、芳香烃类溶剂。含活泼氢或极性大的含氧、氮化合物不宜用于配位聚合。配位聚合对原料纯度要求较高，单体、溶剂、惰性气体中不能含有水、氧等杂质。

3.4.2 配位聚合物的立体规整性

配位聚合所合成的聚合物大多具有良好的立体规整性。我们在第一章绪论部分曾简要介绍过，聚合物的立体异构（stereo-isomerism）有对映异构和顺反异构两种。

对映异构（enantiomers）又称手性异构，由手性中心产生的光学异构体有 R 型（右）

和 S 型（左）两种。丙烯、1-丁烯等 α-烯烃（CH$_2$=CHR）所形成的聚合物含有多个手性中心 C* 原子，每个 C* 连有 H、R 和两个聚合链段。尽管这两个聚合链段不等长，但对旋光活性的影响较小，并不显示光学活性，这种手征中心常称作假手征中心。

顺反异构（cis-trans isomers）是由双键引起的顺式和反式几何异构，两种构型不能互变，如聚异戊二烯。不论哪一类构型，立构规整聚合物多以螺旋状构象存在。聚二烯烃如丁二烯聚合，可以 1,4-或 1,2-加成，可能有顺、反-1,4-和全同、间同-1,2-聚丁二烯四种立体构型异构体，这四种异构体均已制得。1,3-异戊二烯聚合，1,4-加成中有顺、反结构。

聚合物的立构异构现象对聚合物的立体规整性产生影响，进而影响大分子堆砌的紧密程度和结晶度，从而影响密度、熔点、溶解性能、强度、弹性等一系列宏观性能。

例如，使用 Ziegler-Natta 催化剂经配位聚合得到的等规聚丙烯是结晶性聚合物，熔点高达 175℃，拉伸强度可达 35MPa，相对密度约为 0.90，性能接近工程塑料，是目前产量非常大的聚合物。因可耐受 121℃高压灭菌，也广泛用于注射器等医用塑料产品。无规聚丙烯熔点大约 75℃，强度差，用途有限。顺-1,4-聚丁二烯和顺-1,4-聚异戊二烯的 T_g 和 T_m 较低，不易结晶，可作为橡胶使用，而反-1,4-聚二烯烃的 T_g 和 T_m 较高，易结晶，可作为塑料使用。

立构规整度（stereo-regularity）指立构规整聚合物占聚合物总量的百分数。立构规整度可由红外、核磁共振等直接测定，也可由结晶度、密度、溶解度等物理性质来间接表征。

聚丙烯的等规度或全同指数 IIP（isotactic index）可用红外光谱的特征吸收谱带来测定。波数 975cm^{-1} 处是全同螺旋链段的特征吸收峰，而 1460cm^{-1} 处是 CH$_3$ 基团的特征峰，对立体结构并不敏感，由两者的吸收强度（或峰面积）之比可以计算聚丙烯样品的等规度。同理，间规度可用波数 987cm^{-1} 处的特征峰面积来计算。

对于聚二烯烃，常用顺-1,4、反-1,4、全同-1,2、间同-1,2 等的百分数来表征立构规整度。根据红外光谱特征吸收峰（FTIR）的波数位置和核磁共振氢谱（NMR）的化学位移可以定性测定各种立构的存在，由各特征吸收峰面积的积分定量计算这四种立构规整度的比值。

3.4.3 配位聚合催化剂

配位聚合（coordination polymerization）最早用来解释 Ziegler-Natta 引发体系的聚合机理，其本质属于离子聚合。大多数配位聚合以碳阴离子为活性中心，属于配位阴离子聚合机理。

配位聚合催化剂自问世以来，历经半个多世纪的持续改进，性能不断提高，种类也增加了许多。烯烃配位聚合的催化剂有三大类，Ziegler-Natta 催化剂，茂金属催化剂和非茂过渡金属催化剂。其中负载型 Ziegler-Natta 催化剂应用最为广泛，全球大约 90% 的聚丙烯和 50% 的聚乙烯由该类催化剂生产。

（1）Ziegler-Natta 催化剂

以配位聚合聚丙烯 Ziegler-Natta 催化剂为例（表 3-7），催化剂经历了几个不同的发展阶段，催化剂的活性已由最初的几十倍提高到数万倍，若按过渡金属计已达到数百万倍，聚丙烯的等规度已达 98% 以上，生产工艺也得到了简化。

表 3-7 Ziegler-Natta 催化剂发展历程

阶段	年代	催化剂	等规度质量分数/%	催化剂效率/(kgPP/gM)
第一代	1957~1970	$TiCl_3/AlCl_3/AlEt_2Cl$	90~94	0.8~1.2
第二代	1970~1978	$TiCl_3/R_2O/AlEt_2Cl$	94~97	3~5
第三代	1978~1980	$TiCl_4/单酯/MgCl_2/AlEt_3$	90~95	5~15
第四代	1980~1985	$TiCl_4/单酯/MgCl_2/AlEt_3/硅氧烷$	90~98	20~100
第五代	1985~1990	$TiCl_4/二醚/MgCl_2/AlEt_3$	94~98	50~120

最初的 Ziegler 催化剂由 $TiCl_4/Al(C_2H_5)_3$ 组成，Natta 催化剂为 $TiCl_3/Al(C_2H_5)_3$。以后发展到由Ⅳ~Ⅷ族过渡金属化合物和Ⅰ~Ⅲ族金属有机化合物两大组分配合而成的二元体系，统称为 Ziegler-Natta 催化剂系列。其中过渡金属化合物成为主催化剂，Ⅰ~Ⅲ族金属有机化合物称为副（助）催化剂。

主催化剂：Ⅳ~Ⅷ族过渡金属（Mt）化合物，包括 Ti、V、Mo、Zr、Cr 的氯（或 Br、I）化物 $MtCl_n$、氧氯化物 $MtOCl_n$、乙酰丙酮化合物 $Mt(acac)_n$、环戊二烯基（Cp）金属氯化物 Cp_2TiCl_2 等，主要用于 α-烯烃的配位聚合。其中 $MoCl_5$ 和 WCl_6 专用于环烯烃的开环聚合，Co、Ni、Ru、Rh 等的卤化物或羧酸盐组分则主要用于二烯烃的定向聚合。

副催化剂：Ⅰ~Ⅲ族金属有机化合物，如 AlR_3、LiR、MgR_2、ZnR_2 等，R 为烷基或环烷基。其中有机铝应用最多。

Ziegler-Natta 引发体系可分成不溶于烃类（非均相）和可溶（均相）两大类，溶解与否与过渡金属组分和反应条件有关。

① 非均相引发体系

$TiCl_4$-AlR_3（或 AlR_2Cl）在 -78℃ 下溶于庚烷或甲苯，成为暗红色溶液，可引发乙烯聚合，但对丙烯单体的引发活性较低。升温到 -35~25℃，则转变成棕红色沉淀，可有效引发丙烯以及丁二烯的聚合反应。低价氯化钛（或钒），如 $TiCl_3$、$TiCl_2$、VCl_4 等，一般不溶于烃类有机试剂，与 AlR_3 或 AlR_2Cl 混合反应后，仍为（微）非均相，对丙烯聚合有较高的活性。

② 均相引发体系

许多钒系催化剂为均相催化剂。例如，合成乙丙橡胶中的 $VOCl_3/AlEt_2Cl$ 或 $V(acac)_3/AlEt_2Cl$ 体系为均相。卤化钛中的卤素如果被烷氧基、乙酰丙基或 Cp 等有机官能团取代，与 AlR_3 络合后可溶于烃类有机溶剂，一般可引发乙烯聚合，但对丙烯聚合的活性比较差。

20 世纪 70 年代后期，催化效率比第一代 Ziegler-Natta 催化剂提高了数十倍的高效催化剂逐渐应用于聚烯烃合成工业，成为配位聚合的重大突破，目前占据配位聚合产业的主导地位。高效催化剂的主要特征就是催化剂载体化，扩大了催化剂的表面积，增加了有效活性中心的数量。常用的载体为 $MgCl_2$，Mg 的电负性小于 Ti，所以 Mg 的推电子效应可使活性中心 Ti 的电子密度增加，削弱 Ti—C 键，从而有利于烯烃类单体的配位，也利于 Ti—C 键上链增长的单体插入反应，起到了降低链增长活化能，提高催化剂反应活性的目的。

(2) 金属茂催化剂

大多数 Ziegler-Natta 催化剂为非均相催化剂，因活性点的反应活性多种多样，聚合物产品的分子量分布较宽。1980 年，Kaminsky 利用二氯二锆茂（Cp_2ZrCl_2）作主催化剂，使用甲基铝氧烷（简称 MAO，$AlMe_3$ 和水 1∶1 反应产物）作共催化剂，成功开发出可溶解、均相分散的催化剂，对乙烯以及 α-烯烃显示出高聚合活性，且聚合产物的分子量分布也比较窄。此后，茂金属催化剂逐渐发展成为一类重要的配位催化剂。

茂金属配合物是指过渡金属原子与茂环（环戊二烯或取代的环戊二烯负离子）配位形成的过渡金属有机配合物，助催化剂使用最多的是 MAO。茂金属催化剂（metallocene catalyst）的通式为 $LL'MtX_2$。其中 L 和 L′ 是五元环环戊二烯基类配位体（简称茂）；Mt 为 ⅣB 族过渡金属，通常为锆（Zr）；X 为非茂配体，通常是 Cl，也可以是 CH_3。

如图 3-21 所示，茂金属催化剂有普通结构，桥链结构和限定几何构型配位体结构三种形式。单独茂金属催化剂对烯烃聚合基本没有活性，常加甲基铝氧烷 MAO 作共催化剂。茂金属引发聚合机理与 Ziegler-Natta 体系相似，即烯烃分子与过渡金属配位，在增长链端与金属之间插入而增长。

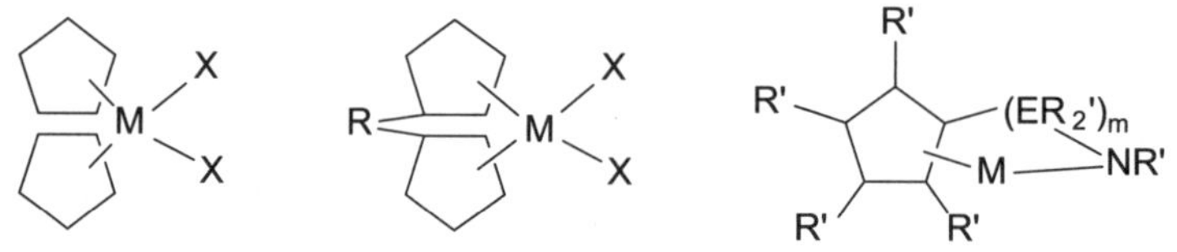

普通结构　　　桥链结构　　　限定几何构型配体结构

图 3-21　金属茂催化剂三种结构

茂金属催化剂均相分散，相比 Ziegler-Natta 催化剂有许多优点：①高活性，金属原子几乎 100% 均可形成活性中心。例如，Cp_2ZrCl_2/MAO 用于乙烯聚合时，活性可高达 10^8 gPE/(gZr·h)，比高效 Zieigler-Natta 催化剂的活性 10^6 gPE/(gZr·h) 要高两个数量级。②活性中心种类单一，可高效控制聚合物的立构规整性。③聚合产物分子量分布窄（1.05～1.8）。④可使用单体更多，例如环烯烃、共轭二烯烃、氯乙烯、丙烯腈等极性单体也可进行配位聚合。茂金属催化剂的缺点有：合成困难，成本高，较难从聚合产物中脱除分离，对氧和水分敏感。

(3) 非茂金属催化剂

非茂金属催化剂是指不含环戊二烯类基团的非茂配体过渡金属元素的化合物，在助催化剂 MAO 存在下引发烯烃聚合。非茂金属催化剂主要为 O、N、S、B、P 等杂原子或其他芳环碳原子与Ⅳ族金属（Ti、Zr、Hf）或Ⅷ族后过渡金属（Ni、Pd、Fe、Co 等）形成的配位体。除具备茂金属催化剂特性（单活性中心、组成均一、聚合物分子量分布窄等）外，还有诸多新特点和独特性能。

非茂金属催化剂的过渡金属元素亲电性弱，耐杂原子的能力强，可使烯烃和极性单体共聚；活性较高，与茂金属催化剂相当；价格低廉，容易制备。

3.4.4 配位聚合的定向机理

配位聚合是一种加聚反应，其链增长机理是先由聚合单体的碳碳双键与配位催化剂中活性中心的过渡元素原子空的 d 轨道进行配位，然后移位，使分子链增长。配位聚合的特点就是 α-烯烃或二烯烃的聚合物的链节排列具有非常规整的立体构型，因此也称作定向聚合。

自 Ziegler-Natta 催化剂问世以来，人们对配位聚合的定向机理进行了广泛研究，并提出了多种机理假设。其中，双金属机理和单金属机理由于实验证据比较充足，得到了普遍公认。无论哪种机理，配位聚合链增长过程可以归纳为：形成活性中心空位，吸附单体定向配位，络合活化，插入增长，类似模板地进行定向聚合，形成立构规整聚合物。下面以 Ziegler-Natta 催化剂引发丙烯和丁二烯的配位聚合为例，简要介绍配位聚合的定向机理。

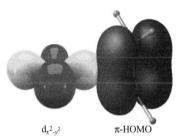

图 3-22 过渡金属 d_{xy} 轨道 /π-LUMO，$d_{x^2-y^2}$ 轨道 /π-HOMO 的重叠形成配位键

(1) Ziegler-Natta 催化剂引发丙烯配位聚合

理解 Ziegler-Natta 催化剂引发丙烯配位聚合的机理，重点在于理解活性中心的化学-物理结构和性质、链增长场所、单体定向机理等问题。

高价态过渡金属离子无配位点与烯类进行 π-络合，但低价过渡金属离子的 d-轨道和烯烃的 π 轨道重叠能形成稳定的 π-络合物。例如，过渡金属 d_{xy} 轨道与烯烃的 π-反键轨道（LUMO）重叠，$d_{x^2-y^2}$ 轨道与烯烃的 π-轨道（HOMO）重叠均可形成配位键（图 3-22）。

① 双金属机理

双金属机理由 Patat-Sinn 和 Natta 分别于 1958 年和 1960 年提出，认为链增长的活性中心是具有 Ti---C---Al 碳桥电子三中心键的络合物。Natta 提出的桥式络合物结构如下：

其中 X 表示卤素原子，P_n 表示聚烯烃增长链。双金属机理的链增长反应历程如下：

烯烃配位

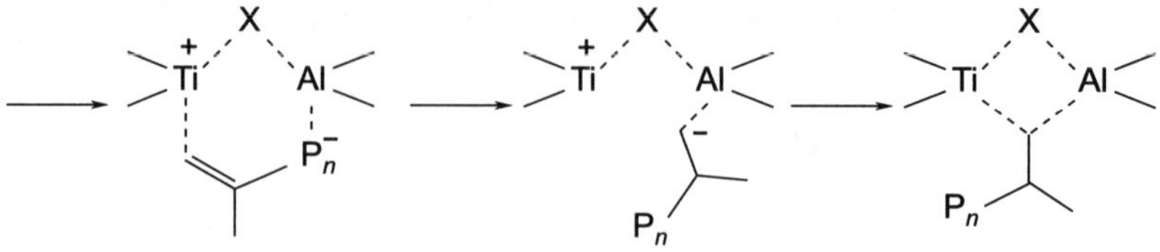

烯烃络合　　　　　　　　　　　　　　　　　　空位恢复

单体先与正电性的 Ti 配位，导致 Ti---C 键松动，极化的单体在 C---Al 键上插入，然后恢复到双金属桥式结构，实现一个链节增长，如此循环下去，便形成聚丙烯大分子。双金属机理认为在非均相催化剂固体表面上形成的双金属活性中心具有不对称性，每一个活性中心只能与丙烯的一种构型络合，达成定向聚合的结果。

后期的实验结果逐渐否定了聚合链在 C---Al 键上增长的观点，因为更多的实验结果表明，聚合物分子链在 Ti---C 键之间插入增长。

② 单金属机理

P. Gossee 和 E. J. Arlman 于 1964 年提出单金属中心机理，其结构如下：

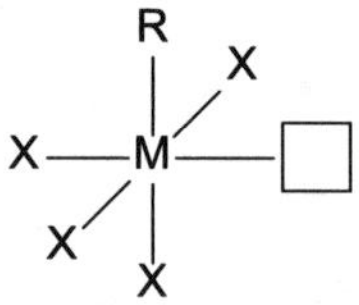

其中，M 表示过渡金属原子，X 表示卤素原子，R 表示烷基，□表示空轨道。
单金属链增长反应机理含活性中心形成、链节增长、空位复原三个步骤。

a. 活性中心形成　　含有空轨道的五配位 Ti 络合物与烷基铝进行 C—Cl 交换。

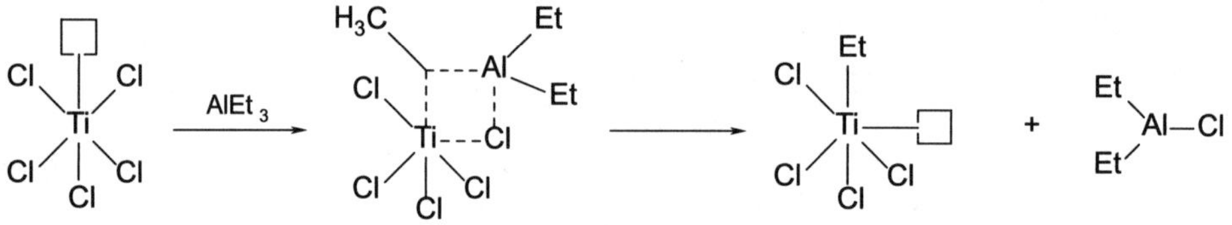

b. 链增长　　丙烯在金属空位配位形成 π-络合物，然后插入进 Ti---C 键。

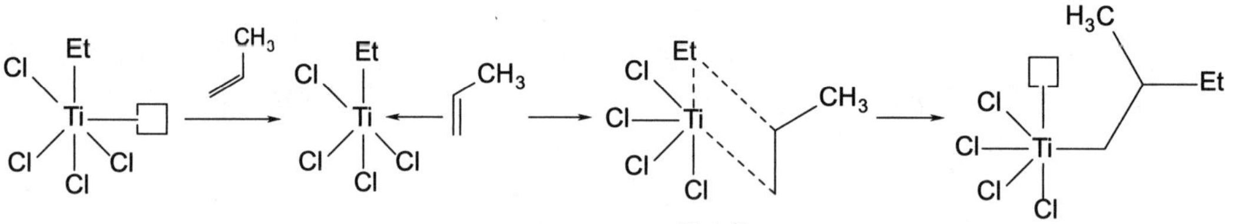

π-络合物

c. 空位复原　　单体插入后空位方位发生改变，使活性中心复原。

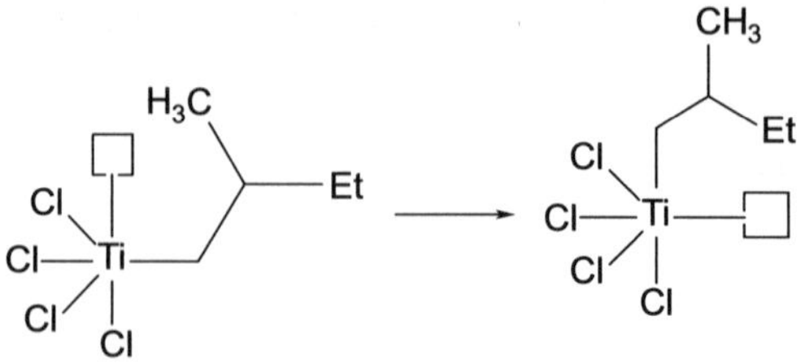

较多实验数据表明，Ziegler-Natta 催化剂引发丙烯配位聚合的活性中心是 Ti---C 键，

层状结晶的 $TiCl_3$ 表面存在 Cl 空位。单金属中心理论较好地解释了丙烯定向聚合的机理，但是仍然存在争议，有待深入研究。

(2) Ziegler-Natta 催化剂引发二烯烃配位聚合

α-烯烃配位聚合理论可以解释共轭二烯烃的部分实验现象，但是共轭二烯烃的聚合反应更为复杂，其配位聚合机理迄今还未解释透彻，其原因有以下三个：

a. 因丁二烯的配位聚合可发生 1,2-加成，也可以发生 1,4-加成，而且 1,4-加成中又会产生顺反结构。

b. 单体有顺反两种构象。例如，丁二烯的 S-顺式占 4%，S-反式占 96%。

c. 增长链端可以是 α-烯丙基，也可以是 π-烯丙基。

式中，左边 α-烯丙基由 Mt 和 CH_2 以 σ 键结合，右边 π-烯丙基则由 Mt 与三个 C 原子形成配位体。

重要的共轭二烯烃的配位聚合机理有两种，分别是烯丙基机理、单体与金属配位机理。

① π-烯丙基机理

该机理认为聚合物的链端结构是控制二烯烃有规聚合的关键。有人把 π-烯丙基可发生对式-同式异构化现象引入到烯丙基机理，提出"烯丙基对式-同式异构化机理"。以丁二烯配位聚合为例：

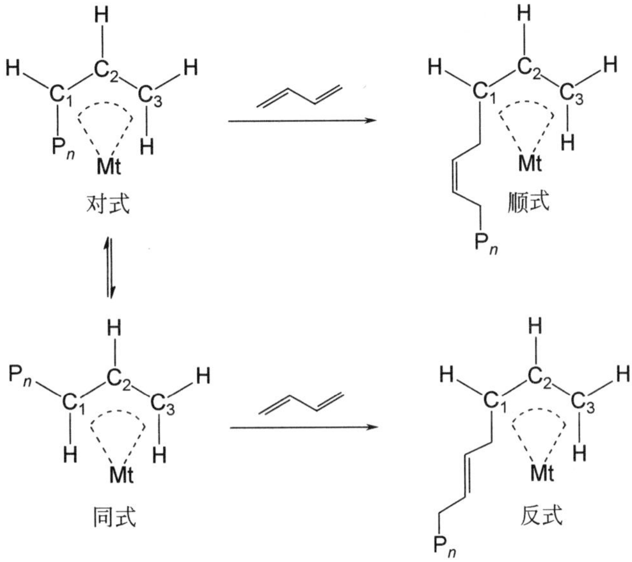

双键配位自然得到对式构型的烯丙基化合物，链增长得到顺式聚合物。当烯丙基异构化

为同式结构，链增长得到反式聚合物。因此，顺式链节含量取决于链增长速度和异构化速度之比。

② 单体与金属配位机理

该机理的基本观点是单体在过渡金属上的配位方式决定单体的加成类型。例如，顺式配位的丁二烯单体，1,4-插入获得顺-1,4-聚丁二烯；反式配位的丁二烯单体，1,4-插入获得反-1,4-聚丁二烯；1,2-插入获得1,2-聚丁二烯。

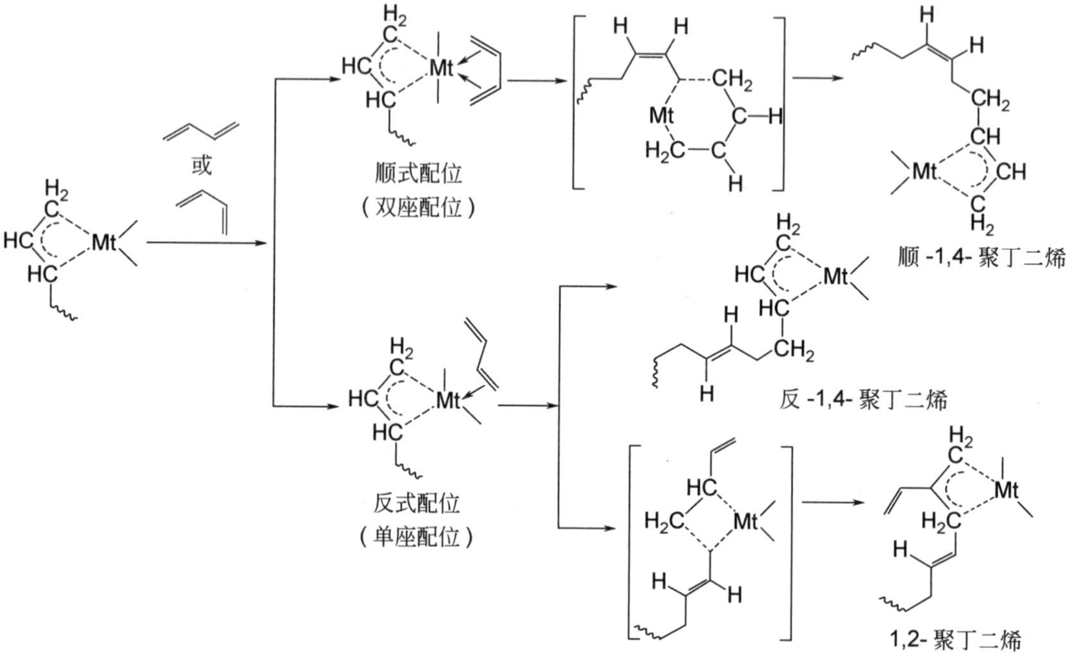

3.4.5 配位聚合的反应历程及反应动力学

配位聚合的机理特征与活性阴离子聚合相似，也存在链引发、链增长、链转移和链终止四个基元反应。和阴离子聚合类似，也具有难终止、难转移的动力学特征。

(1) 配位聚合反应历程

a. 链引发　Ziegler-Natta 催化剂两组分反应后，形成活性种，引发在固体颗粒表面进行。

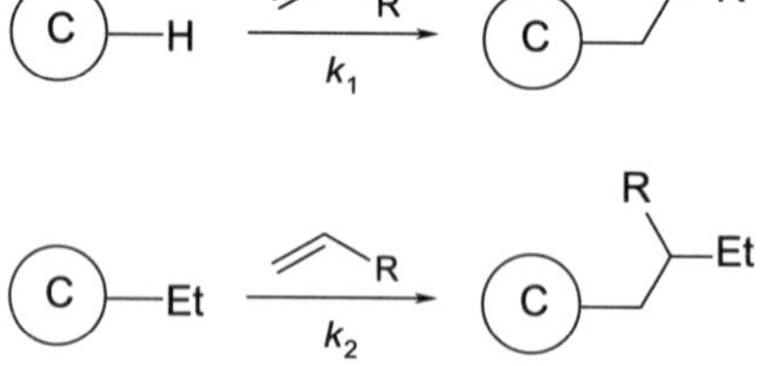

b. 链增长　单体在过渡金属-碳键间插入，实现链增长。

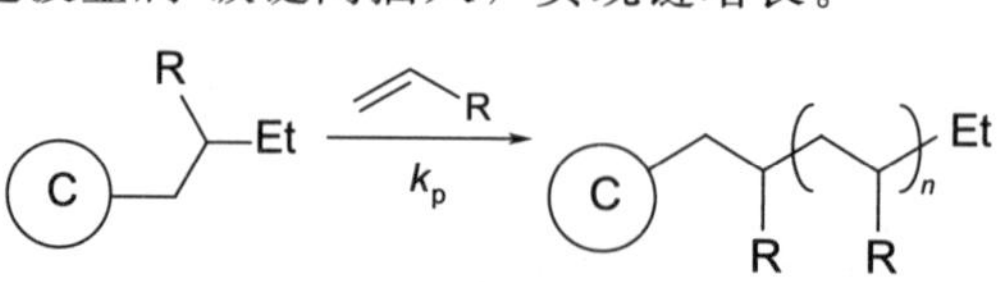

c. 链转移　活性中心可向烷基铝等小分子转移，但转移常数较小。

d. 链终止　配位聚合难终止。水、醇、酸、胺等含活性氢的化合物可以终止配位聚合。聚合前，要除净这些杂质

（2）配位聚合反应动力学

对于均相体系配位聚合，依照阴离子聚合，可写出如下聚合速率方程：

$$R = k_p [C^*][M]$$

式中，$[C^*]$ 和 $[M]$ 分别为催化剂活性中心浓度和单体浓度。

对于载体型非均相聚合体系，吸附过程对聚合速率的影响不容忽视。

以丙烯配位聚合的浆液聚合法为例，主要影响因素如下：

a. 催化剂　由上述聚合速率公式可知，聚合反应速率与催化剂用量成正比，与烷基铝用量无关。但烷基铝用量过大，可能会引起链终止反应，使聚合度降低。

b. 丙烯分压　聚合过程中，丙烯分压越大，意味着丙烯浓度越大，反应速率越快。但是如果丙烯浓度过高，影响聚合热排出，反应产物转移困难。

c. 溶剂　大多采用正庚烷、正己烷等为溶剂。要求对丙烯的溶解度大，传热好，可使无规物和等规物分离等特点。

d. 反应温度　温度对反应速率的影响复杂，高温对络合物活化和链增长有利，但不利于络合物形成和单体吸附。此外，温度上升容易导致平均分子量降低，规整性下降。一般情况下，反应温度控制在70℃以下。

e. 反应压力　压力增高，丙烯溶解度加大，单体浓度增高，聚合速率加快，分子量增大。反应压力过大，对设备性能要求提高，操作困难。聚合压力一般控制在0.7~1.22MPa。

f. 搅拌速率　搅拌有利于单体和催化剂的均匀扩散和充分接触，有利于反应热传递和聚合物颗粒转移。目前，小反应釜的搅拌速率控制在80~120r/min，大反应釜为50~70r/min。

3.4.6　重要配位聚合产物

（1）高密度聚乙烯

高密度聚乙烯（HDPE）在工业上可用 Ziegler-Natta 系和负载型过渡金属氧化物两类催化剂来生产。Ziegler-Natta 催化剂（$TiCl_4/AlCl_3$）可使反应在较温和的温度（60~90℃）和压力（0.2~1.5MPa）下聚合，PE产品的支链少，线形规整，结晶度高达95%，密度达 0.96g/cm^3。

乙烯结构对称，配位聚合无单体定向，产物无立构规整度要求。早期，双组分催化剂活性 PE/Ti 只有 1kg/g，较多的钛组分残留需脱除，使得工艺繁复，成本较高。此后，使用 $MgCl_2$ 作载体，大幅提高了 Ziegler-Natta 的催化剂效率和选择性，添加有机酯化合物作第三组分后，通过络合反应，使链增长速率常数增加，催化剂活性达到 10^3 kgPE/gTi。最近的茂金属催化剂活性高达 5×10^3 kgPE/gZr。残留在聚合物内的过渡金属非常少，不再需要后处理工序。

乙烯配位聚合可选择淤浆聚合和气相聚合两种技术。淤浆聚合多采用烷烃作溶剂，将 Ziegler-Natta 催化剂（$TiCl_4/AlCl_3$）经 $MgCl_2$ 负载，并内、外加酯活化，引发乙烯于50~

80℃、0.18MPa压力下聚合1~4h，即可完成反应。聚合过程中聚乙烯析出，呈淤浆状，故名淤浆聚合。

负载型过渡金属氧化物法以负载型ⅥB族过渡金属（Cr、Mo等）氧化物为催化剂，在温度130~270℃和中等压力1.8~8MPa下，以烷烃为溶剂，也可以使乙烯聚合成高密度聚乙烯。例如，Phillips公司把CrO_3（0.5%~5%）负载于Al_2O_3或SiO_2/Al_2O_3上作催化剂，温度150℃、压力3~5MPa，以二甲苯作溶剂。聚合机理与Ziegler-Natta催化剂配位聚合相似。所得中压聚乙烯支链少，线形规整，与Ziegler-Natta体系的聚乙烯相似，结晶度为90%，有中密度（0.926~0.94g/cm³）和高密度（0.941~0.97g/cm³）两种，分子量约5万。

聚乙烯多用环管反应器和流化反应器聚合。使用流化床时，不用稀释剂，高纯乙烯气体和干粉催化剂一起进入流化床，在2MPa和85~100℃下聚合。气流使聚合物颗粒流化，并帮助散热。聚合物颗粒从流化床底部取出，分离出其中5%的单体回用。

（2）超高分子量聚乙烯

分子量为100万~600万的聚乙烯称为超高分子量聚乙烯（ultra-high molecular weight polyethylene, UHMWPE），超高分子量聚乙烯的高分子量增加大分子链间的相互作用力，受外力时大分子链间的缠结点与吸引点相互作用，达到分散作用力的目的，因此，超高分子量聚乙烯具有极高的拉伸强度、表面硬度、耐磨性、耐蠕变、耐老化和耐溶剂性能。

超高分子量聚乙烯合成方法与高密度聚乙烯类似，多采用Ziegler-Natta催化剂，利用淤浆法工艺可以得到超高分子量聚乙烯。例如，以$TiCl_3/AlEt_2Cl$或$TiCl_4 AlEt_2Cl$为催化剂，以饱和烃（60~120℃馏分）为分散介质，常压下，75~85℃使聚乙烯聚合，便合成得到分子量为100万~500万的超高分子量聚乙烯。

超高分子量聚乙烯熔融状态的黏度高达108Pa·s，流动性极差，需要使用特殊的方法进行加工。以冻胶纺丝-超拉伸技术制备高强度、高模量聚乙烯纤维是20世纪70年代末出现的一种新颖纺丝方法。首先将超高分子量聚乙烯溶解于适当溶剂，制成半稀溶液，经喷丝孔挤出，以空气或水骤冷纺丝溶液，将其凝固成冻胶原丝，形成折叠链片晶。再通过超倍热拉伸使大分子链充分取向和高度结晶，进而使呈折叠链的大分子转变为伸直链，从而制得高强度、高模量纤维。超高分子量聚乙烯纤维是强度最高的化学纤维，又具有较好的耐磨、耐冲击、耐腐蚀、耐光等优良性能，可直接制成绳索、缆绳、渔网和各种织物。超高分子量聚乙烯纤维的复合材料在军事上已用作装甲兵器的壳体、雷达的防护外壳罩、头盔等。

（3）线形低密度聚乙烯

利用配位聚合也可以制备低密度聚乙烯，工业上称之为线形低密度聚乙烯（linear low density polyethylene, LLDPE）。采用Ziegler-Natta引发体系，在100℃、20MPa条件下，乙烯只要与8%~10%的α-烯烃（1-丁烯、1-己烯或1-辛烯）共聚，引入少量侧基（每1000个碳原子10~40个C_2H_5），就足以破坏共聚物的规整性，降低结晶度，改善冲击强度和环境应力开裂。这类共聚物基本上保持线形结构，密度也较低（0.91~0.94g/cm³），故称作线形低密度聚乙烯，性能与高压法低密度聚乙烯（LDPE）相当，用来生产薄膜。但聚合条件比较温和，基建投资和生产成本均较低，因此LLDPE发展迅速，其生产能力近于聚乙烯的1/3。LLDPE的聚合技术与高密度聚乙烯相似，可以采用溶液、淤浆、气相聚合法。

我国2019年聚乙烯消耗量超过3000万吨，其中接近一半为HDPE，LLDPE和LDPE

分别占 30% 和 20%。我国各种 PE 的总产量约为 1500 万吨。各种 PE 产品的主要性能如表 3-8。

表 3-8 工业化聚乙烯品种比较

	LDPE	LLDPE	HDPE(低压法)	HDPE(中压法)	UHMWPE
聚合方法	自由基聚合	Ziegler 配位	Ziegler 配位	负载过渡金属配位	Ziegler 配位
聚合压力/MPa	150~200	20	0.2~1.5	1.8~8	1.0
聚合温度/℃	150~200	100	60~90	130~270	75~85
密度/(g/cm^3)	0.91~0.94	0.91~0.935	0.94~0.965	0.94~0.98	0.92~0.964
分子量/万	10~50	5~20	4~30	4~5	100~700
结晶度/%	55~65	~70	80~95	80~95	55~65
熔点/℃	80~90	110~125	130	136	130~136

(4) 聚丙烯

目前，全球大约 70% 的丙烯来自蒸汽裂解生产乙烯的副产品，大约 30% 来自炼油厂生产汽柴油的副产品，少量由丙烷脱氢制得。丙烯使用 Ziegler-Natta 催化剂经配位聚合，可制成等规聚丙烯和间规聚丙烯。我国 2019 年聚丙烯生产量超过 2400 万吨。商品聚丙烯分子量高，约 15 万~70 万，分子量分布较宽，分布指数约 2~10。

等规聚丙烯熔点高（175℃），拉伸强度高（35MPa），密度低（0.9g/cm^3），耐应力开裂和耐腐蚀，电性能优良，性能接近工程塑料，可制纤维（丙纶）、薄膜、注塑件、管材等。丙烯与少量乙烯或长链 α-烯烃（1-丁烯、1-辛烯等）共聚，可减弱等规聚丙烯的结晶度，改善加工性能，提高冲击强度，降低低温脆性，可以扩大使用范围，如用作热水管材料。

聚丙烯的生产工艺包括淤浆聚合、液相本体聚合、气相聚合等方法。淤浆聚合以己烷或庚烷作溶剂，以 $TiCl_4$/单酯/$MgCl_2$/$AlEt_3$ 作引发体系，温度 50~80℃，压力 0.4~2MPa，以保持丙烯处于溶液状态，丙烯浓度 10%~20%，氢气作分子量调节剂。等规聚丙烯从溶剂中沉析，少量无规聚丙烯留在溶液中，聚合结束后，分离未反应单体。沉析的聚合物经洗涤、干燥、造粒，即成商品。

以液态丙烯进行本体聚合，单体和介质合一，沸腾可带出热量，可以选用环管式或搅拌釜式反应器。聚合结束，聚丙烯也沉析成淤浆状，丙烯蒸发回收后，即得粉状聚丙烯产品，后处理比较简单。

丙烯气相聚合是近期发展起来的技术，原理与乙烯气相聚合相似。选择适当催化剂和聚合条件，丙烯也可聚合成间规聚丙烯。间规聚丙烯结晶度低于等规聚丙烯，但透明性好，具有弹性，抗冲性能好。高间规度的聚丙烯可用 VCl_4 或 V(acac)$_3$/二烷基卤化铝/苯甲醚三元可溶性引发体系，在 -78~-48℃ 下聚合而成。茂金属催化体系也可以用来合成间规聚丙烯。

(5) 乙丙橡胶

乙烯与适量的丙烯进行二元共聚，可形成弹性体，特称作乙丙二元橡胶（EPM）。乙丙二元橡胶主链中无双键，耐老化，但难硫化，只能用过氧化物来交联。引入少量的非共轭二烯烃作第三单体，其共聚物称作乙丙三元橡胶（EPDM），可以常规硫化。乙丙二元橡胶和乙丙三元橡胶统称为乙丙橡胶。

乙丙橡胶具有耐老化、耐臭氧、耐化学品、密度小等优点，因此，车胎除以天然橡胶或丁苯橡胶作主胶外，常配用20%～25%的乙丙橡胶。

二元和三元乙丙橡胶一般含有45%～70%（摩尔分数）的乙烯单体，随着乙烯含量的增加，生胶强度增大，弹性降低，结晶倾向增加。含有70%（摩尔分数）乙烯单体的乙丙橡胶的玻璃化温度约-58℃，有良好的高弹性。乙丙橡胶的数均分子量约4万～20万，重均分子量约20万～40万；三元橡胶中大约每1000个碳原子中含15个双键。三元橡胶中双键达到一定程度，才能保证硫化加工性。

典型三元橡胶中的乙烯和丙烯的比值为60∶40，第三成分约占单体总量的3%。目前常用的第三单体是非共轭二烯烃，通常是桥环结构，而且环中至少有一个双键，如亚乙基降冰片烯、亚甲基降冰片烯、双环戊二烯、1,4-己二烯、环辛二烯等。

乙丙橡胶共聚采用Ziegler-Natta催化剂，大多选用可溶性钒-铝引发体系。制备乙丙橡胶有溶液聚合和悬浮聚合两种技术，与选用的溶剂有关。

① 溶液聚合　早期选用溶液聚合法来合成乙丙橡胶，例如采用$VOCl_3/AlEt_2Cl$或$V(acac)_3/AlEt_2Cl$可溶性引发体系，Al/V物质的量之比为20～40，以己烷或芳烃作溶剂，在0～25℃下进行溶液聚合，以氢气作分子量调节剂。聚合结束后，用醇来沉淀聚合物。

② 悬浮共聚　溶液聚合需要进行溶剂分离和回收，操作复杂，成本增加，因此，现在多改用悬浮法。即将乙烯溶解在液态丙烯中，在50℃下共聚。乙丙共聚物不溶于液态丙烯，析出呈悬浮状，未聚合的残余丙烯容易分离。

乙丙悬浮共聚也可以采用负载型引发剂（载体-Ti-Al、载体-V-Al、载体-Ti-V-Al等体系），使活性进一步提高5～10倍，即为经典钒铝引发剂的50～100倍，达$2×10^4$g共聚物/g钒。$MgCl_2$、$Mg(OH)_2$、$Mg(OH)Cl$等镁化合物是常用的载体，负载的目的是使引发剂充分分散，将引发剂效率从1%提高到90%。

(6) 聚丁二烯

丁二烯是最简单的共轭二烯烃，由石油馏分热裂解和丁烷、丁烯催化脱氢而成。根据聚合机理和引发体系的不同，丁二烯有可能聚合成顺-1,4-、反-1,4-和1,2-聚丁二烯，1,2-结构还可能是无规、全同、间同构型。全同1,2-、间同1,2-和反-1,4-聚丁二烯都呈现塑料性质，而高顺-1,4-聚丁二烯则显示橡胶的高弹性，玻璃化温度-120℃，是橡胶工业中的重点产品。

高顺-1,4-和低顺-1,4-聚丁二烯橡胶中的顺-1,4链节含量不同，聚合条件的主要差异是引发剂和溶剂。高顺-1,4-聚丁二烯橡胶含92%～97%的顺-1,4结构，玻璃化温度为-120℃，橡胶弹性佳，是合成橡胶中的第二大胶种，仅次于丁苯橡胶。镍系、钛系、钴系和稀土系等Ziegler-Natta催化剂都可合成出高顺聚丁二烯。这些体系的聚合工艺条件比较温和：温度30～70℃，压力0.05～0.5MPa，反应时间1～4h。

低顺-1,4-聚丁二烯橡胶含有35%～40%的顺-1,4结构，45%～55%的反式-1,4结构，数均分子量约13万～14万，主要用于塑料改性和专用橡胶制品。其采用丁基锂/烷烃或环烷烃体系，经阴离子溶液聚合而成，聚合温度40～80℃。

熔体流动指数（melt flow index，MFI）是一种表示塑胶材料加工时流动性大小的参数，指在规定的温度和压力下，试样熔体每10min通过标准出料模孔的总质量（g）。数值越大表示该塑胶材料的加工流动性越佳，反之则越差。熔体流动指数测试装置的结构示意如下所示。

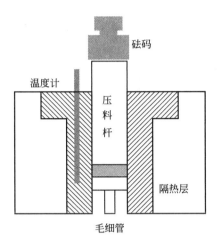

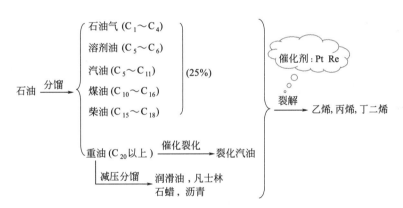

石油裂解是将石油分馏产品在700~1000℃加热分解，使具有长链分子的烃断裂成各种短链的气态烃和少量液态烃，以提供有机化工原料。石油裂解的化学过程比较复杂，生成的裂解气中主要含有乙烯、丙烯、丁二烯等不饱和烃和甲烷、乙烷、氢气、硫化氢等杂质。将裂解产物进行分离，就可以得到所需的多种原料。这些原料在合成纤维工业、塑料工业、橡胶工业等方面有着广泛应用。

裂化（cracking）就是在一定的条件下，将分子量较大、沸点较高的烃断裂为分子量较小、沸点较低的烃的过程。

裂解（pyrolysis）是石油化工生产过程中，以比裂化更高的温度（700~800℃，有时甚至在1000℃以上），使石油分馏产物（包括石油气）中的长链烃断裂成乙烯、丙烯等短链烃的加工过程。

英语读译资料

(1) Polymerization with complex coordination catalysts

Before 1950, the only commercial polymer of ethylene was a highly branched polymer called high-pressure polyethylene, where extremely high pressures were used in the polymerization process. The technique for making linear polyethylene (PE) was discovered by Nobel Laureate Karl Ziegler in the early 1950s. Ziegler prepared high density polyethylene (HDPE) by polymerizing ethylene at low pressure and ambient temperatures using mixtures of triethyl-aluminum and titanium tetrachloride. Another Nobel Laureate, Giulio Natta, used Ziegler's complex coordination catalyst to produce crystalline, stereoregular polypropylene (PP). These catalysts are now known as Ziegler-Natta catalysts. In general, a Ziegler-Natta catalyst is a combination of a transition-metal compound from Groups ⅣB to ⅧB and an organometallic compound of a metal from Groups ⅠA to ⅢA in the periodic table. It is customary to refer to the transition-metal compounds as the catalyst (because reaction occurs at the transition-metal atom site) and the organometallic compound as the co-catalyst.

Here we will use titanium to illustrate the coordination polymerization process. Several exchange reactions between catalyst and co-catalyst occur with Ti(Ⅳ) reduced to Ti(Ⅲ). The extent and kind of stereoregulation can be controlled through a choice of reaction conditions and catalyst/cocatalyst.

The titanium salt is present as a solid. The precise mechanism probably varies a little depending on the catalyst/cocatalyst and reaction conditions. Here we will look at the polymerization of propylene using titanium chloride and triethyl-aluminum. In general, a monomeric molecule is inserted between the titanium atom and the terminal carbon atom in a growing chain. Propagation occurs at the solid titanium salt surface—probably at defects, corners, and edges. The monomer is always the terminal group on the chain. Triethyl-aluminum reacts with the titanium-containing unit producing ethyltitanium chloride as the active site for polymerization.

(2) PE

PE was probably initially synthesized by M. E. P. Friedrich, while a graduate student working for Carl S. Marvel in 1930, when it was an unwanted by-product from the reaction of ethylene and a lithium alkyl compound. In 1932, British scientists at the Imperial Chemical Industries (ICI) accidentally made PE while they were looking at what products could be produced from the high-pressure reaction of ethylene with various compounds. In March 1933, they found the formation of a white solid when they combined ethylene and benzaldehyde under high pressure (about 1400atm pressure). They correctly identified the solid as PE.

They attempted the reaction again, but with ethylene alone. Instead of getting the waxy white solid again, they got a violent reaction and the decomposition of the ethylene. They delayed their work until December 1935 when they had better high-pressure equipment. At 180℃, the pressure inside the reaction vessel containing the ethylene decreased consistently with the formation of a solid. Because they wanted to retain the high pressure, they pumped in more ethylene. The observed pressure drop could not be totally due to the formation of PE, but something else was contributing to the pressure loss. Eventually, they found that the pressure loss was also due to the presence of a small leak that allowed small amounts of oxygen to enter into the reaction vessel. The small amounts of oxygen turned out to be the right amount needed to catalyze the reaction of the additional ethylene that was pumped in subsequent to the initial pressure loss (another "accidental" discovery). The ICI scientists observed no real use for the new material. By chance, J. N. Dean of the British Telegraph Construction and Maintenance Company heard about the new polymer. He needed a material to encompass underwater cables. He reasoned that PE would be water resistant and suitable to coat the wire protecting it from the corrosion caused by the saltwater in the ocean. In July 1939, enough PE was made to coat one nautical mile of cable. Before it could be widely used, Germany invaded Poland and PE production was diverted to making flexible high-frequency insulated cable for ground and airborne radar equipment. PE was produced, at this time, by ICI and by DuPont and Union Carbide for the United States.

PE did not receive much commercial use until after the war when it was used in the manufacture of film and molded objects. PE film displaced cellophane in many applications being used for packaging produce, textiles, and frozen and perishable foods. This PE was branched and had a relatively low softening temperature, below 100℃, preventing its use for materials where boiling water was needed for sterilization.

The branched PE is called low-density HPPE because of the high pressures usually employed for its production, and because of the presence of the branches, the chains are not able to closely pack, leaving voids and subsequently producing a material that had a lower density in comparison with low-branched PE.

Karl Ziegler, director of the Max Planck Institute for Coal Research in Muelheim, Germany, was extending early work on PE attempting to get ethylene to form PE at lower pressures and temperatures. His group found that certain organometallics prevented the polymerization of ethylene. He then experimented with a number of other organometallic materials that inhibited PE formation. Along with finding compounds that inhibited PE formation, they found compounds that allowed the formation of PE under much lower pressures and temperatures. Further, these compounds produced a PE that had fewer branches and higher softening temperatures. Natta, a consultant for the Montecatini Company of Milan, Italy, applied the Zeigler catalysts to other vinyl monomers such as propylene and found that the polymers were of higher density, higher melting, and more linear than those produced by the then classical techniques such as free-radical-initiated polymerization. Ziegler and Natta shared the Nobel Prize in 1963 for their efforts in the production of vinyl polymers using what we know today as solid state stereoregulating catalysts.

(3) Polymers from 1,4-dienes

There are three important 1,4-dienes employed to produce commercially important polymers. These monomers possess a conjugated pi-bond sequence of —C=C—C=C— that readily forms polymers with an average energy release of about 20kcal/mol (80kJ/mol) with the conversion of one of the double bonds into a lower (potential energy wise; generally more stable) energy single bond. For all of these products, cross-linking and grafting sites are available through the remaining double bond.

1,4-Butadiene can form three repeat units: 1,2-, *cis*-1,4-, and *trans*-1,4-. Commercial polybutadiene is mainly composed of *cis*-1,4- isomer and known as butadiene rubber (BR). In general, butadiene is polymerized using stereoregulating catalysts. The composition of the resulting polybutadiene is quite dependent on the nature of the catalyst such that almost total *trans*-1,4-, *cis*-1,4-, or 1,2- units can be formed as well as almost any combination of these units. The most important single application of polybutadiene polymers is its use in automotive tires where over 10^7 t are used yearly in the U.S. manufacture of automobile tires. BR is usually blended with NR or SBR to improve tire tread performance, particularly wear resistance. A second use of butadiene is in the manufacture of ABS copolymers where the stereo-geometry is also important. A polybutadiene composition of about 60% *trans*-1,4-; 20% *cis*-1,4-; and 20% 1,2- configuration is generally employed in the production of ABS. The good low-temperature impact strength is achieved in part because of the low T_g values for the compositions. For instance, the T_g for *trans*-1,4-polybutadiene is about $-14°C$, while that for *cis*-1,4-polybutadiene is about 108°C. Most of the ABS rubber is made employing an emulsion process where the butadiene is initially polymerized forming submicron particles. The SAN copolymer is then grafted onto the outside of the BR particles. ABS rubbers are generally tougher than HIPS rubbers but are more difficult to process. ABS rubbers are used in a number of appliances including luggage, power tool housings, vacuum cleaner housings, toys, household piping, and automotive components such as interior trim.

Another major use of butadiene polymer is in the manufacture of HIPS. Most HIPS has about 4%~12% polybutadiene in it so that HIPS is mainly a PS-intense material. Here, the polybutadiene polymer is dissolved in a liquid along with styrene monomer. The polymerization process is unusual in that both a matrix composition of PS and polybutadiene is formed as well as a graft between the growing PS onto the polybutadiene is formed. The grafting provides the needed compatibility between the matrix phase and the rubber phase. The grafting is also important in determining the structure and size of rubber particles that are formed. The grafting reaction occurs primarily by hydrogen abstraction from the polybutadiene backbone either by growing PS chains or alkoxy radicals if peroxide initiators are employed.

习 题

1. 试列举可溶性和非均相 Ziegler-Natta 催化剂的典型代表，它们对立构规整性有何影响？

2. 丙烯进行自由基聚合、离子聚合及配位阴离子聚合，能否形成高分子聚合物？分析原因。

3. 乙烯和丙烯配位聚合所用 Ziegler-Natta 催化剂两组分有何区别？两组分间有哪些主要反应？钛组分的价态和晶形对聚丙烯的立构规整性有何影响？

4. 丙烯配位聚合中链增长、链转移、链终止等基元反应有何特点？如何控制分子量？

5. 简述丙烯配位聚合时的双金属机理和单金属机理模型的基本论点。

6. 简述茂金属催化剂基本组成、结构类型、提高活性的途径和应用方向。

7. 列举丁二烯进行顺-1,4-聚合的引发体系，并讨论顺-1,4-结构的成因。

英语习题

1. What are the mechanisms for syndiotactic and isotactic placements in propene polymerization? Describe the reaction conditions that favor each type of stereoselective placement.

2. What reaction conditions determine the relative amounts of 1,2-, *cis*-1,4-, and *trans*-1,4-polymerization in the radical and anionic polymerizations of 1,3-butadiene? Indicate the effect of solvent and counterion in these polymerizations. What polymer structures are obtained using traditional Ziegler-Natta and metallocene initiators?

3. What is the catalyst and co-catalyst in the most widely used Ziegler-Natta catalyst? What is the most widely used catalyst for the production of HDPE?

4. Discuss the use of homogeneous versus heterogeneous reaction conditions for the coordination and traditional Ziegler-Natta polymerizations of propene, isoprene, styrene, methyl methacrylate, and *n*-butyl vinyl ether.

第 4 章
开环聚合

开环聚合

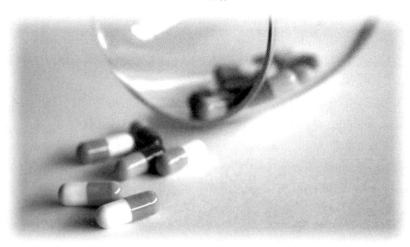

聚己内酯（polycaprolactone，PCL）是一种经开环聚合而成的高分子材料。PCL 具有良好的生物相容性和生物降解性，在人体内可缓慢分解为二氧化碳和水，是一种较为理想的生物医用材料。以 PCL 作为药物缓释载体的药品已有多种面市，成为 PCL 的一个重要用途。此外，PCL 还被用作形状记忆材料、可降解塑料，制备纳米纤维等。

本章目录

4.1 开环聚合单体的结构特征 / 173
4.2 开环聚合机理 / 174
4.3 开环聚合引发剂及溶剂 / 175
4.4 开环聚合高分子材料 / 175
4.5 易位聚合反应 / 180

重点要点

杂环化合物，环张力，阳离子开环聚合，阴离子开环聚合，三氧六环，己内酰胺，环酯，丙交酯，环酐，环碳酸酯，环硫醚，易位聚合，聚硅氧烷，聚磷氮烯。

许多杂环化合物在特定引发剂作用下,能够开环聚合形成线形聚合物。开环聚合物大多属于杂链高分子(图4-1),与缩聚物相似,但是开环聚合属于连锁聚合反应。

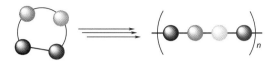

图 4-1 开环聚合反应

开环聚合在高分子合成方法中占有重要地位,这主要因为它的适用单体范围大,聚合物结构具有较强的设计灵活性。一些看似无法聚合的小分子经环化生成环状单体后,通过开环聚合可以得到相应高分子。开环聚合得到的聚合物结构往往和逐步聚合类似,其链增长反应速率常数也和逐步聚合反应类似,要比烯烃单体聚合低几个数量级。此外,与烯烃单体的连锁聚合相比,开环聚合容易出现解聚平衡反应,获得高分子量相对较难。

4.1 开环聚合单体的结构特征

决定环状单体能否开环聚合的关键是热力学因素。环状单体的分子结构决定环张力、聚合热乃至聚合自由焓。环状结构中原子间共价键的变形大,意味着环张力和聚合热大,聚合自由焓为负,环稳定性低,则容易开环聚合。图4-2给出了一些已经被证实可以进行开环聚合的单体。

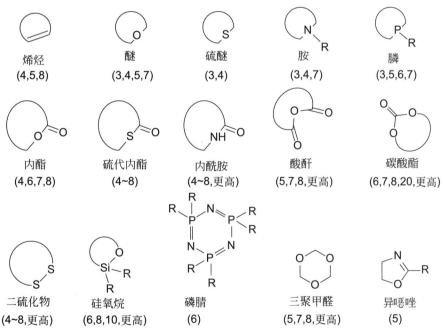

图 4-2 可开环聚合环状化合物
括号内数字为环单元个数

由这些单体可以总结出一些容易开环聚合单体的分子结构特点。这些开环聚合单体都是杂环化合物，包括环醚、环缩醛、内酯、内酰胺、环胺、环硫醚、环酸酐等。一般而言，杂环化合物开环聚合相对容易，环烷烃的键极性小，不易开环聚合。此外，开环聚合反应的活性与环大小密切相关。

环的张力有两类：一类是键角变形引起的角张力，另一类是氢或取代基间斥力造成的构象张力。三元环和四元环的张力较大，易开环聚合。五元环和六元环张力较小，相对稳定，较难开环聚合。

4.2 开环聚合机理

离子聚合是开环聚合的主要方式，但是也有比较特殊的环状化合物（含有不饱和双键侧基）按自由基机理开环聚合。离子开环聚合有两类机理，即阴离子开环聚合和阳离子开环聚合，它们的链增长反应通式如图 4-3 所示。

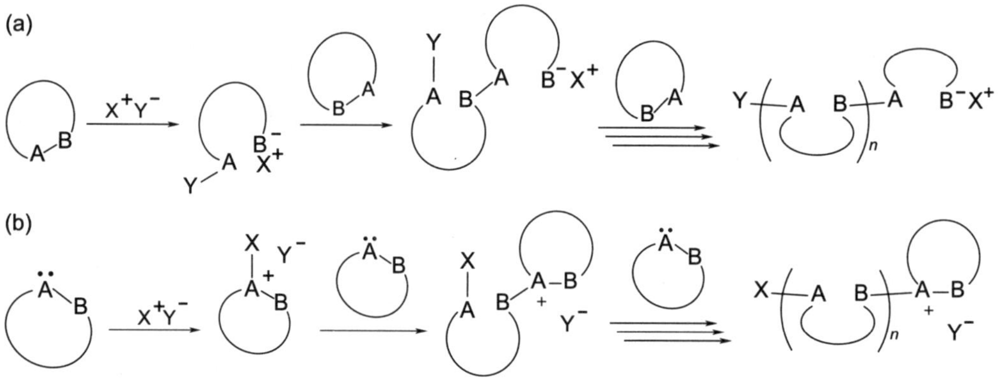

图 4-3 阴离子开环聚合机理（a）和阳离子开环聚合机理（b）

一种方式是亲核性引发剂进攻单体，使环结构断裂，形成离子对，位于聚合物链末端的阴离子活性物种不断与单体反应，实现链增长；另一种方式是环状单体内的亲核性官能团与亲电性引发剂反应，形成络合中间体，成为阳离子活性物种，第二个单体的亲核单元与该阳离子活性物种反应，使环结构断裂，实现链增长。

用于烯烃单体聚合的阴离子引发剂，如碱金属的烷氧化物（如醇钠）、氢氧化物、氨基化物、有机金属化合物、碱土金属氧化物等，也可用于引发阴离子开环聚合。烯烃单体聚合的常用阳离子引发剂，如质子酸和路易斯酸等，也能够引发一些环状单体的阳离子聚合反应。

还有一些开环聚合反应是通过活化单体来进行聚合的。这类反应的阴离子或阳离子活性物种位于单体，而不是聚合链末端。链增长反应通过中性的聚合物末端和活性化的单体反应。图 4-4 为质子化单体的开环聚合反应。

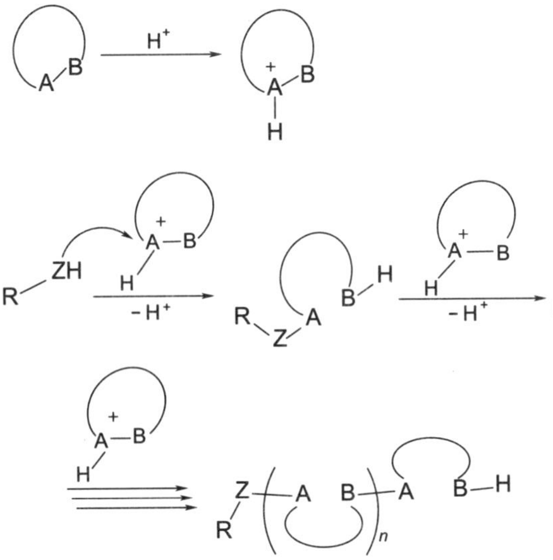

图 4-4　质子化单体的开环聚合机理

4.3　开环聚合引发剂及溶剂

大部分用于烯烃单体阳离子聚合的引发剂也可用于开环聚合。如：
- 质子酸　高氯酸、三氟甲基磺酸等。
- 路易斯酸　BF_3、$AlCl_3$、PF_5、$FeCl_3$、$TiCl_4$、SbF_5、$SnCl_4$ 等。
- 其他　CH_3I、$CF_3SO_3CH_3$、$(C_6H_5)_3CPF_6$、$(CH_3CH_2)_3OSbF_6$ 等。

阴离子开环聚合引发剂也与烯烃单体阳离子聚合引发剂相似。如：
- 钠金属、钠/萘体系等。
- 醇盐　CH_3CH_2ONa、$(CH_3)_3COK$ 等。
- 胺类化合物　三乙胺、苯胺等。

适合阳离子开环聚合的有机溶剂有四氯化碳、氯仿、甲苯等。阴离子开环聚合比较典型的溶剂有正己烷、环己烷等非极性烷烃类溶剂，乙醚、四氢呋喃等醚类溶剂。单体为低黏度液体情况下，也可进行本体聚合。

4.4　开环聚合高分子材料

环氧乙烷、环氧丁烷、四氢呋喃、二氧五环、三氧六环可以开环聚合得到，但三氧六环另归为缩醛类。环氧乙烷是高活性单体，乙醇钠（sodium ethoxide）可引发聚合。烷基氧负离子进攻环氧乙烷中的碳原子，开环聚合成线形聚合物。该体系属于活性阴离子聚合，较难终止，需加入质子酸使活性链失活。

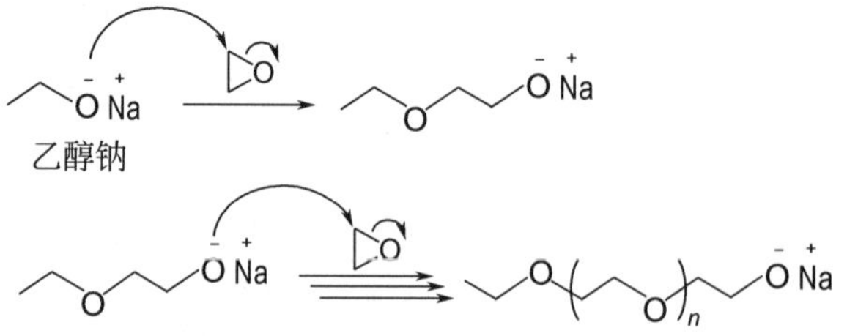

以乙二醇为起始剂，环氧乙烷开环聚合为聚乙二醇（polyethylene glycol，PEG），分子量不等（200～5000）。同理，环氧丙烷也可开环聚合为聚丙二醇（polypropylene glycol，PPG）。分子量低于一万的 PEG 和 PPG 都是水溶性高分子，无毒且具有较好的生物相容性，常用于化妆品、油墨、乳化剂、润滑油和增塑剂等。

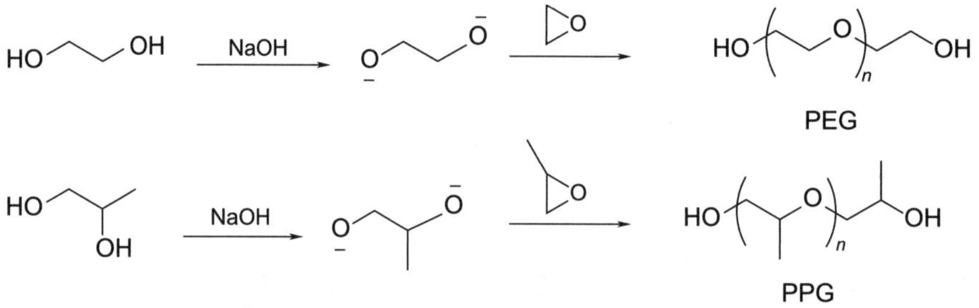

环氧树脂中添加少量叔胺，可引发环氧树脂固化。其固化机理为典型的阴离子型开环聚合。相应固化产物可用作胶黏剂、涂料等。

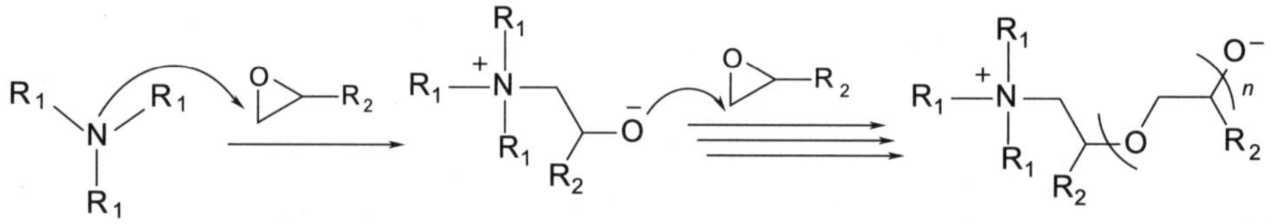

三氧六环是甲醛的三聚体，经三氟化硼-水体系引发进行阳离子开环聚合，形成聚甲醛（polyoxymethylene，POM）。引发反应是 $H^+(BF_3OH)^-$ 与三氧六环单体形成的氧鎓离子，其与第二个单体反应，开环转化为碳阳离子活性种，三聚甲醛单体与活性种反应使链增长。聚甲醛是一种工程塑料，强度、韧性和耐疲劳性较好，可在 180～220℃下模塑成形，用于塑料齿轮等零部件生产。

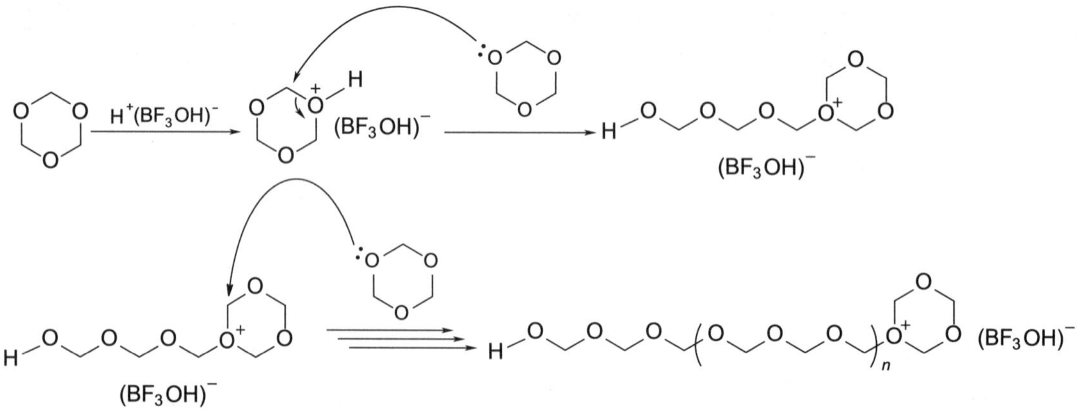

内酰胺的聚合反应可由酸、碱或水引发。由己内酰胺（ε-caprolactam）合成尼龙-6 的聚合反应已经工业化。尼龙-6 的性质和尼龙-66 相似，只是其熔点 T_m 相对较低（223℃）。工业化生产的聚酰胺中，尼龙-6 和尼龙-66 占 90% 以上。水引发的己内酰胺聚合反应，也称水解聚合，是尼龙-6 的主要生产方法。阴离子引发的聚合反应主要用于模具内直接成型的聚合反应。阳离子引发的内酰胺聚合反应转化率和收率较低，产品没有实际应用价值。

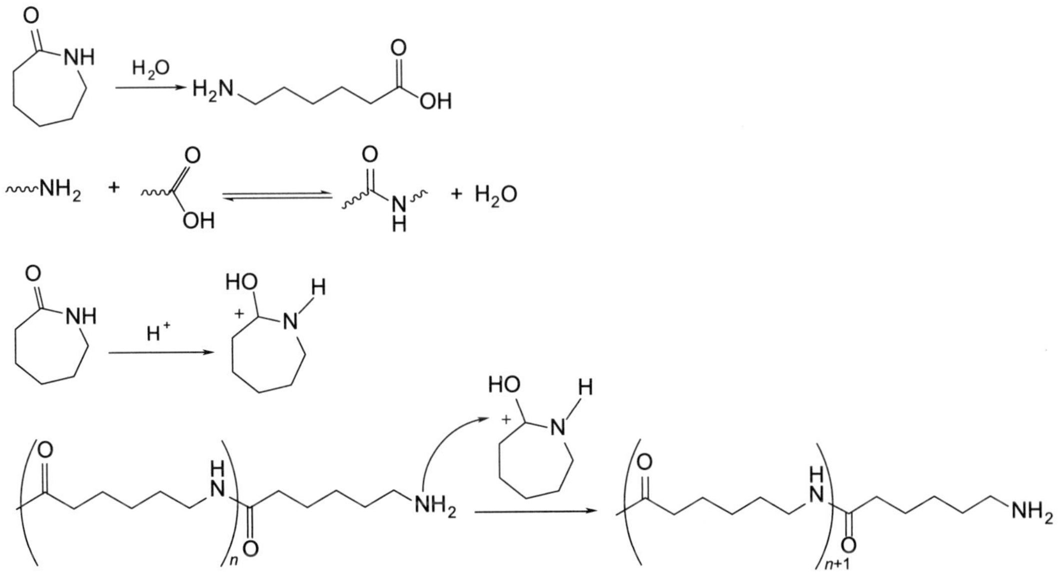

水解聚合是将己内酰胺与 5%～10% 的水共混后，在 250～270℃下加热 24h 后完成聚合反应。因产物中含有约 10% 未反应的单体和低聚物，聚合物产品切片后需要用热水浸取，除去杂质。最后在 100～120℃和 130Pa 的减压条件下，将水分降至 0.1% 以下，即可得到尼龙-6 产品。

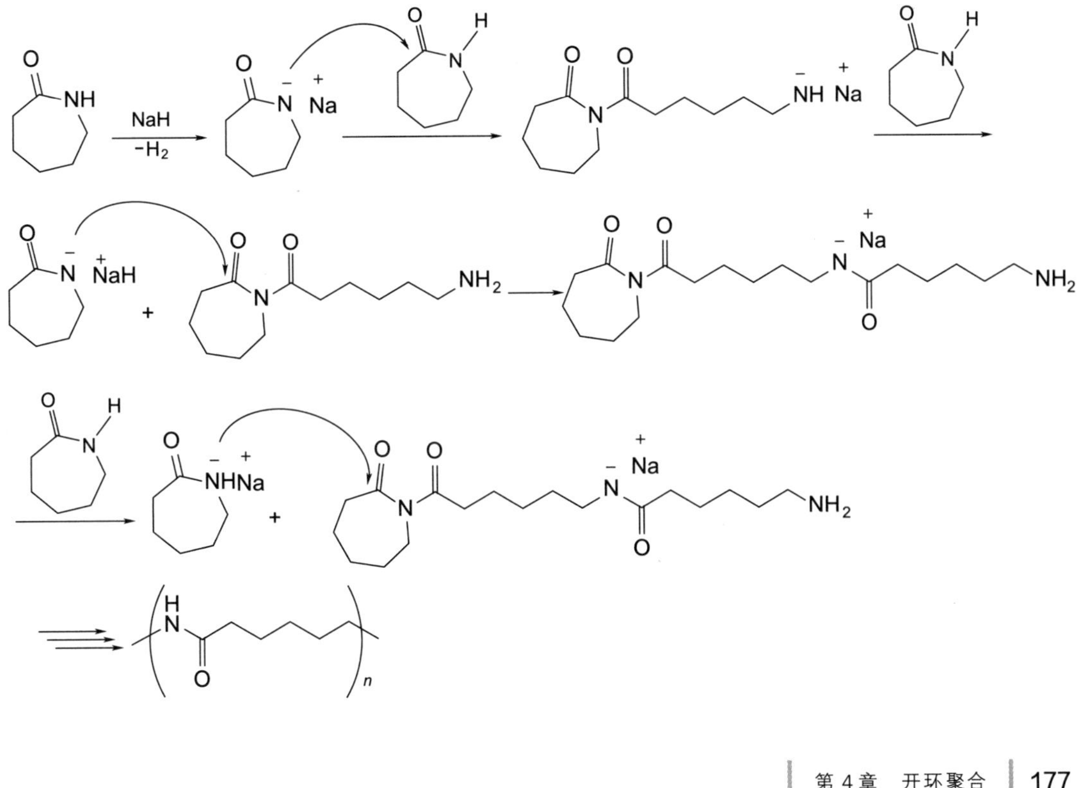

主要的聚合反应机理为：己内酰胺水解生成氨基酸；氨基酸中的羧酸使单体质子化，该反应中的质子化几乎只发生在羰基氧上。该氨基亲核进攻质子化单体，使单体开环形成新的氨基端。链增长反应依次展开，形成线形聚合物。水解聚合是阳离子开环聚合的一个特例。工业上添加等物质的量的羧酸铵盐（由二元胺和二元酸制备），可增加聚合反应速度。

另一方面，碱金属、金属氢化物等强碱可引发己内酰胺的阴离子开环聚合反应，其反应机理较环醚的开环聚合更为复杂。以氢化钠引发为例，氢化钠中的氢夺取内酰胺单体的酰胺氢原子形成氢气，并使单体阴离子化。该阴离子单体亲核进攻另一单体，形成活性极高的伯胺阴离子，再进一步生成伯胺，并活化另一单体。活化单体具有极强的亲核性，容易与该伯胺化合物中的酰基碳反应，阴离子中心转化在酰胺氮原子上面，并可再转向另一内酰胺单体，使其阴离子化，自身则成为以内酰胺为端基的伯胺聚合物。此后，依此机理进行链增长反应。

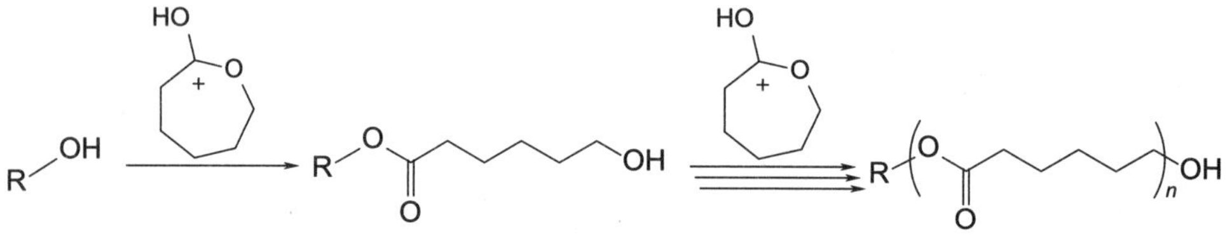

聚己内酯（polycaprolactone，PCL）由 ε-己内酯开环聚合而成。PCL 具有良好的生物相容性、良好的有机高聚物相容性，以及良好的生物降解性，可用作细胞生长支持材料。此外，PCL 还具有良好的形状记忆温控性质，广泛应用于药物载体、增塑剂、可降解塑料、纳米纤维纺丝、塑性材料的生产与加工领域。

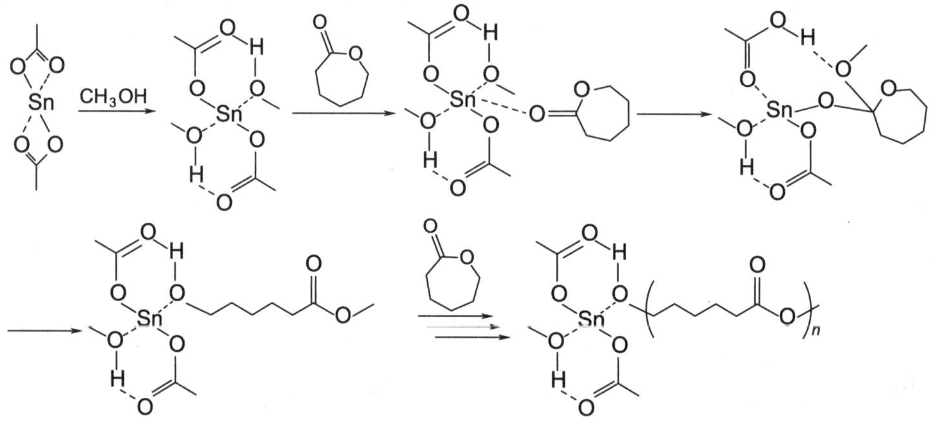

ε-己内酯在胺、醇、羧酸等活性氢存在下可开环聚合，得到不含重金属离子的聚合物，但是一般需要 200℃ 以上的反应温度，且产物分子量较低。

质子酸、Lewis 酸、烷基化试剂等可引发 ε-己内酯的开环聚合，但伴随的副反应较多，分子量不高。碱金属烷基化合物能引发阴离子开环聚合，但是产物容易发生链转移和酯交换反应，不利于得到高分子量聚合物。一些过渡金属络合物引发的配位聚合反应条件温和，可得分子量高、分布窄的聚合物，是目前研究最多的一种方法。醋酸亚锡/甲醇体系引发 ε-己内酯开环聚合反应机理如下，首先甲醇和乙酸亚锡配位，接着 ε-己内酯和锡配位，甲醇转化成甲氧负离子，并亲核进攻单体的酰氧键使单体开环。

α-羟基丙酸（乳酸）易二聚成六元环丙交酯，再进一步开环聚合成相应的高分子量线形聚乳酸（PLA）。聚乳酸具有良好的生物相容性和生物降解性，易水解成 α-羟基酸，在人体代谢过程中排出，可用于制造缓释药物胶囊、手术后无需拆除的缝合线等。

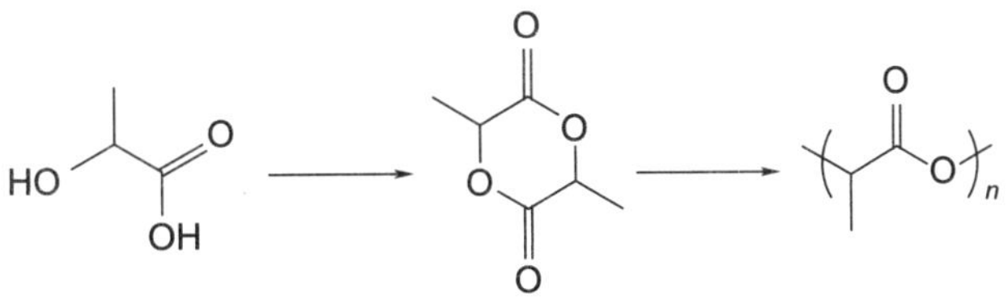

和 PCL 类似，丙交酯开环聚合引发剂有阴离子型、阳离子型、过渡金属配位型等多种。商业用聚乳酸多采用如辛酸亚锡等有机金属化合物引发合成，但残留的痕量金属会限制其在生物材料方面的应用。不含金属的有机小分子引发剂，如吡啶衍生物、氮杂环卡宾化合物、胍、脒等强碱性引发剂最近也被证实可引发丙交酯的开环聚合。

聚二甲基硅氧烷主链由硅和氧相间连接而成，硅上连有两个甲基。聚二甲基硅氧烷的起始原料是二氯二甲基硅烷，由单质硅与氯代烷或氯苯直接反应而成。二氯二甲基硅烷缩聚成八元环四聚体，再开环聚合成聚二甲基硅氧烷，俗称硅胶。

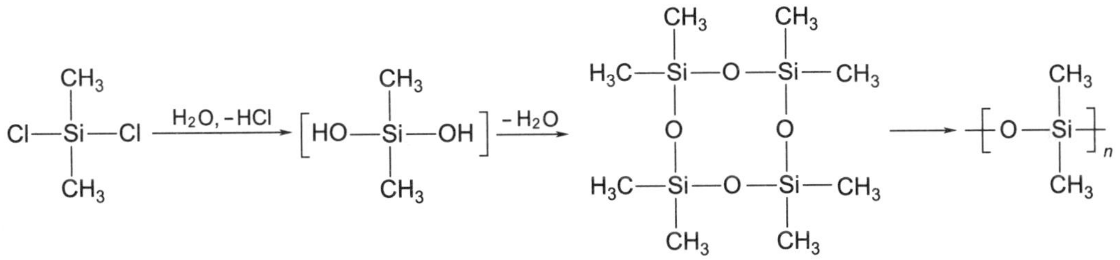

聚二甲基硅氧烷主链结构中氧和硅原子相连，硅原子有两个侧基，氧的键角较大，侧基间相互作用较小，容易绕 Si—O 单键内旋，使聚合物分子链具有良好的柔性，是高分子中最柔顺的一员。此外，它还有良好的耐高温、耐化学品、耐氧化、疏水、电绝缘等性质。低分子线形聚二甲基硅氧烷可用作硅油、皮革制品表面处理剂、洗发剂以及护肤用品等；高分子量聚二甲基硅氧烷具有橡胶特性，用作包装材料、美容手术填充材料等。

聚磷氮烯又名聚磷腈，主链为磷氮双键结构，磷原子上有两个侧基。P=N 键能很大，因此稳定。氮磷键角大，又无侧基，使得高分子主链柔性较大。聚磷氮烯的侧基具有多样灵活的可设计性，可以在较大范围内调整聚合物的性能。目前，聚磷氮烯用作低温弹性体、生物材料、聚合物药物、水凝胶、液晶材料、阻燃纤维、半导体材料等。

制备聚磷氮烯的起始原料为五氯化磷与氯化铵，可形成环状二聚体。230～300℃下减压，即可开环聚合成透明聚二氯磷氮烯，分子量可达 200 万。

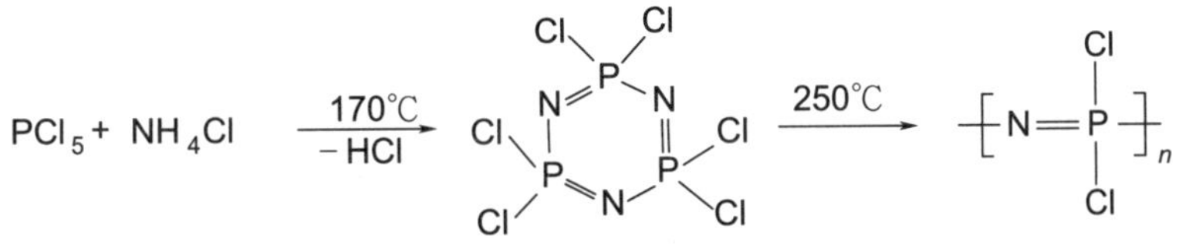

聚二氯磷氮烯分子链高度柔顺，T_g 为 -63℃，低温下呈橡胶态，其应力-松弛的弹性体行为比天然橡胶还好，俗称"无机橡胶"。位于聚二氯磷氮烯侧基的 P—Cl 键较弱，易水解成粉末，无法直接使用。但聚二氯磷氮烯侧基氯原子被有机基团取代后，就成为很稳定的半无机高分子。例如聚二氯磷氮烯与醇钠或酚钠反应，形成烷氧或芳氧衍生物；与胺类反应，形成氨基衍生物；与有机金属化合物（如格氏试剂、烷基锂）反应，可引入烷基、芳基。

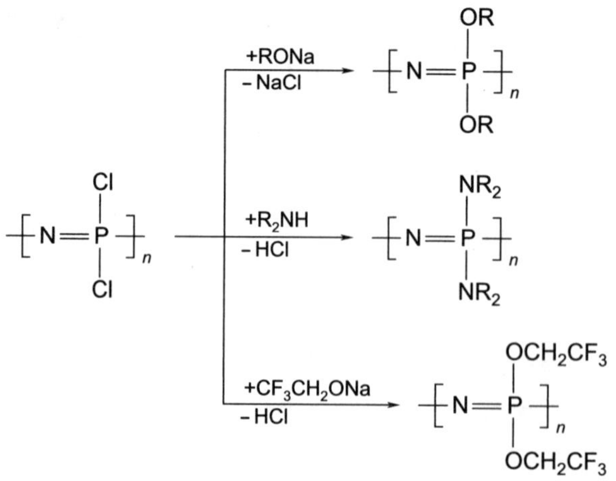

$$\left[N = P \begin{matrix} Cl \\ | \\ Cl \end{matrix} \right]_n \xrightarrow[-NaCl]{+RONa} \left[N = P \begin{matrix} OR \\ | \\ OR \end{matrix} \right]_n$$

$$\xrightarrow[-HCl]{+R_2NH} \left[N = P \begin{matrix} NR_2 \\ | \\ NR_2 \end{matrix} \right]_n$$

$$\xrightarrow[-HCl]{+CF_3CH_2ONa} \left[N = P \begin{matrix} OCH_2CF_3 \\ | \\ OCH_2CF_3 \end{matrix} \right]_n$$

当引入氟乙氧基后，则成为微结晶的软塑料，T_g低（−66℃），熔点高（242℃），类似聚乙烯，能溶纺、成膜，疏水性能与聚四氟乙烯、聚硅氧烷相当。

4.5 易位聚合反应

开环易位聚合（ring-opening metathesis polymerization，ROMP）反应起源于20世纪50年代中期的烯烃易位反应（metathesis reaction）。烯烃易位反应是指两种烯烃互相交换双键两端的基团，从而生成如下所示两种新烯烃的反应。最初认为该反应是一种以金属环丁烷为中间体的特异反应，但近年来，研究者们证明了很多烯烃易位反应的中间体，使得烯烃易位反应得到了广泛的应用。

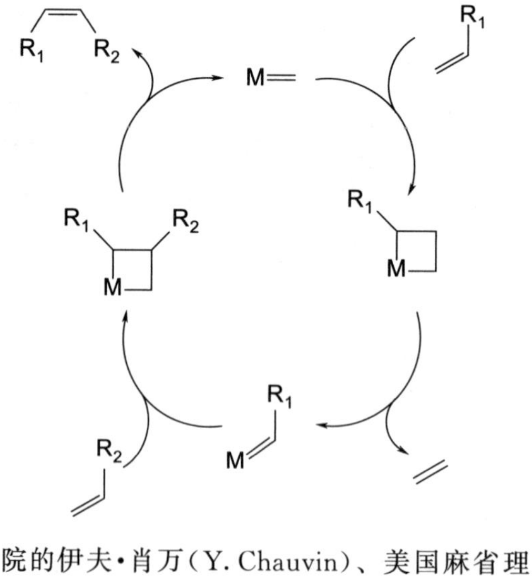

2005年，法国石油学院的伊夫·肖万（Y. Chauvin）、美国麻省理工学院的罗伯特·格拉布（Robert H. Grubbs）和加利福尼亚州加州理工学院的理查德·施罗克（Richard R. Schrock）三位科学家因烯烃易位反应获得诺贝尔化学奖。现在，越来越多的结构明确、

稳定高效的催化剂被合成，使得烯烃易位反应能够和传统的碳碳键的合成方法相媲美。因此，与烯烃易位反应相关的研究已成为化学界极为重要的课题。

开环易位聚合（ROMP）于1967年首次报道。ROMP反应不是简单的链烯烃双键断裂的加成聚合，也不是内酰胺或者环醚等杂环的开环聚合，而是双键不断易位，链不断增长，而单体分子上的双键仍保留在生成的聚合物大分子中的反应。

此外，非环二烯也可以通过双键易位生成聚合物，同时释放出低分子烯。

能够进行ROMP反应的环烯烃一般要求环张力较大，烯烃结构两端没有体积较大的取代基团。单体可以是单环烯，如环丁烯、环辛烯、环戊烯和环庚烯等，也可以是双环烯，如降冰片烯及其衍生物，还可以是三环烯。单环烯与双环烯共聚可以制得高度交联的体形聚合物（图4-5）。

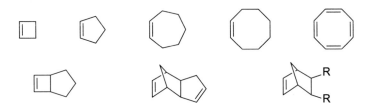

图 4-5　可进行 ROMP 反应的典型聚合物

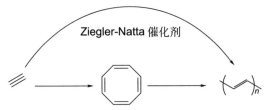

导电高分子聚乙炔（polyacetylene）通常由乙炔采用 Ziegler-Natta 催化剂通过配位聚合制备，但是分子量较难控制。使用镍催化剂合成环状烯烃后，再经 ROMP 也可制备，而且分子量比较容易控制。

英语读译资料

(1) Ring-opening polymerization

The ring-opening polymerization of cyclic compounds is the third of the three general classes of polymerization processes.

In contrast to condensation reactions, it does not result in the loss of a small molecule. Moreover, it does not involve a loss of multiple-bonding enthalpy, whereas the loss of unsaturation is a powerful driving force for vinyl polymerization. It should be noted that ring-opening polymerization is the principal method for the synthesis of inorganic polymers.

Cyclic organic compounds that have been polymerized include cyclic ethers, lactones, lactams, and imines. In addition to these classical ring-opening polymerizations, a number of non-classical reactions have been developed that fall outside the established categories. These include zwitterion polymerizations and "no catalyst" reactions, various cycle-polymerizations, polymerization of bicyclic monomer, and the unusual polymerization of *para*-xylene.

Two general types of mechanisms have been proposed for ring-opening polymerizations. In the first, the catalyst is presumed to attack the ring initially, with concurrent or subsequent cleavage. The resultant ionic or zwitterionic end group then attacks another ring with concurrent ring cleavage, and so on. The alternative mechanism supposes that an initial ring cleavage does not occur. Instead, the primary interaction of the catalyst with the cyclic monomer generates a coordination intermediate, which then functions as the true initiating species. In many cases, the distinction between these two pathways is difficult to establish.

(2) Polysiloxanes

Polysiloxanes, also referred to as silicones, possess an unusual combination of properties that are retained over a wide temperature range ($-100\sim250$℃). Silicones are very stable to high temperature, oxidation, and chemical and biological environments, and possess very good low-temperature flexibility because of the low T_g values (-127℃ for dimethylsiloxane), a consequence of the long Si—O bond (1.64Å compared to 1.54Å for C—C) and the wide Si—O—Si bond angle (143° compared to 109.5° for C—C—C). The Si—O bond is stronger than the C—C bond (450kJ/mol vs. 350kJ/mol) and siloxanes are very stable to high temperature, oxidation, chemical and biological environments, and weathering, and also possess good dielectric strength, and water repellency. Fluid applications include fluids for hydraulics, antifoaming, water-repellent finishes for textiles, surfactants, greases, lubricants, and heating baths. Resins are used as varnishes, paints, molding compounds, electrical insulation, adhesives, laminates, and release coatings. Elastomer applications include sealants, caulks, adhesives, gaskets, tubing, hoses, belts, electrical insulation such as automobile ignition cable, encapsulating and molding applications, fabric coatings, encapsulants, and a variety of medical applications (antiflatulents, heart valves, encasing of pacemakers, prosthetic parts, contact lenses, coating of plasma bottles to avoid blood coagulation). Silicone elastomers differ markedly from other organic elastomers in the much greater effect of reinforcing fillers in increasing strength properties.

The higher-MW polysiloxanes are synthesized by anionic or cationic polymerizations of cyclic siloxanes. The most commonly encountered polymerizations are those of the cyclic trimers and tetramers, for example, for octamethylcyclotetrasiloxane. The anionic polymerization of cyclic siloxanes can be initiated by alkali metal hydroxides, alkyls, and alkoxides, silanolates such as potassium trimethylsilanoate, $(CH_3)_3SiOK$, and other bases. Both initiation and propagation involve a nucleophilic attack on monomer analogous to the anionic polymerization of epoxides.

Cationic polymerization has been initiated by a variety of protonic and Lewis acids. The cationic process is more complicated and less understood than the anionic process. Polymerization under most reaction conditions involves the presence of a step polymerization simultaneously with ROP. This appears to be the only way to reconcile the observed (complicated) kinetics for the overall process. Initiation consists of protonation of monomer followed by subsequent reaction with monomer to form the tertiary oxonium ion. Propagation for the ring-opening chain polymerization follows in the same manner with nucleophilic attack of monomer on the tertiary oxonium ion.

(3) Hydrolytic polymerization of lactams

The polymerization of lactams (cyclic amides) can be initiated by bases, acids, and water. Initiation by water, referred to as hydrolytic polymerization, is the most often used method for industrial polymerization of lactams. Anionic initiation is also practiced, especially polymerization in molds to directly produce objects from monomer. Cationic initiation is not useful because the conversions and polymer molecular weights are considerably lower.

Nylon-6 is produced commercially, accounting for almost one-third of all polyamides. Nylon-11 and Nylon-12 are specialty polyamides, which find use in applications requiring higher moisture resistance and hydrolytic stability.

Hydrolytic polymerization of ε-caprolactam to form nylon 6 is carried out commercially in both batch and continuous processes by heating the monomer in the presence of $5\% \sim 10\%$ water to temperatures of $250 \sim 270\text{°C}$ for periods of 12h to more than 24h. Several equilibria are involved in the polymerization. These are hydrolysis of the lactam to ε-aminocaproic acid, step polymerization of the amino acid with itself, and initiation of ring-opening polymerization of lactam by the amino acid.

Hydrolytic polymerization is simply a special case of cationic polymerization. The overall rate of conversion of E-caprolactam to polymer is higher than the polymerization rate of ε-aminocaproic acid by more than an order of magnitude. Step polymerization of ε-aminocaproic acid with itself accounts for only a few percent of the total polymerization of ε-caprolactam. Ring-opening polymerization is the overwhelming route for polymer formation. Polymerization is acid-catalyzed as indicated by the observations that amines and sodium ε-aminocaproate are poor initiators in the absence of water and the polymerization rate in the presence of water is first-order in lactam and second-order in COOH end groups.

Although step polymerization of ε-aminocaproic acid with itself is only a minor contribution to the overall conversion of lactam to polymer, it does determine the final degree of polymerization at equilibrium since the polymer undergoes self-condensation. The final degree of polymerization is dependent in large part on the equilibrium water concentration. Most of the water used to initiate polymerization is removed after about $80\% \sim 90\%$ conversion in order to drive the system to high molecular weight. Molecular weight control at a desired level also requires control of the initial water and monomer concentrations and the addition of small but specific amounts of a monofunctional acid. The MWD is essentially the Flory most probable distribution except for the presence of monomer and cyclic oligomers. The final product in the industrial polymerization of ε-caprolactam contains about 8% monomer and 2% cyclic oligomer. About one-half of the cyclic fraction is the cyclic dimer with the remaining one-half consisting mostly of the trimer and tetramer, although small amounts of cyclics up to the nonamer have been identified. Cyclic oligomers are formed when any of the propagations discussed above occur intramolecularly (both for cationic and hydrolytic polymerizations). The monomer and cyclic oligomer are removed in the industrial production of poly (ε-caprolactam) by extraction with hot water (or vacuum). The water content of the final product is lowered to less than about 0.1% by vacuum drying at 100~200℃ and about 1torr pressure.

习 题

1. 以辛基酚为起始剂，甲醇钾为引发剂，简述环氧乙烷开环聚合的机理。辛基酚用量对聚合速率、聚合度、聚合度分布有何影响？

2. 以甲醇钾为引发剂聚合得到的聚环氧乙烷分子量可以高达 30000~40000，但在同样条件下，聚环氧丙烷的分子量却只有 3000~4000，为什么？说明两者聚合机理有何不同？

3. 环氧丁烷、四氢呋喃开环聚合时需选用阳离子引发剂，环氧乙烷、环氧丙烷聚合时却多用阴离子引发剂，而丁硫环则可阳离子聚合，也可阴离子聚合，为什么？

4. 甲醛和三聚甲醛均能聚合成聚甲醛，但实际上多选用三聚甲醛作单体，为什么？在较高的温度下，聚甲醛很容易连锁解聚成甲醛，提高聚甲醛的热稳定性有哪些措施？

5. 己内酰胺可以由中性水和阴、阳离子引发聚合，为什么工业上很少采用阳离子聚合？阴离子开环聚合的机理特征是什么？如何提高单体活性？什么叫乙酰化剂，有何作用？

6. 为什么半无机和有机高分子的主链多限于 C、N、O、P、S、Si 等主族元素，而其他元素受到限制？

7. 合成聚硅氧烷时，为什么选用八甲基环硅氧烷作单体，且使用碱作引发剂？如何控制聚硅氧烷的分子量？

8. 聚硅氧烷和聚磷氮烯都是具有低温柔性和高弹性的半无机聚合物，试说明其结构有何相似之处。聚硅氧烷多由分子量和交联来改变品种，较少更改侧基；相反，聚磷氮烯却通过侧基的变换来改变品种，较小调节分子量和交联。试说明原因。

9. 简述尼龙-6 的生产工艺以及分子量的调控方法。

10. 70℃下用甲醇钠引发环氧丙烷聚合，环氧丙烷和甲醇钠的浓度分别为 0.80mol/L 和 2.0×10^{-4} mol/L，有链转移反应，试计算 80% 转化率时聚合物的数均分子量。

英语习题

1. Discuss the effect of ring size on the tendency of a cyclic monomer toward ring-opening polymerization.

2. When termination is by chain transfer during a cationic chain polymerization, what is the relationship of average DP and the kinetic chain length?

3. Anionic polymerization of propylene oxide is usually limited to producing a relatively low-molecular-weight polymer. Discuss the reasons for this occurrence.

4. The polymerization of an epoxide by hydroxide or alkoxide ion is often carried out in the presence of an alcohol. Why? How is the degree of polymerization affected by alcohol? Discuss how the presence of alcohol affects both the polymerization rate and molecular weight distribution.

5. Explain the following observations: a. a small amount of epichlorohydrin greatly increases the rate of the polymerization of tetrahydrofuran by BF_3 even though epichlorohydrin is much less basic than tetrahydrofuran; b. the addition of small amounts of water to the polymerization of oxetane by BF_3 increases the polymerization rate but decreases the degree of polymerization.

6. An equilibrium polymerization is carried out with an initial concentration of 12.1 mol/L tetrahydrofuran, $[M^*] = 2.0 \times 10^{-3}$ mol/L, and $K_p = 1.3 \times 10^{-2}$ L/(mol·s). Calculate the initial polymerization rate if $[M]_c = 1.5$ mol/L. What is the polymerization rate at 20% conversion?

7. What are the roles of an acylating agent and activated monomer in the anionic polymerization of lactams?

8. Consider the equilibrium polymerization of ε-caprolactam initiated by water at 220℃. For the case where $[I]_0 = 0.352$, $[M]_0 = 8.79$ and $[M]_c = 0.484$, the degree of polymerization at equilibrium is 152, calculate the values of K_i and K_p at equilibrium.

9. Discuss by means of equations the occurrence of backbiting, ring-expansion reactions in the polymerizations of cyclic ethers, acetals, and amines.

第 5 章
天然高分子

天然高分子（一）　天然高分子（二）

  甲壳素的生物合成量每年约 1000 亿吨，是一种取之不尽用之不竭的可再生自然资源。甲壳素广泛存在于微生物、菌菇类植物的细胞壁，以及昆虫表皮、软体动物骨骼内。虾、螃蟹等甲壳类动物的甲壳内甲壳素尤其丰富。壳聚糖是甲壳素中大部分重复单元脱乙酰基的产物，与纤维素化学结构相近，仅仅在重复糖环单元 C2 位上取代基不同，纤维素 C2 位是羟基，壳聚糖 C2 位是氨基。壳聚糖分子中的氨基官能团比羟基化学反应性强，能够较容易地进行多种化学修饰反应，因此被认为是较纤维素更有潜力的天然原材料。

本章目录

5.1 　多糖　/ 187
5.2 　蛋白质　/ 196
5.3 　核酸　/ 199
5.4 　木质素　/ 201
5.5 　天然橡胶　/ 205

重点要点

  醛糖，酮糖，D 型糖，α-吡喃葡萄糖，β-吡喃葡萄糖，糖环，纤维素，淀粉，壳聚糖，甲壳素，海藻酸，环糊精，普鲁兰，氨基酸，多肽，蛋白质，蛋白质四级结构，蚕丝，丝素，DNA，RNA，木质素，天然橡胶。

天然高分子材料，如棉、麻、丝、毛、木材、皮革、生漆等，自古以来都是重要的生产原材料。20世纪中叶，合成高分子材料发展迅猛，涤纶、聚乙烯、聚丙烯等高分子产品开始大规模生产，导致天然高分子材料的地位下降。21世纪以来，随着能源及环保意识的增强，适合社会可持续发展的天然高分子材料再次被重视。天然高分子可再生，且生产量极大。例如，全球纤维素的生物产量高达千亿吨，超过现有石油总储量。有效利用天然高分子材料是构建可持续发展社会的重要途径。

天然高分子材料可分为五大类：多糖、蛋白质、核酸、木质素以及生漆、天然橡胶和虫胶等植物分泌树脂。其中，前三类天然高分子最为重要。纤维素和淀粉都属于多糖类聚合物，分别承担植物的支撑增强和动植物的储存作用。蛋白质决定生命体的结构和性状，主宰细胞体内的绝大多数化学反应，如基因表达、电子传递、神经传递、学习和记忆相关反应等。核酸类物质，包括 DNA 和 RNA，在细胞内起到储存和表达遗传信息、合成蛋白质、复制细胞的功能。

5.1 多　　糖

5.1.1 糖的基础知识

单糖（monosaccharide）是多糖（polysaccharide）的基本单元，不能再进一步水解。多糖含有 20 个以上的单糖结构单元。含有 3~20 个单糖的短链糖称作寡糖（oligosaccharide）。单糖有醛糖（aldoses）和酮糖（ketoses）之分。甘油醛是最简单的醛糖，具有一个手性碳，因此有两个异构体（图 5-1）。将甘油醛按费歇尔投影式展开，醛基在顶端时，手性碳上的羟基在右侧的为 D 型糖，反之为 L 型糖。

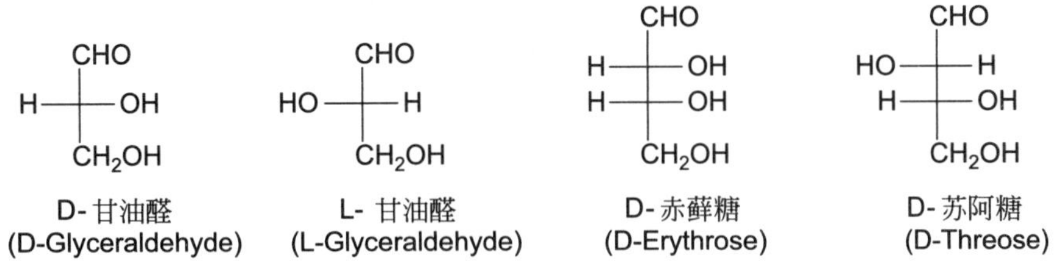

CHO	CHO	CHO	CHO
H—OH	HO—H	H—OH	HO—H
CH$_2$OH	CH$_2$OH	H—OH	H—OH
		CH$_2$OH	CH$_2$OH
D-甘油醛	L-甘油醛	D-赤藓糖	D-苏阿糖
(D-Glyceraldehyde)	(L-Glyceraldehyde)	(D-Erythrose)	(D-Threose)

图 5-1　甘油醛及 D 型 4 碳糖

每增加一个手性碳，异构体数量增加一倍。单糖分子立体异构体的数目与分子中手性碳原子的个数有关。含有 n 个手性碳的单糖，其立体异构体应该有 2^n 个。

六碳糖具有 4 个手性碳，有 16 个异构体，其中一半为 D 型糖（图 5-2）。D-葡萄糖就是六碳 D 型糖中的一员。D-葡萄糖是自然界中产量最大，资源最丰富的单糖。

单糖中的一个羟基和醛基缩合，形成半缩醛结构成环。形成五元环的称为呋喃糖，形成六元环的称为吡喃糖。以 D-葡萄糖为例，它的 C5 羟基和醛基缩合，生成六元环的半缩醛，

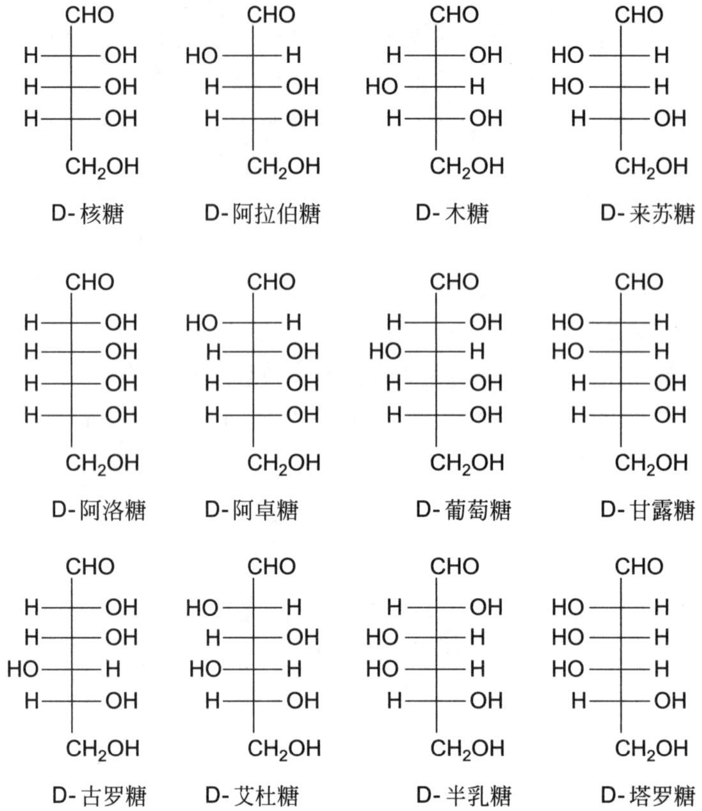

图 5-2　D 型 5 碳糖及 6 碳糖的分子结构

也可以是 C4 羟基和醛基缩合，生成五元环的半缩醛。D-葡萄糖的吡喃糖六元环结构更加稳定（图 5-3）。

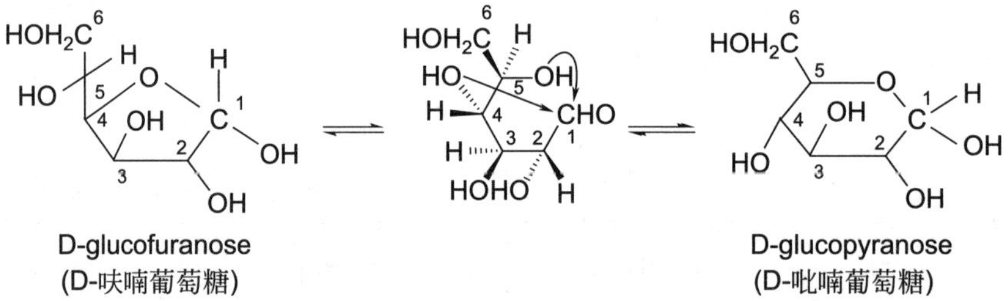

图 5-3　D-葡萄糖的成环

半缩醛缩合反应在 1 位碳上形成羟基，使 1 位碳成为手性碳，因此又形成两个异构体，称为 α-异构、β-异构。就 D-葡萄糖而言，1 位碳羟基和 5 位羟甲基位于环异侧的为 α-异构，位于同侧的为 β-异构（图 5-4）。α-吡喃葡萄糖是淀粉的结构单元，β-吡喃葡萄糖是纤维素的结构单元。

5.1.2　多糖

重要的多糖聚合物有纤维素、淀粉和甲壳素等，也有糖原、半纤维素、海藻酸盐等产量

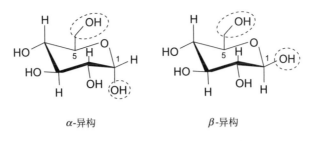

图 5-4 D-葡萄糖的 α-异构和 β-异构

相对较小的聚合物。

纤维素和淀粉的重复单元都是葡萄糖。它们的最大区别是糖苷键的连接方式不同。淀粉是通过 α-1,4-糖苷键连接，纤维素是通过 β-1,4-糖苷键连接。甲壳素也是通过 β-1,4-糖苷键连接，和纤维素结构相似，其主要结构单元是 N-乙酰化-D-氨基葡萄糖（N-acetyl-D-glucosamine），可以看作是纤维素葡萄糖单元中 2 位碳羟基被乙酰基取代的结果。甲壳素脱除乙酰基，就成为壳聚糖。

(1) 纤维素

天然高分子中纤维素年生产量最大，应用历史也最悠久。棉花是自然界中纤维素含量最高的纤维，其纤维素含量为 90%～98%。木材是纤维素最主要的来源，木材中除纤维素外，还含有木质素和半纤维素。木材中的纤维素溶解后再生纺丝，可得再生纤维素纤维（图 5-5）。

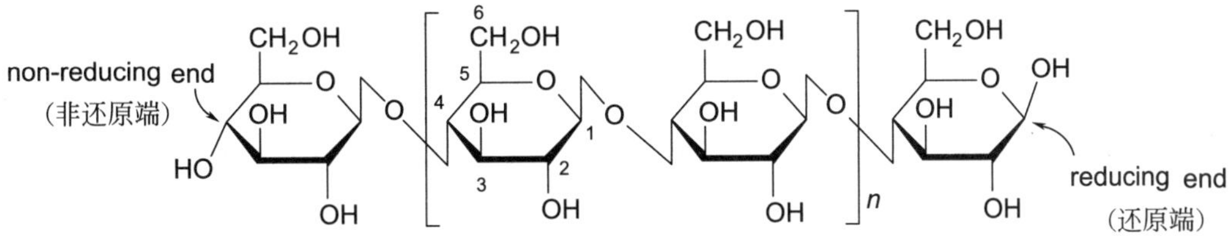

图 5-5 纤维素分子式

① 纤维素的晶型

纤维素是由 β-1,4-糖苷键连接的，由脱水-D-吡喃葡萄糖单元构成的天然高分子化合物，两个相邻单元互成 180°交错。纤维素的重复单元称为纤维素二糖（cellobiose）。纤维素分子链的一端是半缩醛，具有还原性，为还原端；另一端是糖环的羟基，为非还原端。天然纤维素中葡萄糖单元的数量也就是聚合度（DP）较高。棉花的聚合度达 10000，分子量约 150 万；亚麻的聚合度更高（36000），分子量可达 590 万。纤维素中每个葡萄糖单元分子结构中含有三个羟基，在分子链内及链间形成大量氢键，使得纤维素分子链处于伸展状态，并形成片状结晶网络，受热时不能塑化熔融，也不容易溶解。

纤维素具有四种常见晶型，称作纤维素Ⅰ、纤维素Ⅱ、纤维素Ⅲ、纤维素Ⅳ。绝大多数天然纤维素具有纤维素Ⅰ型结晶结构。近期又发现，纤维素Ⅰ型又可细分为纤维素Ⅰα 和纤维素Ⅰβ 晶型，分别属于三斜晶系和单斜晶系。纤维素Ⅰα 晶型为亚稳态，加热退火处理可转化为更为稳定的纤维素Ⅰβ 晶型。纤维素Ⅱ型和纤维素Ⅲ型属单斜晶系，纤维素Ⅳ属正交晶系。纤维素各种晶型的晶胞参数如表 5-1 所示。

表 5-1 各种纤维素晶型的参数差异

晶型参数	纤维素Ⅰα（三斜）	纤维素Ⅰβ（单斜）	纤维素Ⅱ（单斜）	纤维素Ⅲ（单斜）	纤维素Ⅳ（正交）
$a/Å$	6.72	7.78	8.10	10.25	8.03
$b/Å$	5.96	8.20	9.05	7.78	8.13
$c/Å$	10.40	10.38	10.31	10.34	10.34
$α/°$	118.1	90	90	90	90
$β/°$	114.8	90	90	90	90
$γ/°$	80.4	96.5	117.1	122.4	90

纤维素Ⅰ经浓碱液浸泡处理可转化为纤维素Ⅱ，大多再生纤维素的主要晶型为纤维素Ⅱ。纤维素Ⅱ较纤维素Ⅰ更加稳定，无法逆转成纤维素Ⅰ。纤维素Ⅰ和纤维素Ⅱ在有机胺类或氨水浸泡膨胀后可转化为纤维素Ⅲ。纤维素Ⅳ由纤维素Ⅲ转化生成。

不同种类的纤维素具有不同的结晶度。海藻纤维素的结晶度可高达90%，细菌纤维素的结晶度在40%～70%，棉花的结晶度在60%左右，各种麻纤维的纤维素结晶度在40%～60%。

② 天然纤维素

天然纤维素的结构复杂多样，以棉纤维为例，棉纤维具有扁平的带状结构，界面类似腰果仁形状，直径大约为20μm。棉纤维的横截面由许多同心层组成，可分为初生层、次生层、中腔三个部分（图5-6）。

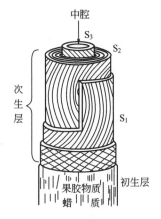

图 5-6 棉纤维的形态结构

初生层在棉纤维伸长期形成，它的外皮是一层极薄的蜡质与果胶，表面有螺旋状条纹。表皮层下面是初生层，由网状的原纤组成，厚度为0.1～0.2μm。次生层在棉纤维加厚期沉积而成，几乎都是纤维素，是棉纤维的主体层，一般含三个层次。棉纤维生长停止后遗留的空腔称为中腔，中腔内留有少数原生质和细胞核残余。

③ 再生纤维素

由于数目众多的氢键的存在，纤维素的致密结晶结构很难被常规溶剂所溶解。纤维素既不溶于水也不溶于普通的有机溶剂。如果能将纤维素溶解，会为纤维素的加工成型带来巨大变革。但是，能够溶解纤维素的溶剂很少，主要有离子液体、混合碱溶液、有机/无机溶剂等少数几个体系。

黏胶法曾经是生产再生纤维素的主要方法，它通过化学反应将纤维素溶解并再生。纤维素在强碱性条件下与二硫化碳反应生成纤维素黄酸酯（xanthate），成为黏胶液。黏胶液在酸性凝固浴中再生，可制得纤维或膜状产品。

$$\text{Cell—OH} + \text{NaOH} + \text{CS}_2 \longrightarrow \underset{\text{纤维素黄酸酯}}{\text{Cell—O}-\overset{\text{S}}{\underset{\|}{\text{C}}}-\text{S}^-\text{Na}^+} \xrightarrow{\text{H}_2\text{SO}_4} \underset{\substack{\text{黏胶纤维}\\(\text{再生纤维素})}}{\text{Cell—OH}}$$

黏胶纤维（viscose fiber），别名莫代尔，是以木纤维或棉短绒为原料，经碱化、老化、磺化等工序制成可溶性纤维素黄原酸酯，再溶于稀碱液制成黏胶，经湿法纺丝而制成的再生纤维素纤维。

将纤维素浆粕制成黏胶需要经历两个化学反应。首先将浆粕与碱液作用，生成碱纤维素，再与二硫化碳反应生成纤维素黄酸酯，使纤维素溶解而制得黏胶纺丝液。在纺丝工段，黏胶通过喷丝孔形成细流进入含酸凝固浴，纤维素黄酸酯分解再生成水化纤维素。成型后的水化纤维素纤维经水洗、脱硫、酸洗等后处理加工除去附在纤维表面的硫酸及其盐类和部分硫，再经上油和干燥处理即可出厂。黏胶法由于废水排放量大、所用二硫化碳容易分解、生成有毒和有恶劣气味的物质，环境问题较多，近些年来，黏胶法产量逐渐减少。

N-甲基吗啉-N-氧化物（NMMO）是一种脂肪族环状叔胺氧化物，可以和纤维素上的羟基形成配合物，破坏纤维素分子间的氢键，使纤维素溶解。使用 NMMO 溶解再生纤维素已经实现工业化生产。因 NMMO 溶剂价格高，再生纤维素生产过程中非常重视 NMMO 的循环利用，目前的再生纤维素生产工艺里，NMMO 的回收利用率高达 99%。

氢氧化钠/尿素（NaOH/Urea）水溶液溶解体系需要在低温下进行，因为温度越低，碱液对纤维素的溶胀作用越大。尿素在碱液中可以破坏纤维素分子之间的氢键，有利于促进纤维素溶解。

氯化锂/二甲基乙酰胺（LiCl/DMAc）溶解体系在近年发展迅速。Li^+ 在 DMAc 的羰基和氮原子之间发生配位，游离出的 Cl^- 与纤维素羟基结合，使纤维素和 DMAc 之间形成较纤维素分子间氢键更强的氢键作用，使纤维素溶解。该溶剂体系价格较高，回收困难，至今未能实现工业化生产。

离子液体被公认为是有机溶剂的理想替代溶剂。1-丁基-3-甲基咪唑氯盐（[BMIM]Cl）是最早用于溶解纤维素的离子液体。加热到 100～110℃时，纤维素在该离子液体中的溶解度可达 10%。同样条件下，1-烯丙基-3-甲基咪唑氯盐（[AMIM]Cl）溶解纤维素的能力更强。向纤维素的离子液体溶液中加入水、丙酮或乙醇等溶剂，可使纤维素析出，获得再生纤维素。

④ 纤维素衍生物

纤维素分子链中每个葡萄糖单元分别在 C2、C3、C6 位置上有 3 个活泼羟基，可进行多种化学反应，获得纤维素衍生物（图 5-7）。其中 C6 上羟基与 C2、C3 相比，空间位阻最小，具有更好的活性。纤维素 C6 上的伯羟基可氧化成醛基或羧基，C2 和 C3 的仲羟基经高碘酸盐氧化可形成二醛或羧基。过度氧化会导致纤维素分子链断裂，分子量下降。四甲基哌啶氮

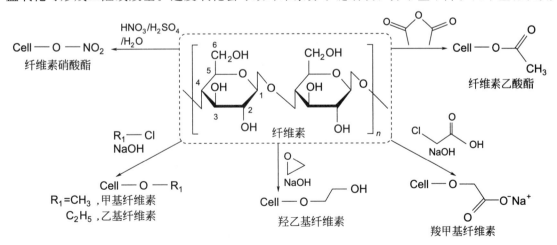

图 5-7　纤维素衍生物

氧化物（TEMPO）作为氧化剂可将 C6 羟基氧化为羧基，增加纤维表面的亲水性。

纤维素在强酸性条件下被酯化，生成相应的纤维素酯。常见的纤维素酯有纤维素硝酸酯、纤维素乙酸酯等。纤维素的羟基也可和卤代烷、缩水甘油等醚化剂在碱性条件下醚化，生成纤维素醚。主要的纤维素醚衍生物有羧甲基纤维素、乙基纤维素、甲基纤维素、羟乙基纤维素等。

(2) 淀粉

淀粉（starch）是一种自然界植物体内合成的生物多糖类聚合物，分布在植物种子、根块和茎叶等部位。在生物体中以颗粒形态存在，粒径因植物种类而异，一般在数百纳米至 $200\mu m$ 范围（图 5-8，表 5-2）。

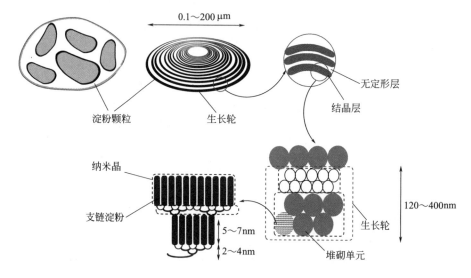

图 5-8 淀粉颗粒结构

表 5-2 淀粉颗粒粒径

来源	淀粉颗粒粒径/μm	来源	淀粉颗粒粒径/μm
小麦(小)	2～3	玉米	5～20
小麦(大)	22～36	稻米	3～8
马铃薯	15～75		

淀粉由直链淀粉和支链淀粉构成。淀粉颗粒中 15%～35% 为直链淀粉。它是一种线形多糖，结构单元是 α-1,4-D-葡萄糖。支链淀粉是一种高度分支的分子，以 1,4-糖苷键连接的 D-葡萄糖分子链为骨架，含有大约 5% 的 1,6-糖苷键连接的支链。直链淀粉和支链淀粉的质量比因物种不同而异，对淀粉的物理和生物特性有较大影响。直链淀粉降解性相对较强，能够溶于水，也能形成强韧的薄膜。支链淀粉较稳定，仅能在水中分散，形成的膜强度较弱。

直链淀粉溶于热水，遇碘呈深蓝色。显色反应的机理是碘分子进入直链淀粉的螺旋结构中，并与淀粉形成蓝色的配合物。支链淀粉不溶于水，热水中则溶胀而成糊状。支链淀粉遇碘呈现紫色。

淀粉颗粒在加热条件下大量吸水，结晶结构被破坏直至消失，大部分直链淀粉溶解，悬浮液黏度增加并最终转变为黏稠糊浆，这种现象称为淀粉糊化。淀粉糊化分为可逆吸水、不可逆吸水、淀粉粒解体三个阶段。在可逆吸水阶段，水分子仅仅吸附在淀粉颗粒非结晶区，体积略有膨胀，干燥可复原；不可逆吸水阶段，水分子进入淀粉微晶间隙，不可逆地大量吸水，结晶区减少；最后在淀粉粒解体阶段，结晶区完全消失，淀粉分子形成凝胶或溶解。

淀粉发生糊化的温度称为糊化温度，淀粉种类不同，糊化温度也不同。糊化温度对淀粉及其衍生物的性质有重要影响。

经含水加热糊化，失去结晶结构的淀粉分子链在低温下静置，部分分子链通过氢键作用重新有序排列，重新形成结晶，这种现象称为老化。直链淀粉的链状结构空间障碍小，易于老化，支链淀粉的分支部分也易于老化。－4℃的储存温度可促进淀粉老化，60℃以上则不易老化。迅速冷却，淀粉分子来不及重新取向，也可减少老化。

构成淀粉的葡萄糖环结构单元有三个醇羟基，故淀粉的吸水溶胀性优于纤维素，所以淀粉更容易进行化学改性处理。

过氧化氢、过氧乙酸、高锰酸钾、高碘酸及过硫酸等氧化剂可以用来氧化淀粉。其中，工业上应用最广泛的氧化剂是次氯酸钠。部分经特殊处理的氧化淀粉，经口服对尿毒症有一定的疗效。

醚化淀粉包括非离子型醚化淀粉和离子型醚化淀粉。生产规模较大的非离子型淀粉有羟乙基淀粉和羟丙基淀粉。代表性的羟乙基淀粉醚化剂是氯乙醇，具有安全性高、易操作的优点。羟乙基淀粉目前主要用于造纸、纺织和医药工业中。羧甲基淀粉是阴离子型淀粉醚衍生物，使用氯乙酸作为醚化剂经羧甲基化反应制备。羧甲基淀粉溶于冷水，是食品工业中常见的增黏剂。目前，羧甲基淀粉广泛用于医药、石油、日用化工、纺织以及造纸工业。

酯化淀粉衍生物主要包括乙酸酯淀粉、磷酸酯淀粉、尿素淀粉、黄原酸酯淀粉、烯基琥珀酸酯淀粉等，在食品、医药及化妆品等行业具有较大应用价值。

(3) 环糊精

淀粉经特殊酶解处理可得环状低聚糖，称为环糊精（cyclodextrin，CD）。环糊精有 α、β、γ 三种类型，分别由 6、7、8 个葡萄糖环单元连接而成，呈中空杯状结构（表 5-3，图 5-9）。环糊精具有一定的水溶性，环糊精杯形内部的疏水性比外壁强，可包埋大小合适的客体分子。环糊精的这种特殊的主客体超分子吸附作用广泛用作各种吸附材料，医药领域中的增溶缓释，以及用于掩盖食物及饮料中的异味。

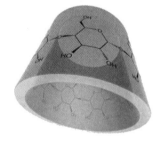

图 5-9 环糊精的杯形结构

表 5-3 环糊精的重要参数

	α-环糊精	β-环糊精	γ-环糊精
葡萄糖单元数	6	7	8
分子量	973	1135	1297
溶解度/(g/100mL)	14.5	1.85	23.2
空腔直径/nm	0.5	0.6	0.8
空腔高度/nm	0.8	0.8	0.8

(4) 普鲁兰多糖

普鲁兰（Pullulan）多糖是一种微生物多糖，由出芽短梗霉发酵所得，在医药、食品、石油、化工等行业具有广阔的应用前景。普鲁兰多糖的重复单元是麦芽三糖，重复单元间以 α-1,6-糖苷键连接，聚合度为 100～5000，没有分支结构，属于直链多糖。

普鲁兰多糖无毒性，可用于食品、日用化学和医药工业，它能迅速溶解于冷水或温水中，溶解速度快，溶液呈中性，可与水溶性高分子如羧甲基纤维素、海藻酸钠和淀粉等互溶，不溶于乙醇、氯仿等有机溶剂。普鲁兰多糖成膜性优良，容易形成强韧且气密性良好的薄膜。普鲁兰多糖在动物消化器官内较难消化，可保持数小时不分解。

普鲁兰多糖已广泛应用于食品、纺织、制药、造纸等领域。例如，普鲁兰多糖无需增塑剂就可形成透明韧性膜。普鲁兰多糖膜阻气性良好，抗油脂，易水溶，适用于食品包装材料。因普鲁兰多糖不易消化，可用于低热量食品和饮料的添加剂。普鲁兰多糖用作食品增稠剂，可改善口感。在制药工业，添加少量的普鲁兰多糖，可适当提高软胶囊的柔韧性。

(5) 甲壳素与壳聚糖

甲壳素的结构单元是乙酰氨基葡萄糖环，由 β-1,4-糖苷键相连。甲壳素一般以结晶微纤的形式存在，直径为 2.5～2.8nm。甲壳素广泛存在于蟹、虾、昆虫的甲壳中，年生物生产量在纤维素之后，居第二位。因主要产自海洋，也称作海洋纤维素。壳聚糖是甲壳素脱除大部分乙酰基的产物。脱乙酰度（degree of deacetylation）是脱乙酰基葡萄糖胺数占总结构单元数的比例，是影响壳聚糖溶解性等性能的重要参数。甲壳素、壳聚糖、纤维素三者化学结构相近，仅仅是糖环单元中 C2 位置上的取代基不同，纤维素 C2 位是羟基，甲壳素、壳聚糖 C2 位分别是乙酰氨基和氨基（图 5-10）。壳聚糖分子中的氨基官能团比羟基化学反应性强，更容易进行多种化学修饰反应，因此被认为是较纤维素更有潜力的天然原材料。壳聚糖生物降解性和生物相容性较好，有一定的抑菌作用。也有报道认为分子量为数千的壳聚糖具有抗癌作用。

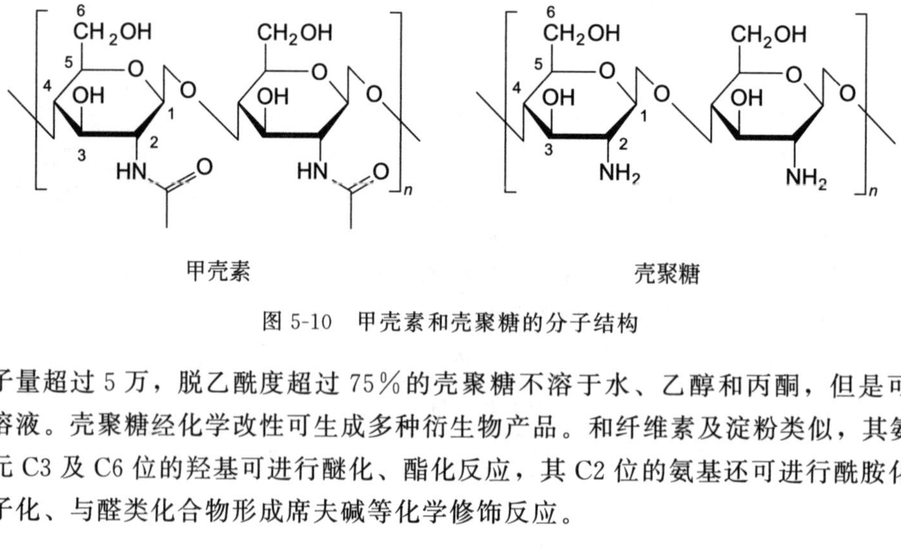

图 5-10 甲壳素和壳聚糖的分子结构

分子量超过 5 万，脱乙酰度超过 75% 的壳聚糖不溶于水、乙醇和丙酮，但是可以溶于稀酸水溶液。壳聚糖经化学改性可生成多种衍生物产品。和纤维素及淀粉类似，其氨基葡萄糖环单元 C3 及 C6 位的羟基可进行醚化、酯化反应，其 C2 位的氨基还可进行酰胺化、季铵盐阳离子化、与醛类化合物形成席夫碱等化学修饰反应。

5.1.3 其他多糖

海藻酸（alginic acid）是由海洋褐藻植物中提取的天然高分子，是存在于海藻细胞壁和细胞间质的一种聚阴离子多糖。海藻酸是一种嵌段线形聚合物，其结构单元为 β-D-甘露醛酸和 α-L-古罗糖醛酸，在一个高分子链中，可能只含有一种结构单元，也可能含两种结构单元（图 5-11）。海藻酸钠盐可用作食品添加剂，如乳制品增稠剂等。海藻酸钠水溶胶还可用于医药领域的药物载体材料、微胶囊材料、创伤敷料等。在纺织印染业中，海藻酸钠也可以用作印染糊料，以改善印染质量。

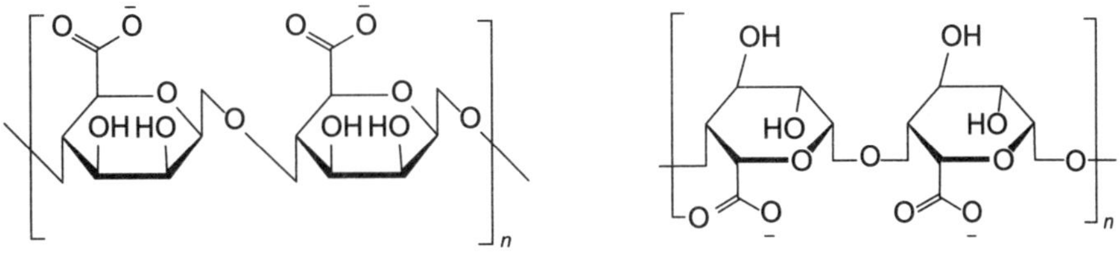

图 5-11 海藻酸结构单元分子式

魔芋是天南星科魔芋属多年生草本块茎植物，其主要成分是魔芋葡甘聚糖，由 D-葡萄糖和 D-甘露糖按 1∶1.6 或 1∶1.9 的物质的量比，通过 β-1,4-糖苷键结合构成的复合多糖（图 5-12）。在主链甘露糖碳 3 位连接有支链结构。大约每 32 个糖环单元中含有 3 个左右的支链。大约每 19 个糖环含有 1 个乙酰基团。魔芋葡甘聚糖是一种具有较强吸水性的高分子化合物，吸水后体积可膨胀 80~100 倍，食后易被消化吸收。魔芋葡甘聚糖可用作造纸、印刷胶液、橡胶、陶瓷、摄影胶片的胶黏添加剂，在纺织工业中用作毛、麻、棉纱的浆料，丝绸双面透印的印染糊料和后处理的柔软剂，等等。

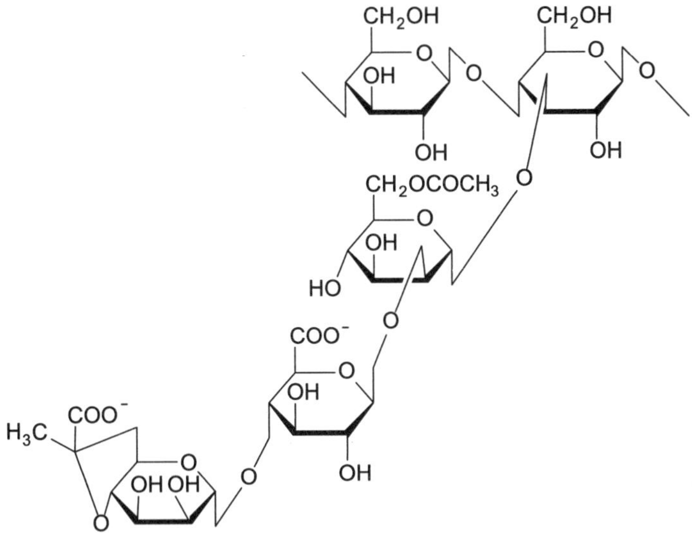

图 5-12 魔芋葡甘聚糖分子结构

黄原胶也是一种水溶性生物高分子聚合物。黄原胶由 5 个糖环连接构成重复单元，其主链结构和纤维素类似，由 D-葡萄糖以 β-1,4-糖苷键相连，每两个葡萄糖环中的一个 C3 位连接一个短侧链。侧链由甘露糖-葡萄糖醛酸-甘露糖连接而成。黄原胶在水溶液中呈现阴离子聚合物的特征。黄原胶可用于食品、饮料行业的增稠剂、乳化剂和成型剂，还可以用作医药业中的药物缓释材料。

5.2 蛋白质

5.2.1 氨基酸

蛋白质是重要的天然高分子材料，蚕丝、毛、动物胶、酪素等都是传统的工业原料。蛋白质的结构单元为α-氨基酸。自然界中共发现有700多种氨基酸，其中有20种为人体必需氨基酸。它们都是氨基连在α碳上的有机酸，α碳原子为手性碳原子，构成生物蛋白质的氨基酸均为L-型。20种氨基酸的差异体现在R取代基的不同（图5-13）。

R为非极性脂肪链的氨基酸有五个。它们在肽链中构建疏水孔，也在细胞膜中起重要作用。第一个是甘氨酸，R为一个氢原子，也是最简单的氨基酸。其余四个氨基酸分别是丙氨酸、缬氨酸、亮氨酸、异亮氨酸，它们的R取代基分别是甲基、异丙基、异丁基、丁基。含有羟基官能团的有两个，分别是丝氨酸和苏氨酸。含硫氨基酸两个，分别是半胱氨酸和蛋氨酸。半胱氨酸含巯基，可以与另一个半胱氨酸形成双硫键，是肽链中常见的交联点。R为羧基的氨基酸有天冬氨酸和谷氨酸。含酰氨基官能团的氨基酸有天冬酰胺和谷氨酰胺。碱性氨基酸有赖氨酸和精氨酸。含苯环结构的氨基酸有苯丙氨酸和酪氨酸。色氨酸、脯氨酸和组氨酸则含有杂环结构。

氨基酸分子含有氨基和羧基，属于两性化合物，在特定pH值下呈中性。氨基酸呈中性时的pH值称为等电点（pI）。等电点下的氨基酸水溶性差。20种标准氨基酸的等电点各不相同，等电点可用于分离和鉴别氨基酸。

5.2.2 蛋白质分级结构

两个氨基酸的羧基和氨基进行缩合反应，生成肽。在生物体内，肽的合成需要酶的催化作用。实验室内则需将氨基或羧基进行保护处理。数目众多的氨基酸单元通过肽键成为多肽，就是生物体内的各种蛋白质。蛋白质具有四级结构，分别是氨基酸的序列，大分子构象，大分子空间形态，以及肽链和肽链之间的超分子作用。

氨基酸序列实际上是氨基酸取代基构型的排列顺序。蛋白质骨架结构一致，都是酰胺肽键，决定蛋白质性质的是氨基酸取代基的种类及排列方式。

蛋白质大分子的构象为二级结构，主要有α-螺旋和β-折叠两种构象。带有较大侧基的线形肽链，呈右手螺旋式盘绕，称α-螺旋。约3~6个氨基酸单元构成一个螺旋周期，螺旋之间羰基和氨基形成氢键，使构象稳定。螺旋结构赋予蛋白质高弹性，拉伸时可伸长。羊毛角蛋白、肌蛋白、胶原蛋白都具有α-螺旋结构。氨基酸取代基体积较小时，线形蛋白质分子链回折成多个链段，形成β-折叠。两条或多条肽链也可能顺向（如羽毛）或逆向（如丝蛋白）平行排列成β-折叠片状结构。β-折叠片状结构伸长率较小，但拉伸强度较高。肽链间氢键不足时，蛋白质分子链也能形成无规线团结构。

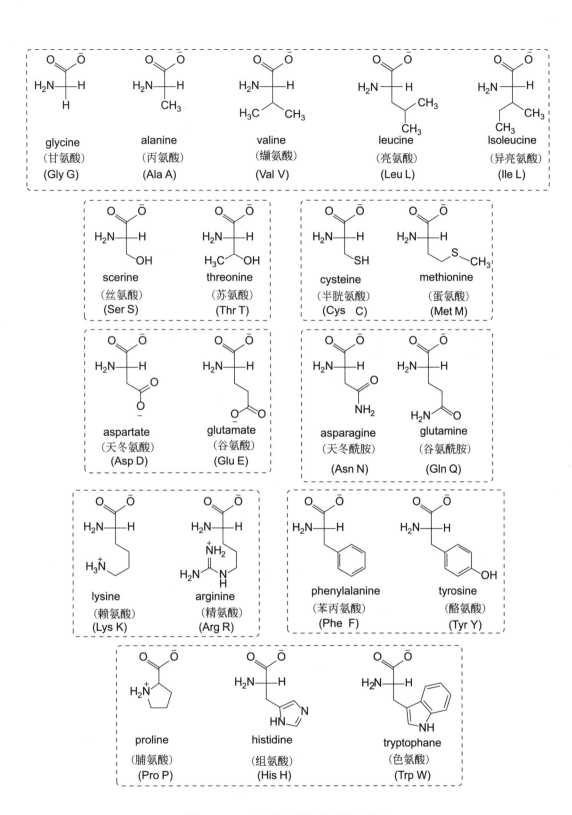

图 5-13 20 种人体必需氨基酸的结构式

蛋白质分子链在二级结构基础上进一步盘绕或折叠，相邻的二级结构彼此作用，形成规则的二级结构聚集体，产生特定空间结构，这就是蛋白质的三级结构。三级结构是指一条多肽链形成一个或多个紧密球状单位或结构域，如双硫键。二级结构还不能完全描述蛋白质大分子的空间构象状况时，就用三级结构来描述。由半胱氨酸形成的双硫键是稳定三级结构的唯一共价键形式，其他类型包括离子键、氢键和范德华力。

四级结构表示复数个具有独立的三级结构的蛋白质分子链相互作用而成的蛋白质分子链的聚集态结构，属于高分子和高分子之间的超分子结构（图5-14）。

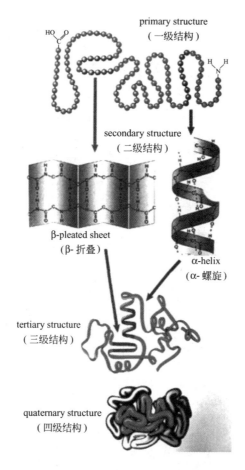

图5-14 蛋白质的四级结构

5.2.3 重要天然蛋白质

羊毛属于角蛋白。3条蛋白质链为一组，盘绕成2nm的α-螺旋原纤维，2条原纤维为内核，9条原纤维围绕该内核缠绕成8nm的微纤维，许多根微纤维构成大约1μm的粗羊毛纤维。羊毛蛋白含大量半胱氨酸（含硫量约为6%），原纤维之间有双硫键交联。在碱性条件下，双硫键可分解为硫醇，经可逆反应又可恢复成双硫交联。羊毛织物水洗后收缩起皱，经湿烫，又能平直挺括，这与双硫键的可逆反应有关。

蚕丝中含量较高的氨基酸有甘氨酸、丙氨酸、丝氨酸等，占总量的80%以上。多肽链

大多呈逆向平行 β-折叠片状结构。蚕丝由单丝经丝胶黏合而成，单丝纤维由微原纤聚集而成。丝织物洗后收缩发皱，是因水分进入单丝非晶区，破坏氢键和离子键。丝织物经熨烫干燥，可恢复光滑平整有弹性。

蚕丝主要由内层的丝素蛋白和外层的丝胶蛋白两部分组成。丝素蛋白中含量较高的氨基酸有甘氨酸、丙氨酸、丝氨酸等，占总量的 80% 以上。丝素蛋白多肽链呈逆向平行 β-折叠片状结构。丝胶是一种可溶于水的糖蛋白，经沸水煮后可与丝素蛋白分离。丝素蛋白可溶于浓酸、高浓度盐、盐-有机溶剂等，不溶于水、稀酸和碱类溶剂。

丝素蛋白的结晶度约为 50%，非晶区具有一定的水溶性，由无规线团和 α-螺旋结构组成，赋予蚕丝较好的弹性。结晶区呈水不溶性，以反平行 β-折叠片状结构为主，属单斜晶系，链段排列整齐，结构稳定。结晶区给予蚕丝较好的机械强度和弹性模量。

丝素蛋白可在特殊的盐溶液中膨胀，最终形成黏稠的液体，经透析除盐可得到丝素蛋白溶液，然后可再生成凝胶、薄膜、多孔材料等不同形态，可用作创面覆盖材料、生物传感器等。

胶原蛋白存在于皮、骨、腱、韧带中，几乎占了动物蛋白质的三分之一。胶原蛋白经水解，可制备明胶。将动物皮、骨加压蒸煮，其中胶原蛋白轻度水解成水溶性的动物胶，俗称明胶。明胶是食品、医药、照相工业的重要原料。

酪素主要来自牛奶，也可从大豆中提取。牛奶酪素由脱脂奶经发酵后提取，产率约为 3%。酪素可热压成板材或挤出成棒材，浸在 60℃ 左右的 5% 甲醛溶液中，数天后酪素蛋白中侧氨基交联，再用 100℃ 甘油或油类处理软化，就成为酪素塑料。其性能与酚醛树脂相似，可制成纽扣、服饰等塑料制品。酪素碱溶液在 50℃ 下纺丝，经酸浴固化，再由甲醛交联，也可制成酪素纤维。

5.3 核　酸

核酸是第三大类生物高分子，存在于所有生物活细胞中，参与生物过程和生命活动，肩负着遗传的特殊功能，如储存和传递遗传信息，导向和控制蛋白质和酶的模板合成、新细胞的生长和分裂等。

如图 5-15 所示，核酸由含氮杂环碱、戊糖和磷酸组成。杂环碱和戊糖结合成核苷，核苷与磷酸缩合成核苷酸（核苷的磷酸酯），核苷酸共缩聚成聚核苷酸（即核酸的学名）。

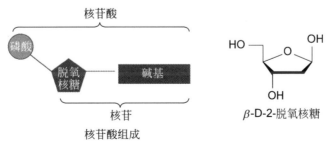

图 5-15　核苷酸结构

核酸可分为核糖核酸 RNA 和脱氧核糖核酸 DNA 两种，分别含有 D-核糖或 2′-D-脱氧核糖。这两种糖都是呋喃戊糖。

DNA 中出现有两种嘧啶碱和两种嘌呤碱。分别是 A、T、C、G。嘧啶碱有胸腺嘧啶和胞嘧啶，嘌呤碱有腺嘌呤和鸟嘌呤两种。图 5-16 是四种杂环碱的结构。

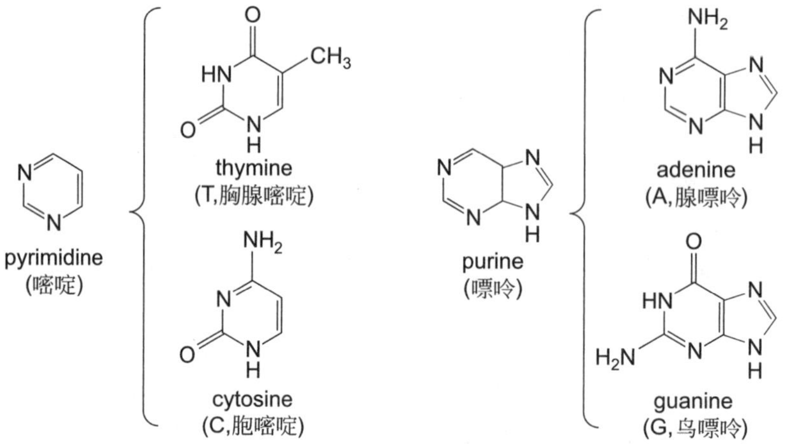

图 5-16 核苷酸以及 DNA 中杂环碱的分子结构

碱基与脱氧核糖磷酸酯链接，形成四种核苷酸，就是 DNA 高分子中的结构单元。嘌呤和嘧啶之间通过氢键配对，形成碱基对，且 A 只和 T 配对、C 只和 G 配对，这种碱基之间的一一对应的关系叫作碱基互补配对原则（图 5-17）。

DNA 结构单元，四种核苷酸　　碱基互补配对原则

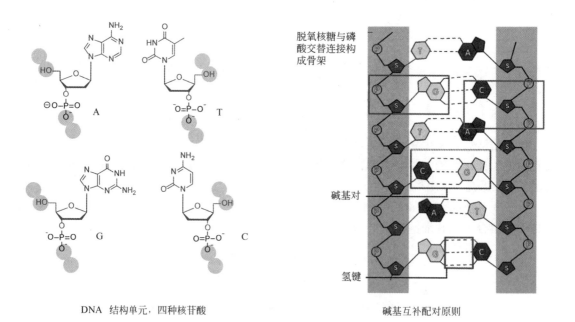

图 5-17 DNA 分子的结构

两条反向平行的脱氧核苷酸长链盘旋成 DNA 双螺旋结构。DNA 在分子结构上具有两组碱基互补配对，三种基本物质（磷酸、脱氧核糖、碱基），四个碱基（A、T、G、C），五

种元素（C、H、O、N、P）这样的分子结构特点（图 5-18）。另外，DNA 还有一些与其他高分子不同的特点。例如，它是等规有序聚阴离子，具有自组装能力，可选择性吸附化学物质，生物相容性良好等特点。

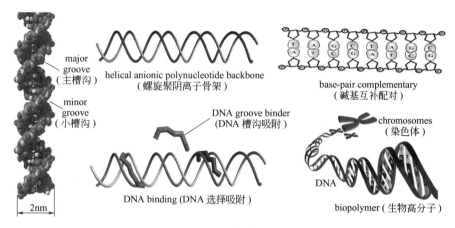

图 5-18　DNA 双螺旋的结构特点

DNA 的这些结构特点也被用于功能材料的制备。DNA 是一种聚阴离子，磷酸基在 DNA 双螺旋骨架上有序排列，可结合金属离子、阳离子表面活性剂、聚阳离子等阳离子物质，制备各种功能性的高分子材料。DNA 的碱基对与碱基对之间有一个大约 0.34nm 的空隙，DNA 双螺旋还会形成一宽一窄，深度大约相同的凹槽。具有平板形状或呈月牙形状的有机小分子可以分别插入到碱基对之间的空隙或吸附在凹槽中。DNA 的吸附对小分子结构有很高的选择性。一些致癌物质往往容易被 DNA 吸附。利用选择性吸附，DNA 可以用作环保材料以排除有毒污染物。异种 DNA 在人体内几乎无排异现象，所以，也可以被用作生物医用材料。此外，基于碱基互补配对原则，DNA 具有较强的自组装能力。利用 DNA 两条单链的自组装能力，通过灵活设计 DNA 碱基序列，可自组装为在纳米尺度上高度有序的纳米材料。

5.4　木　质　素

1838 年法国农学家 Payen 使用硝酸和碱处理木材，并用乙醇和乙醚洗涤时得到不溶性的纤维状的纤维素。被硝酸溶解的部分，含碳量更高，被称为"真正的木质材料"（the ture woody materials）。1857 年，Schulze 分离出了这种可溶性组分，并称之为"lignin"，中文称作"木质素"（图 5-19）。

木质素（lignin）是一类由苯丙烷单元通过醚键和碳-碳键连接的非晶态高聚物，广泛分布于木质化植物细胞壁中。它与纤维素、半纤维素通过共价键连接在一起，在细胞壁内起保护细胞形状的作用，是植物细胞内力学结构材料。木质素是植物界中仅次于纤维素的第二丰富的天然高分子，全球随植物生长而来的木质素年产量估算为 1500 亿吨。木质素在木材中的含量为 20%～40%，禾本科植物中木质素的含量一般比木材中含量低，约为 15%～25%。

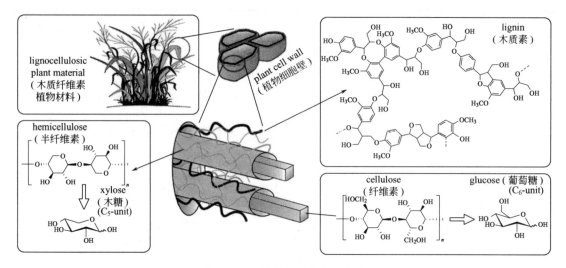

图 5-19 木质素分子结构及分布

木质素中出现频率较高的结构单元有三个,分别是紫丁香基丙烷结构单元,愈创木基丙烷结构单元和对羟基苯丙烷结构单元。它们的含量因植物种类而异(表 5-4)。在木质素的生物合成过程中,这些结构单元,经过酶解过程脱氢,产生苯氧基团,再经随机偶联反应,生成三维高分子交联结构。

表 5-4 木质素含量及三种主要单体的分子结构

单元结构	木质素含量/%	主要结构单元/%		
		p-hydroxyphenyl (对羟基苯丙烷)	guaiacyl (愈创木基丙烷)	styringyl (紫丁香基丙烷)
软木类 (softwood)	27~40	—	90~95	5~10
硬木类 (hardwood)	20~25	—	50	50
禾本科 (grasses)	15~25	5	75	25

芳香酚结构单元主要通过醚键和碳碳共价键连接(图 5-20),因此,木质素分子结构中存在的官能团主要包括羟基、羰基、羧基、甲氧基等。木质素的物理性质和化学性质主要取决于酚羟基的含量,当酚羟基的含量升高时,其溶解性能增大,化学反应性也进一步增强。木质素的物理性质不仅取决于来源,也取决于分离提取的方法,因而具有多变性和复杂性。

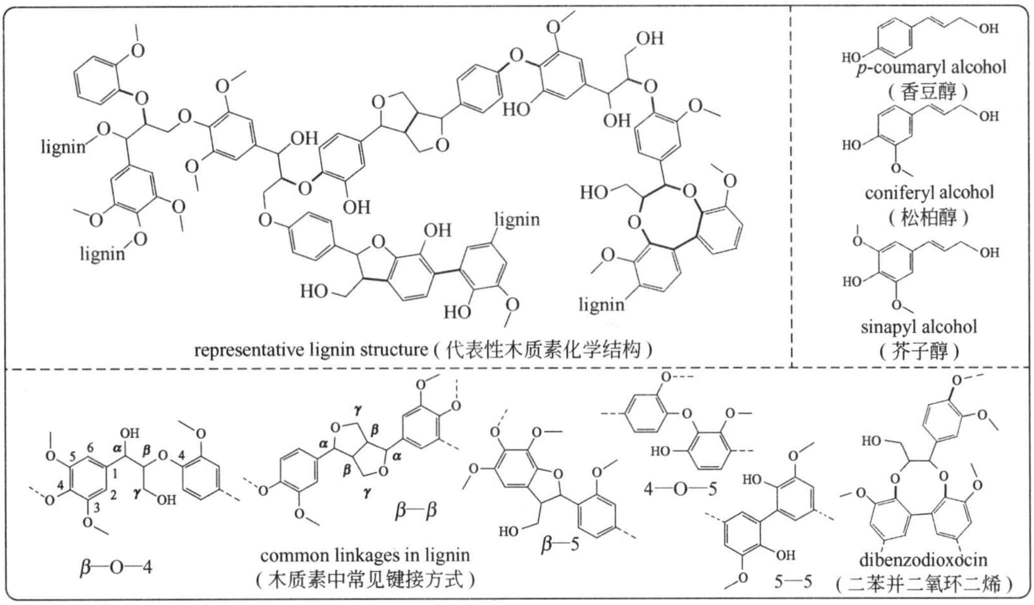

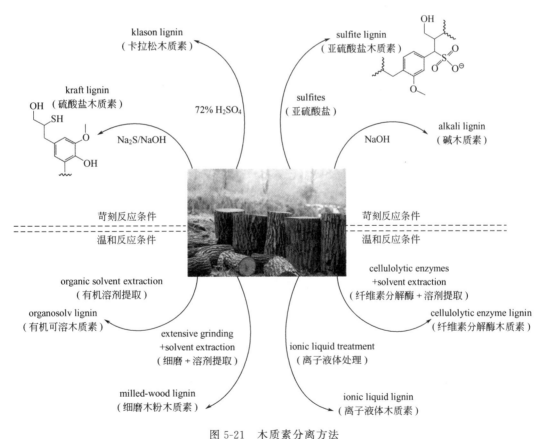

图 5-20　木质素分子结构中的主要连接方式

图 5-21　木质素分离方法

一般而言，木质素的玻璃化温度为127~193℃，具有良好的热稳定性，235℃开始失重，300℃仅失重2%，除木质素磺酸盐外，大部分木质素不溶于水。

从植物体中分离木质素有两种方法，一种是将木质素作为不溶性部分过滤分离，另一种是直接溶解木质素分离（图5-21）。例如，用65%~72%的硫酸或42%的盐酸处理脱脂植物原料，使纤维素和半纤维素等多糖溶解，并以稀酸补充水解，保留下来的残渣即为酸木质素。可溶性木质素一般采用无机试剂、酸性有机试剂及中性溶剂等将木质素可溶化处理后，与多糖等其他物质分离。例如，使用亚硫酸钙、镁、钠或铵的酸性亚硫酸盐溶液，将木质素磺化，变为水溶性的木质素磺酸盐而溶出，或者使用碱分离得到碱木质素；采用甲醇、乙醇、丁醇等并加入少量的无机酸分离木质素；也可以采用甲醇、二氧六环、乙醇和丙酮等中性溶剂分离木质素。因大部分木质素来源于造纸工业，木质素也可按制浆造纸的方法分类，包括磺化木质素、碱木质素等。

木质素经化学改性后可应用于多个工业领域。例如：木质素磺酸镁可作为混凝土减水剂广泛应用于建筑业；木质素经化学改性，添加到天然橡胶、溴化丁基橡胶和丁苯橡胶等橡胶中，可增加轮胎中橡胶与帘线黏合性，提高外胎耐磨性能；木质素采用酚醛树脂和聚乙烯改性后，可用于胶合板、玻璃纤维板、刨花板和无机绝热材料的生产与加工，木质素酚醛树脂、木质素脲醛等树脂已完全实现工业化生产；此外，木质素在特殊催化剂下热解，可制备多种芳香族绿色平台化合物，能够用于进一步合成高性能树脂（图5-22）。

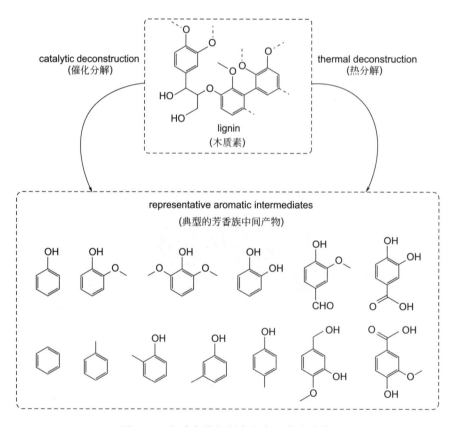

图5-22 木质素热解制备绿色平台化合物

5.5 天然橡胶

植物分泌树脂是指某些植物渗出的无定形半固体或固体有机物质。从植物分泌树脂中可提炼出有实用价值的天然高分子,如天然橡胶、生漆、虫胶等。以下简单讲述天然橡胶的加工与应用。

普通的天然橡胶(natural rubber,NR)是一种以聚异戊二烯为主要成分的天然高分子化合物。天然橡胶中聚异戊二烯含 97% 以上的顺-1,4-加成结构,含 1% 的反-1,4-加成结构和约 2% 头-尾连接的 3,4-加成结构单元。天然橡胶中的聚异戊二烯呈线形结构,无定形,玻璃化温度为 $-73℃$,分子量约 20 万~50 万。天然橡胶的分子链上有醛基,每条大分子链上平均有一个,正是醛基发生缩合或与蛋白质分解产物发生反应形成的支化、交联,使得橡胶在贮存中黏度增加。天然橡胶的大分子末端一般为二甲基烯丙基,另一端为焦磷酸酯基。

常用的天然橡胶取自三叶橡胶树,橡胶树分布在亚洲、非洲、大洋洲、拉丁美洲的 40 多个国家和地区。种植面积较大的国家有:印度尼西亚、泰国、马来西亚、中国、印度、越南、尼日利亚、巴西、斯里兰卡、利比里亚等。我国橡胶树园区主要分布于海南、广东、广西、福建、云南,台湾也有种植,其中海南为主要种植区。橡胶树所产生的胶乳含 60% 左右的水分,35% 聚异戊二烯,5% 蛋白质、类脂类有机物、无机盐等。橡胶园的生产过程包括,割胶(割开树皮,牛奶状胶乳不断流出)、收胶(收集容器内的鲜胶乳)、胶乳净化(通过离心分离等方法除去杂质)、胶乳凝固(加酸使胶乳凝固)、压薄、压绉(机械滚压脱除部分水分)、造粒(经锤磨机造粒)、干燥(烘干、脱除水分和挥发性物质)、压实包装(加压制成紧密橡胶块)。天然橡胶经过塑炼,在剪切力和氧的双重作用下降解,分子量降低。通过硫化交联后制成橡胶制品(图 5-23)。

橡胶树割胶

天然橡胶原料压制块

聚异戊二烯结构式

图 5-23 天然橡胶来源及化学结构

天然橡胶在常温下有较高弹性,略有塑性,低温时结晶硬化,有较好的耐碱性,但不耐强酸,不溶于水、低级酮和醇类,但在非极性溶剂如三氯甲烷、四氯化碳等中能溶胀。天然橡胶中含有较多不饱和双键，具有烯类化合物的反应特性。在天然橡胶的各类化学反应中，最重要的是氧化裂解反应和硫化反应。前者是生胶塑炼加工以及橡胶老化的原因，后者则是生胶进行硫化加工的理论依据。天然橡胶的氯化、环化、氢化等反应，可应用于天然橡胶的改性方面。天然橡胶由于分子量大、分子量分布宽、分子链易于断裂，因此很容易进行塑炼、混炼、压延、压出、成型等。

天然橡胶具有优良的回弹性、绝缘性、隔水性及可塑性等，经过适当处理后还具有耐油、耐酸、耐碱、耐热、耐寒、耐压、耐磨等特性，适合制备生产多种产品。例如，雨鞋、暖水袋、轮胎、工业用传送带、排灌胶管、密封垫片、防震设备等产品都采用天然橡胶原料。

英语读译资料

(1) Polysaccharides

Polysaccharides are complex carbohydrates in which tens, hundreds, or even thousands of simple sugars are linked together through glycoside bonds. Because they have only the one free anomeric —OH group at the end of a very long chain, polysaccharides are not reducing sugars and don't show noticeable mutarotation. Cellulose and starch are the two most widely occurring polysaccharides.

Cellulose consists of several thousand D-glucose units linked by 1,4-β-glycoside bonds like those in cellobiose. Different cellulose molecules then interact to form a large aggregate structure held together by hydrogen bonds.

Nature uses cellulose primarily as a structural material to impart strength and rigidity to plants. Leaves, grasses, and cotton, for instance, are primarily cellulose. Cellulose also serves as raw material for the manufacture of cellulose acetate, known commercially as acetate rayon, and cellulose nitrate, known as guncotton. Guncotton is the major ingredient in smokeless powder, the explosive propellant used in artillery shells and ammunition for firearms.

Potatoes, corn, and cereal grains contain large amounts of starch, a polymer of glucose in which the monosaccharide units are linked by 1,4-α-glycoside bonds like those in maltose. Starch can be separated into two fractions: amylose, which is insoluble in cold water, and amylopectin, which is soluble in cold water. Amylose accounts for about 20% by weight of starch and consists of several hundred glucose molecules linked together by 1,4-α-glycoside bonds.

Amylopectin accounts for the remaining 80% of starch and is more complex in structure than amylose. Unlike cellulose and amylose, which are linear polymers, amylopectin contains 1,6-α-glycoside branches approximately every 25 glucose units.

(2) Silk

Silk is a protein fiber that is woven into fiber from which textiles are made, including clothing and high-end rugs. It is obtained from the cocoon of silkworm larvae. While most silk is harvested from commercially grown silkworms, some is still obtained from less well-established sources. Silk was first developed as early as 6000BC. The Chinese Empress Xi Ling-Shi developed the process of retrieving the silk filaments by floating the cocoons on warm water. This process and the silkworm itself were monopolized by China until about 550AD when two missionaries smuggled silkworm eggs and mulberry seeds from China to Constantinople (Istanbul). First reserved to use by the Emperors of China, its use eventually spread to the Middle East, Europe, and North America, but now its use is worldwide. The history of silk and the silk trade is interesting.

The early work focused on a particular silkworm, Bombyx mori, which lives on mulberry bushes. There are other silkworms each with its on special properties, but in general, most silk is still derived from the original strain of silkworm. Crystalline silk fiber is about four times stronger than steel on a weight basis.

In the silk fibroin structure, almost every other residue is glycine with either alanine or serine between them, allowing the sheets to fit closely together. While most of the fibroin exists as beta sheets, regions that contain more bulky amino acid residues interrupt the ordered beta structure. Such disordered regions allow some elongation of the silk. Thus, in the crystalline segments of silk fibroin, there exists directional segregation using three types of bonding: covalent bonding in the first dimension, hydrogen bonding in the second dimension, and hydrophobic bonding in the third dimension. The polypeptide chains are virtually fully extended; there is a little puckering to allow for optimum hydrogen bonding. Thus, the structure is not extensible in the direction of the polypeptide chains. By comparison, the less specific hydrophobic forces between the sheets produce considerable flexibility. The crystalline regions in the polymers are interspersed with amorphous regions in which glycine and alanine are replaced by amino acids with bulkier pendent groups that prevent the ordered arrangements to occur. Furthermore, different silk worm species spin silks with differing amino acid compositions and thus, with differing degrees of crystallinity.

(3) Paper

Paper comes in many forms with many uses. The book you are reading is made from paper, we have paper plates, paper napkins, newspapers and magazines, cardboard boxes, in fact the amount of paper items is probably more than twice, by weight, that of all the synthetic polymers combined. About 30% paper is writing and printing paper. The rest is mainly used for tissues, toweling, and packaging. If you rip a piece of ordinary paper, not your book page please, you will see that it consists of small fibers. Most of these cellulosic fibers are randomly oriented, but a small percentage are preferentially oriented in one direction because the paper is made from a cellulose-derived watery slurry with the water largely removed through use of heated rollers that somewhat orient the fibers.

Modern paper is made from wood pulp, largely cellulose, which is obtained by the removal of lignin from debarked wood chips by use of chemical treatments with sodium hydroxide, sodium sulfite, or sodium sulfate. Newsprint and paperboard, which is thicker than paper, often contains a greater amount of residual lignin.

Wood is almost entirely composed of cellulose and lignin. In the simplest paper making scheme, the wood is chopped, actually torn, into smaller fibrous particles as it is pressed against a rapidly moving pulp stone. A stream of water washes the fibers away dissolving much of the water-soluble lignin. The insoluble cellulosic fibers are concentrated into a paste called pulp. The pulp is layered into thin sheets and rollers are used to both squeeze out much of the water and to assist in achieving paper of uniform thickness. This paper is not very white. It is also not very strong. The remaining lignin is somewhat acidic (lignin contains acidic phenolic groups that hydrolyze to give a weakly acidic aqueous solution) that causes the hydrolytic breakdown of the cellulose. Most of the newsprint is of this type or it is regenerated, reused paper.

Pulping processes are designed to remove the nonsaccharide lignin portion of wood which constitutes about 25% of the dry weight. The remaining is mostly cellulose with about 25% hemicellulose (noncellulose cell wall polysaccharides that are easily extracted by dilute aqueous base solutions). Pulping procedures can be generally classified as semichemical, chemical, and semimechanical. In semimechanical pulping, the wood is treated with water or sulfate, bisulfite, or bicarbonate solution that softens the lignin. The pulp is then ground or shredded to remove much of the lignin giving purified or enriched cellulose content. The semichemical process is similar but digestion times are longer and digesting solutions more concentrated giving a product with less lignin but the overall yield of cellulose-intense material is lowered by 70%~80%. Further, some degradation of the cellulose occurs.

Most paper is produced by the chemical process where chemicals are employed to solubilize and remove most of the lignin. While overall yields are lower than the other two main processes, the product gives good quality writing and printing paper. Three main chemical processes are used. In the soda process extracting solutions containing about 25% sodium hydroxide and 2.4% sodium carbonate are used. In the sulfite process the extracting solution contains a mixture of calcium dihydrogen sulfite and sulfur dioxide. The sulfide process utilizes sodium hydroxide, sodium monosulfide, and sodium carbonate in the extracting solution.

After the chemical treatment, the pulped wood is removed, washed, and screened. Unbleached, brown-colored paper is made directly for this material. Most whiten or bleached paper is made from treatment of the pulp with chlorine, chlorine dioxide, hypochlorite, and/or alkaline extraction. In general, sulfate pulped paper is darker and requires more bleaching and alkaline extraction to give a "white" pulp.

The sulfide process, also called the kraft process (the term "kraft" comes from the Swedish word for strong since stronger paper is produced), is more commonly used. The kraft process is favored over the sulfite treatment of the paper because of environmental considerations. The sulfite process employs more chemicals that must be disposed of particularly mercaptans, RSHs, which are quiteodorous. Research continues on reclaiming and recycling pulping chemicals.

If pure cellulose was solely used to make paper, the fiber mat would be somewhat water soluble with only particle surface polar groups and internal hydrogen bonding acting to hold the fibers together. White pigments such as clay and titanium dioxide are added to help "cement" the fibers together and to fill voids producing a firm, white writing surface. This often occurs as part of an overall coating process.

Most paper is coated to provide added strength and smoothness. The coating is basically an inexpensive paint that contains a pigment and a small amount of polymeric binder. Unlike most painted surfaces, most paper products are manufactured with a short lifetime in mind with moderate performance requirements. Typical pigments are inexpensive low-refractive index materials such as plate-like clay and ground natural calcium carbonate. Titanium dioxide is used only when high opacity is required. The binder may be a starch or latex or a combination of these. The latexes are usually copolymers of styrene, butadiene, acrylic, and vinyl acetate. Other additives and coloring agents may also be added for special performance papers. Resins in the form of surface coating agents and other special surface treatments (such as coating with polypropylene and polyethylene) are used for paper products intended for special uses such as milk cartons, ice cream cartons, light building materials, and drinking cups. The cellulose supplies the majority of the weight (typically about 90%) and strength with the special additives and coatings, providing special properties needed for the intended use.

(4) The xanthate viscose process for the production of rayon

Cellulose is sometimes used in its original or native form as fibers for textile and paper, but often it is modified through dissolving and reprecipitation or through chemical reaction. The xanthate viscose process, which is used for the production of rayon and cellophane, is the most widely used regeneration process. The cellulose obtained by the removal of lignin from wood pulp is converted to alkali cellulose. The addition of carbon disulfide to the latter produces cellulose xanthate.

While terminal hydroxyl and aldehyde groups, such as present in cellobiose, are also present in cellulose, they are not significant because they are only present at the ends of very long chains. The hydroxyl groups are not equivalent. For instance, the pK_a values of the two ring hydroxyl groups are about 10 and 12, which is about the same as the hydroxyl groups on hydroquinone and the first value about the same as the hydroxyl on phenol. The pK_a value of the nonring or methylene hydroxyl group is about 14, same as found for typical aliphatic hydroxyl groups.

In the cellulose-regenerating process, sodium hydroxide is initially added such that approximately one hydrogen, believed to be predominately a mixture of the hydroxyl groups on carbons 2 and 3, is replaced by the sodium ion. This is followed by treatment with carbon disulfide forming cellulose xanthate, which is eventually rechanged back again, regenerated, to cellulose.

The orange-colored xanthate solution, or viscose, is allowed to age and is then extruded as a filament through holes in a spinneret. The filament is converted to cellulose when it is immersed in a solution of sodium bisulfite, zinc(II) sulfate, and dilute sulfuric acid. The tenacity, or tensile strength, of this regenerated cellulose is increased by a stretching process that reorients the molecules so that the amorphous polymer becomes more crystalline. Cellophane is produced by passing the viscose solution through a slit die into an acid bath. Since an average of only one hydroxyl group in each repeating glucose unit in cellulose reacts with carbon disulfide, the xanthate product is said to have a degree of substitution (DS) of 1 out of a potential DS of 3.

习 题

1. 从分子结构和主要性能上分别分析纤维素和淀粉，直链淀粉、支链淀粉和糖原，纤维二糖和麦芽糖，$α$-D-葡萄糖和$β$-D-葡萄糖的差别。

2. 简述纤维素的再生方法，比较它们的优缺点。

3. 半纤维素与纤维素结构有何区别？写出木糖的结构式。

4. 壳聚糖和藻酸的结构特征和用途与纤维素有何关系和区别？

5. 写出构成蛋白质的氨基酸通式，对带烷基、苯环、羟基、羧基、氨基、含硫的氨基酸各举一例。

6. 有哪些天然蛋白质属于纤维蛋白？具有哪些结构特征？

7. 蛋白质二级结构中的 $α$-螺旋和 $β$-折叠有何区别？氨基酸残基的结构对二级结构有何影响？

8. 说明毛和丝织品洗涤后起皱和熨烫后平整的原因，提出免烫丝织品处理途径的设想。

9. 简述核糖、核苷、核苷酸、核酸的相互关系，嘧啶碱和嘌呤碱、核糖和脱氧核糖在结构上有何区别？

10. 在生产黏胶纤维的过程中，如果发现纺丝黏胶液黏度过大，需要如何解决？

11. 论述改善黏胶纤维生产中废水排放和有害气体排放的途径。

12. 分析再生海藻酸纺丝的可能性，提出海藻酸纺丝的具体方案，分析纺丝生产过程对环境产生的影响。

13. 描述利用双螺旋 DNA 的基于碱基互补配对原则的高度自组装能力制备高度有序的纳米材料的方法。

英语习题

1. Why is starch digestible by humans? Why is cellulose not digestible by humans?

2. How does cellobiose differ from maltose?

3. Why is cellulose stronger than amylose?

4. How does the monosaccharide hydrolytic product of celllulose differ from the hydrolytic product of starch?

5. What is paper made from?

6. How many hydroxyl groups are present on each anhydroglucose unit in cellulose?

7. Which would be more polar——tertiary or secondary cellulose acetate?

8. Why would you expect chitin to be soluble in hydrochloric acid?

9. Which is more apt to form a helix: amylose or amylopectin?

10. Why is amylopectin soluble in water?

11. How do the configurations differ for gutta percha and NR?

12. The formation of what polymer is responsible for tanning?

13. Will the tensile force required to stretch rubber increase or decrease as the temperature is increased?

14. Does a stretched rubber band expand or contract when heated?

15. List three requirements for an elastomer.

16. Why is there an interest in the cultivation of guayule?

17. Are the polymerization processes for synthesis and natural *cis*-polyisoprene similar or different?

18. What does the presence of C_5H_8 units in NR indicate?

19. Why does a rubber band become opaque when stretched?

20. What is the most important contribution to retractile forces in highly elongated rubber?

21. What is present in so-called vulcanized rubber compounds?

22. When a rubber band is stretched, what happens to its temperature?

23. Why is lignin sometimes referred to as being a two-dimensional polymer?

24. If the annual production of paper is more than 100 million tons, how much lignin is discarded from paper production annually?

25. What are some of the obstacles in using polymer-intensive plants as feed stocks for the preparation of fuels such as ethanol?

26. Since there are many plants that give a rubber-like latex, what are the impediments to their use as replacements for natural rubber?

27. If you were beginning an industrial research project aiming at obtaining useful products from lignin, what might be some of the first areas of research you might investigate?

第 6 章
聚合物的化学反应

聚合物的化学反应

近年来,芯片制造技术成为我国优先发展的工业领域。光刻胶是微电子技术中微细图形加工的关键材料之一。普通光刻胶在显影过程中存在衍射、反射和散射,降低了光刻胶图形的分辨率。在数纳米尺度下的光刻技术中,光刻胶对微电路的质量有较大影响。

本章目录

6.1 聚合物侧基反应 / 214
6.2 分子量增大较多的聚合物反应 / 218
6.3 降解与老化 / 221

重点要点

侧基反应,接枝反应,交联反应,扩链反应,降解反应,高分子材料老化,光降解,光氧化降解,光热降解,氧化降解。

聚合物的化学反应是提高聚合物性能，制备功能聚合物的一个重要途径。许多小分子的有机反应也可以应用于聚合物的侧基反应，但是聚合物的化学反应具有和有机小分子不同的特点。有机小分子在溶剂中有着较好的运动性，在分散介质中呈均匀分散状态，但是聚合物侧基因为连接在大分子链内，决定其渗透扩散性能的运动性必然受限，在微观的局部环境下也无法达到分散均匀。

高分子链特有的凝集态会对其侧基的化学反应有较大影响。由于反应试剂很难渗透进入高分子结晶区，处于结晶态的高分子链就难有反应机会。处于非晶态的高分子链相对更容易进行反应。同样，小分子反应试剂也不容易扩散至处于玻璃态的聚合物链段，在高弹态下的高分子或处于溶胀状态下的高分子链反应速率更快。溶解状态的高分子链常呈无规线团卷曲状，高分子链卷曲中心的基团浓度一般较高，而线团外部的浓度较低。当小分子试剂有利于线团内部富集时，反应速率增加。

一些聚合物所特有的化学因素也会影响聚合物的侧基反应。聚合物侧基成对反应时，残存在已转化侧基中间的未反应基团成为孤立基团，其反应受限。这就是聚合物化学反应的"概率效应"。例如，聚氯乙烯与锌粉共热脱氯成环，按概率计算，环化程度只能达到87%左右，有13%的氯原子被孤立隔离在两环之间，不能继续反应。

聚合物侧基的化学反应也常常呈现"邻近基团效应"。聚合物侧基或反应后新基团均可能会通过位阻效应、电子效应以及静电作用等影响邻近基团的反应活性。体积较大的侧基会因位阻效应使邻近基团的反应活性降低，基团转化程度受限。例如聚乙烯醇与三苯基乙酰氯

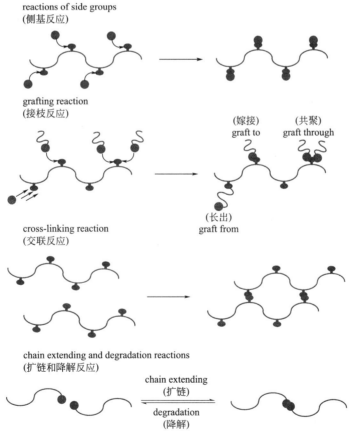

图 6-1 聚合物的化学反应类型

反应，乙酰化度达 50% 时，反应就会变得十分缓慢。也有一些条件下，大分子上基团活性受邻近基团影响而增强。例如，聚甲基丙烯酸甲酯在稀碱液中水解，因羧基阴离子易与相邻酯基形成六元环酐，再开环成羧基，表现出自动催化效应。

聚合物侧基反应属于高分子和小分子的化学反应，一般不会引起聚合物的深度变化。如图 6-1 所示，聚合物的化学反应还有使分子量增大的接枝、扩链、交联等反应，其中部分是高分子和高分子的反应。聚合物的降解反应会使聚合度变小。

6.1 聚合物侧基反应

6.1.1 高分子侧羟基反应

聚乙烯醇（PVA）是维尼纶纤维的原料，也可用作胶黏剂和分散剂。单体乙烯醇不存在，聚乙烯醇由聚乙酸乙烯酯制备。聚乙酸乙烯酯可由乙酸乙烯经自由基聚合而得。在碱性条件下，聚乙酸乙烯酯在甲醇溶剂中醇解成聚乙烯醇，醇解前后聚合度几乎不变。聚乙酸乙烯酯也可在碱性水溶液中经皂化反应生成聚乙烯醇。在醇解过程中，乙酸根转变成羟基的摩尔分数称作醇解度（DH）。纤维用聚乙烯醇的 DH>99%，悬浮聚合分散剂用聚乙烯醇的 DH 为 80%。当聚乙烯醇的 DH<50% 时，呈现油溶性。聚乙烯醇纤维生产需要的 DH>99%，经加热溶解于水，纺丝拉伸而得。聚乙烯醇纤维中的结晶区不溶于热水，但非晶区亲水溶胀，需与甲醛反应，通过形成缩醛交联，降低亲水性。因此，维尼纶纤维生产工艺由聚乙酸乙烯酯醇解、聚乙烯醇纺丝拉伸、缩醛交联三个主要工序构成。此外，聚乙烯醇与醛类试剂反应形成的缩甲醛或缩丁醛衍生聚合物是优良的玻璃胶黏剂，也可用作电绝缘膜和涂料（图 6-2）。

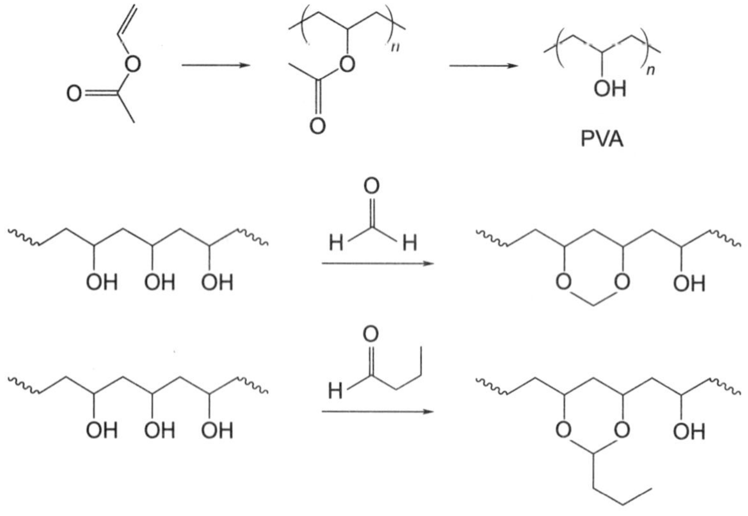

图 6-2 聚乙烯醇的主要化学反应

聚乙烯醇的羟基和肉桂酰氯反应生成肉桂酸酯侧基（图 6-3）。在光照条件下，肉桂酸酯的双键结构发生加成反应，形成交联，使聚合物固化。所以，该聚合物作为感光性高分子用于激光制版印刷行业。

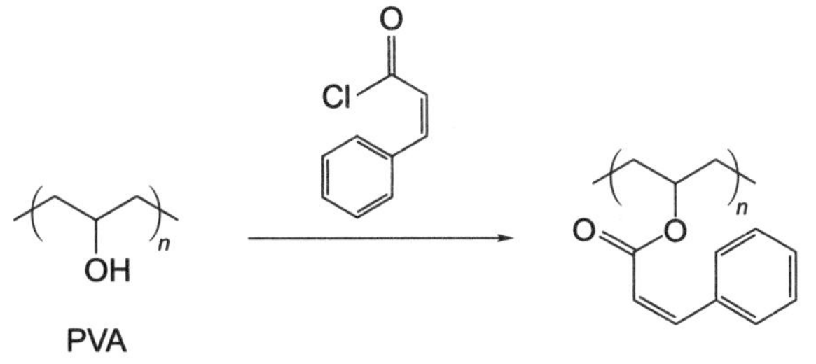

图 6-3　聚乙烯醇和肉桂酰氯反应

6.1.2　高分子侧氨基反应

聚丙烯腈水解可形成聚丙烯酰胺，聚丙烯酰胺经 Hofmann 反应可转化成聚乙烯胺（图 6-4）。聚乙烯胺的氨基具有较高的化学反应活性，可灵活设计并转化为酰胺、磺酰胺、脲等官能团（图 6-5）。高分子侧氨基也可转化成叠氮、异氰酸酯、硫代异氰酸酯等更高活性的官能团（图 6-6）。

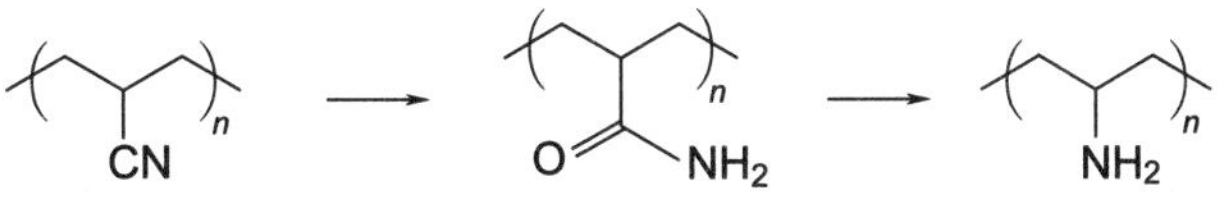

图 6-4　聚丙烯腈水解及其 Hofmann 反应

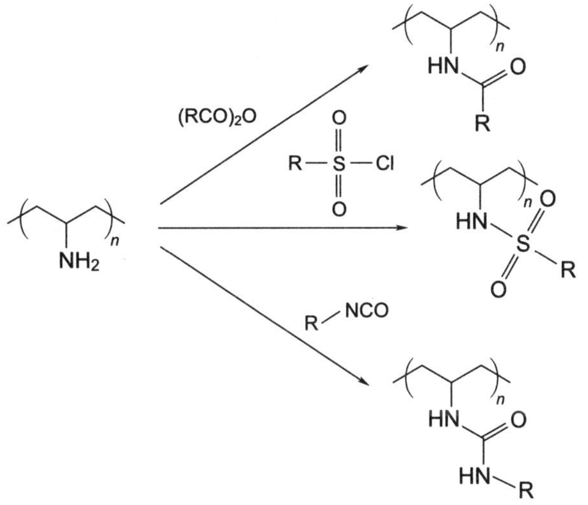

图 6-5　聚乙烯胺转化成酰胺、磺酰胺、脲等反应

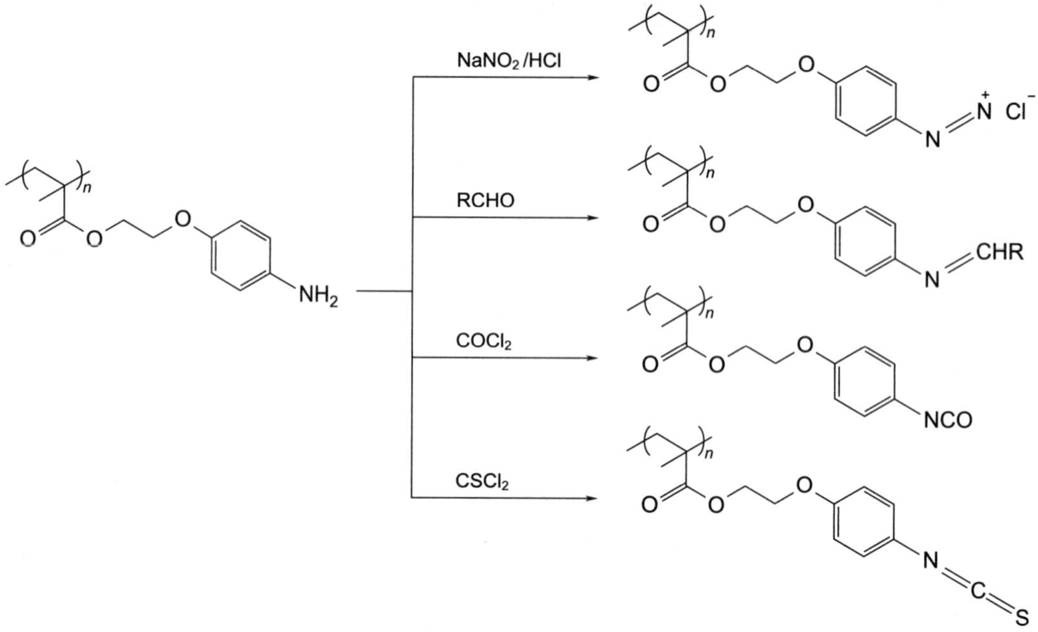

图 6-6　高分子侧氨基转化成叠氮、异氰酸酯、硫代异氰酸酯等的反应

6.1.3　高分子侧酰基反应

聚丙烯腈水解形成的聚丙烯酰胺进一步水解形成聚丙烯酸。聚丙烯酸钠盐也可由丙烯酸钠单体在水溶液中聚合。分子量数千的聚丙烯酸钠具有胶体保护作用，可与水中微量的钙、镁离子形成配合物，可有效避免碳酸钙的形成，用于锅炉水以及热交换用水的防垢剂。分子量数百万的聚丙烯酰胺较长的分子链可以将水中的微小颗粒架桥凝聚，形成絮凝沉降，是一种常用的水处理絮凝剂。

聚甲基丙烯酸和卤代烷烃类化合物在碱性条件下反应，可得较高转化率的聚甲基丙烯酸酯类。此外，聚甲基丙烯酸经脱水反应形成聚酸酐，再与水、醇、胺等试剂反应，也可形成聚甲基丙烯酸酯和聚甲基丙烯酰胺等高分子（图 6-7）。

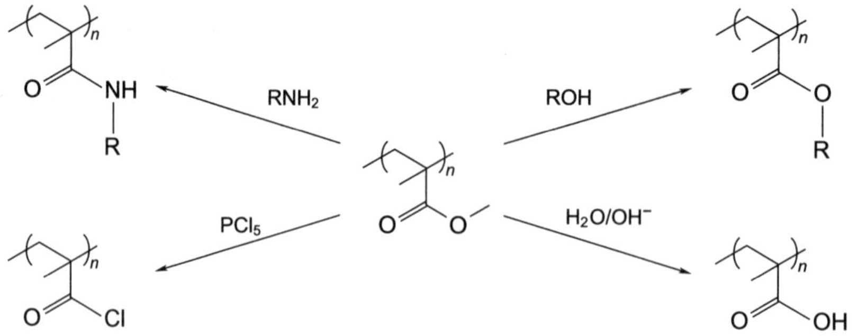

图 6-7　高分子侧酰基反应

聚甲基丙烯酸甲酯或聚丙烯酸甲酯的甲酯侧基可通过水解、醇解、胺解等反应引入多种功能性侧基。

6.1.4 芳香环侧基反应

聚苯乙烯中的苯环具有较高的反应活性，可通过亲电取代反应，如烷基化、氯化、磺化、氯甲基化、硝化等，引入多种官能团。例如，苯乙烯和二乙烯基苯自由基共聚得到的体形聚合物经磺化反应引入磺酸根基团，即成阳离子交换树脂。与氯代二甲基醚反应，则可引入氯甲基，进一步引入季铵基团，即成阴离子交换树脂（图 6-8）。

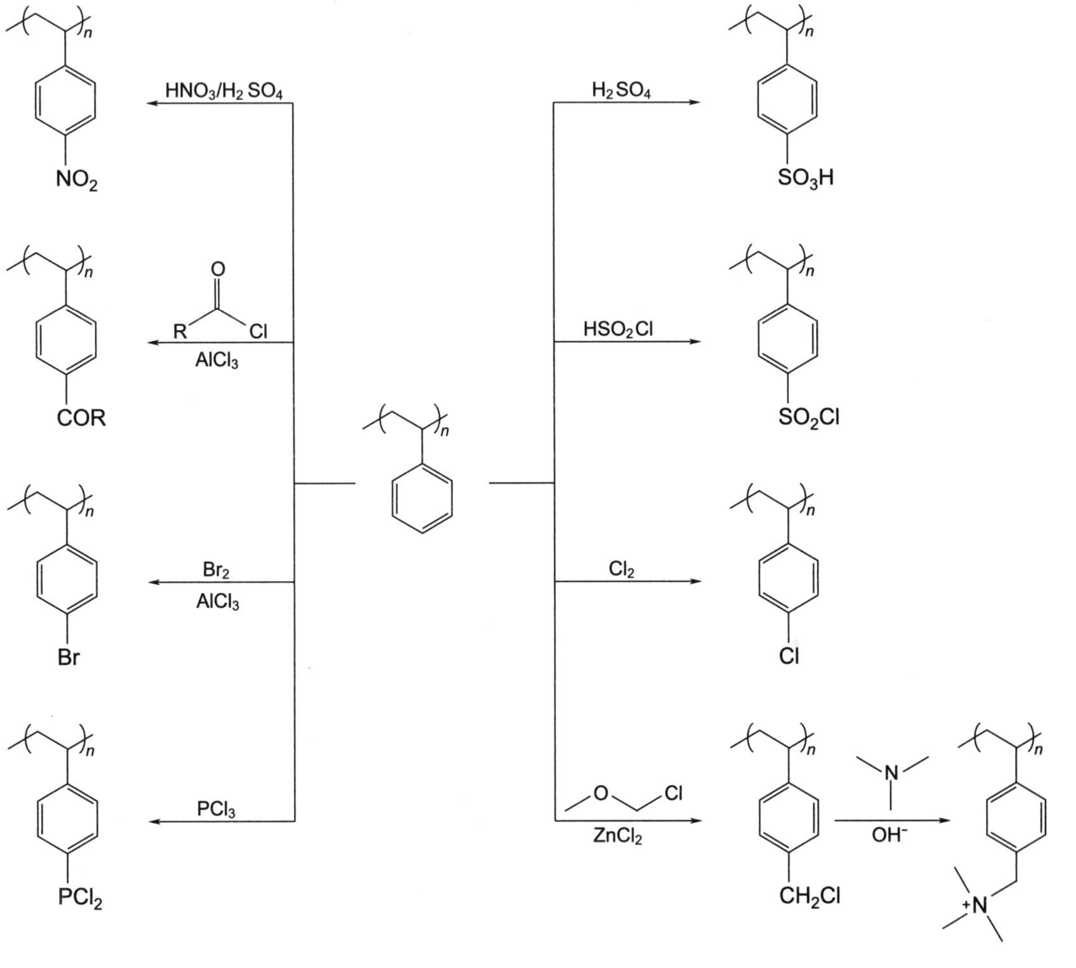

图 6-8 聚苯乙烯的化学反应

6.1.5 环化反应

聚丙烯腈经环化可制成碳纤维（图 6-9），其工艺含三个主要工序：200~300℃下预氧化，800~1900℃下炭化，2500℃下石墨化，析出其他所有元素，形成碳纤维。碳纤维是高强度、高模量、耐高温的石墨态纤维，与合成树脂复合成为高性能复合材料，是航空航天制造领域的重要原材料。

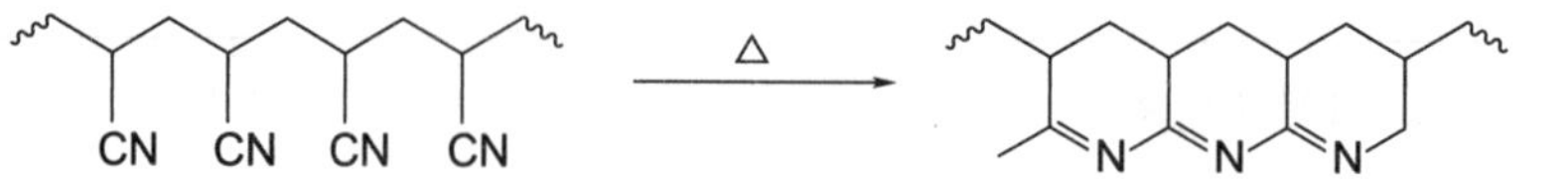

图 6-9 聚丙烯腈环化反应

6.1.6 氯化反应

聚乙烯、聚丙烯等聚烯烃的氯化反应如图 6-10 所示,通过取代氢原子引入氯原子,进而改善高分子材料的性能。聚乙烯耐酸碱,但易燃。通过氯化反应,引入部分氯侧基,形成氯化聚乙烯(CPE),可改善阻燃性能。

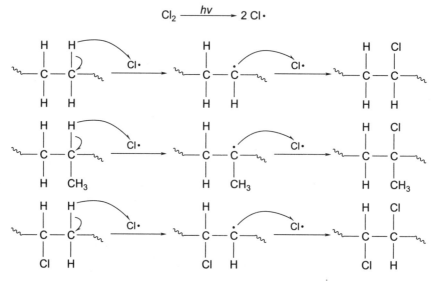

图 6-10 聚烯烃的氯化反应

聚丙烯中侧甲基的叔氢原子,更容易被氯原子取代。聚丙烯经氯化,可增加极性,提高黏结性能。氯化聚丙烯可用作聚丙烯的附着力促进剂。

聚氯乙烯可在水悬浮液中进一步氯化。当氯含量由 56.8% 提高到 65% 左右时,热变形温度由 80℃ 提高至近 120℃,溶解性能、耐候、耐腐蚀、阻燃等性能也相应改善。氯化聚氯乙烯可用于热水管、涂料、化工设备等方面。

6.2 分子量增大较多的聚合物反应

6.2.1 接枝聚合

通过接枝聚合,可将物性迥异的高分子链连接在一起,得到具有特殊性能的高分子材料。接枝共聚物的性能取决于主链和支链的结构和长度，以及支链数。依据接枝链段的接入反应方式，可分长出（graft from）接枝、嫁接（graft to）接枝、聚合物单体共聚（graft through）接枝三大类（图 6-11）。例如，淀粉的两个侧羟基可与 Ce^{4+} 首先形成配位体，再经分解，在淀粉主链上产生自由基活性点，然后接枝聚合丙烯酸钠盐等亲水性单体，可合成高吸水性树脂。该接枝聚合由于自由基键接在主链上，可削弱均聚副产物的形成，有效提高接枝效率。还有许多侧基反应可用来合成接枝共聚物，例如在聚苯乙烯的苯环上引入异丙基，氧化成氢过氧化物，再分解成自由基，而后引发单体聚合，长出支链，形成接枝共聚物。

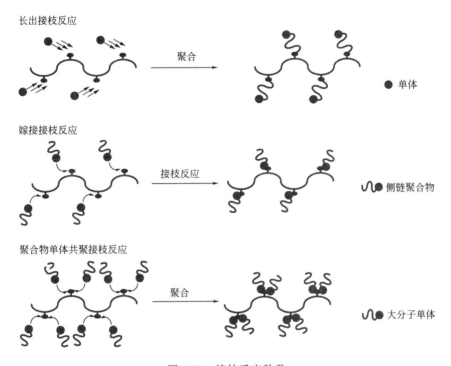

图 6-11 接枝反应种类

嫁接法接枝聚合需要预先合成主链和支链高分子，再通过主链中活性基团和支链活性端基的反应，将支链嫁接到主链上。例如，预先分别制备含氯甲基侧基的聚苯乙烯主链和具有羧基端基的聚苯乙烯支链，两者缩合，可得接枝共聚物。

高分子链端基为可聚不饱和双键结构的低聚物单体，与普通乙烯基单体共聚，就可形成接枝共聚物。其中，聚合物单体的长侧基成为支链，而乙烯基单体就成为主链。这一方法可避免链转移法的效率低和混有均聚物的缺点。

ABS 是一种通用热塑性工程塑料，由丁二烯（B）、丙烯腈（A）、苯乙烯（S）接枝共聚合成，材料性能可由三种组分的配比及物理形态调控，主要用于汽车和电子电器制造。ABS 结合了聚丙烯腈耐热、耐候，聚丁二烯抗冲击强度高，聚苯乙烯刚性较好的优点，具有刚性好、耐热耐低温、抗冲击强度高、易于加工、产品尺寸稳定和表面光泽好等优点。ABS 的接枝聚合工艺较成熟，这里简单介绍乳液聚合接枝法和本体聚合接枝法（图 6-12）。

乳液聚合接枝法是接枝聚合生产 ABS 最早的方法。首先乳液聚合丁二烯得聚丁二烯乳胶，然后在此乳胶中继续通过乳液聚合接枝共聚苯乙烯和丙烯腈单体，经凝聚、脱水、干燥得到 ABS 成品。本体聚合接枝法先经乳液聚合制备聚丁二烯乳胶，继续在乳液状态下接枝

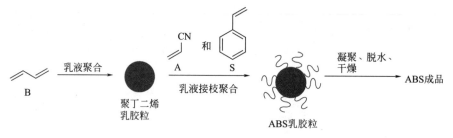

图 6-12　ABS 乳液聚合接枝法

少量苯乙烯和丙烯腈单体凝聚脱水后，用苯乙烯和丙烯腈单体溶解，再进行本体接枝聚合，造粒得 ABS 成品（图 6-13）。

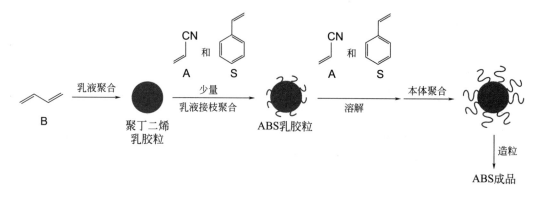

图 6-13　ABS 本体聚合接枝法

6.2.2　扩链

分子链端基为高反应活性官能团的预聚物，经化学反应使预聚物端基反应进行键接，分子量成倍增加，这一合成方法称为扩链。扩链聚合中，带有端基的预聚物称为遥爪预聚物，一般是分子量不超过一万的低聚物。例如，遥爪聚丁二烯（$M_w = 3000 \sim 6000$），室温呈液态，称液体橡胶，在成型过程中经端基反应，扩链成高聚物（图 6-14）。

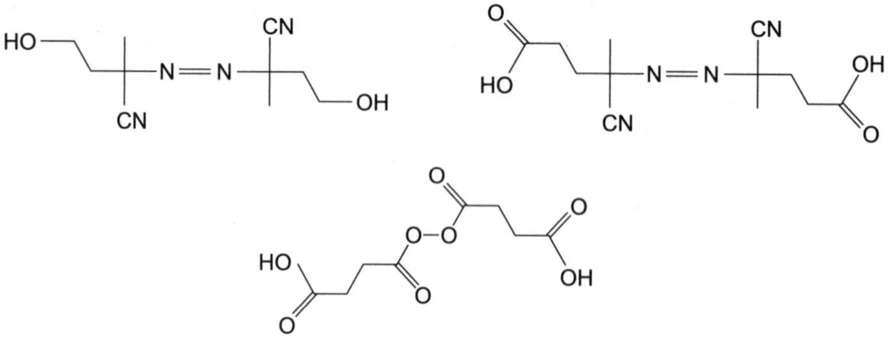

图 6-14　带官能团端基的自由基聚合引发剂

遥爪预聚物可灵活设计合成。例如，用带官能团端基的偶氮或过氧化类引发剂，引发丁二烯、异戊二烯、苯乙烯、丙烯腈等聚合，经偶合终止，即成带官能团端基的预聚物。

6.2.3 交联

通过在高分子链之间形成共价键交联，可提高高分子材料的使用性能。例如，天然橡胶等生胶通过硫化交联后才会表现出弹性，塑料经适当交联反应可提高强度和耐热性，棉、丝织物使用甲醛交联后可改善防褶皱性能。

以橡胶硫化交联为例，聚丁二烯等高分子主链上含双键结构，可与单质硫反应形成硫碳共价键。单质硫的化学结构为八个硫原子构成的八元环，可在适当条件下开环和双键反应形成锍离子（sulfonium）。锍离子夺取聚二烯烃中的氢原子，形成烯丙基碳阳离子。碳阳离子先与单质硫反应，再与大分子双键加成，产生交联，继续与大分子反应，再生出大分子碳阳离子。如此反复，形成大网络结构（图 6-15）。

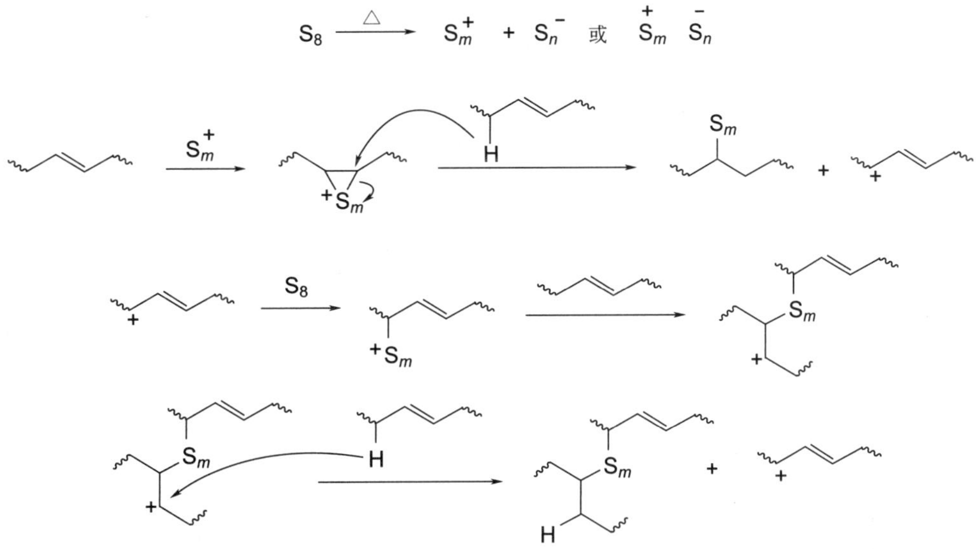

图 6-15 橡胶硫化反应

单质硫的硫化速率慢，硫的利用率低。橡胶工业中需要添加硫化促进剂。例如，苯并噻唑二硫化物可与硫结合，形成多硫化物，进一步与碳碳双键反应形成交联。形成的多硫交联逐步脱硫变短，直至单硫原子交联，从而提高了硫的利用效率。大多数情况下，硫化反应还需要添加氧化锌和硬脂酸等活化剂，以提高硫化速度和效率。

在体形缩聚中也涉及交联反应。例如，酚醛树脂模塑粉受热，交联成热固性制品；环氧树脂用二元胺交联固化；皮革用甲醛鞣制则是蛋白质氨基酸的交联过程；蚕丝（聚酰胺）用甲醛交联处理，可获得免熨防皱的效果。

6.3 降解与老化

聚合物的降解泛指能将聚合度变小的化学反应。高分子材料在使用过程中，受环境影响，如热、力、超声波、光和辐射等物理因素，氧、水、微生物等化学因素，都可以引发降解。高分子材料的降解又往往是这些环境因素的综合作用，例如，热/氧、光/氧、水解/生物降解等都是常见的降解复合模式。

对高分子材料而言，高分子链的解聚会使材料性能劣化，如变软发黏、变脆发硬、机械强度降低等。高分子材料性能随时间劣化的现象，称为高分子材料老化，发生在高分子材料设计、合成及使用中。

6.3.1 热降解

聚合物的热稳定性主要与键能有关，也受邻近基团的电子效应和位阻效应影响。例如，C—C 键的热稳定性次序如下：

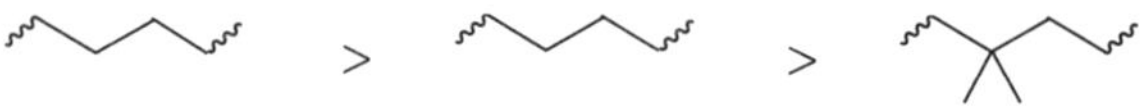

同理，C—H 键的热稳定性次序为：

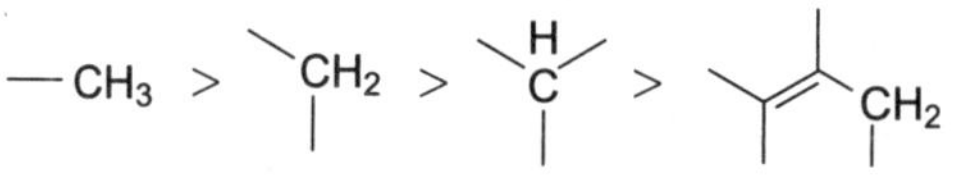

苯环等芳香共轭体系，比烷烃中相应的键都要稳定。热分解半衰期温度（T_h）可用来说明聚合物热稳定性与结构的关系。聚四氟乙烯中碳氟键能特别高，T_h 达 509℃，说明热稳定性好。甲基、亚甲基、苯环也比较热稳定，例如聚乙烯（$T_h=404℃$）、聚苯乙烯（$T_h=364℃$）。因位阻效应，聚异丁烯（$T_h=348℃$）、聚甲基丙烯酸甲酯（$T_h=327℃$）和 α-甲基苯乙烯（$T_h=286℃$）的热稳定性变差。由于 C—Cl 的键能较小，聚氯乙烯（$T_h=260℃$）容易热解。

聚合物热降解有解聚、无规断链、侧基脱除三种类型。解聚是聚合（链增长）的逆反应，一般是受热形成端基自由基，然后按连锁机理，"拉链"式地脱除单体。

例如，聚甲基丙烯酸甲酯的解聚从链端开始，分子量逐渐降低，在 270℃ 全部解聚成单体。甲基丙烯酸甲酯聚合时多以歧化方式终止，数目较多的大分子为碳碳双键端基，容易在烯丙基碳碳单键处断裂，产生两个自由基：三级碳自由基和稳定的烯丙基自由基。后者夺取三级碳自由基上甲基中的氢，转化成单体，而失氢的三级碳自由基继续解聚。该反应反复进行，形成拉链式解聚。

聚乙烯热解时，无规断链为主要方式。聚乙烯断链后形成的自由基活性高，经分子内"回咬"转移而断链，形成低分子物。聚乙烯的热解气态产物以丙烯为主，其他有甲烷、乙烷、丙烷，以及一些饱和烃和不饱和烃，乙烯量较少。聚苯乙烯在 350℃ 热解，无规断链和解聚同时进行，产生约 40% 单体，少量甲苯、乙苯、甲基苯乙烯，以及二、三、四聚体。聚氯乙烯、聚氟乙烯、聚乙酸乙烯酯、聚丙烯腈等受热时，在温度相对不高时，主要以脱除侧基为主。聚氯乙烯接近 120℃ 时开始脱氯化氢，颜色变黄。200℃ 下脱氯化氢更快，形成共轭双键结构生色基团，聚合物颜色变深，强度变差。聚氯乙烯受热脱氯化氢属于自由基机理，大致分三步反应。

① 聚氯乙烯分子链中含烯丙基氯等薄弱结构，易分解产生氯自由基。

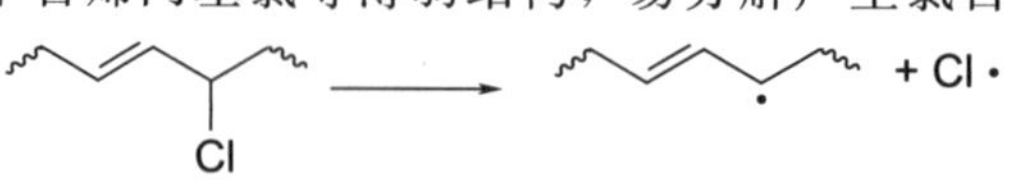

② 氯自由基夺取主链氢原子，形成氯化氢和链自由基。

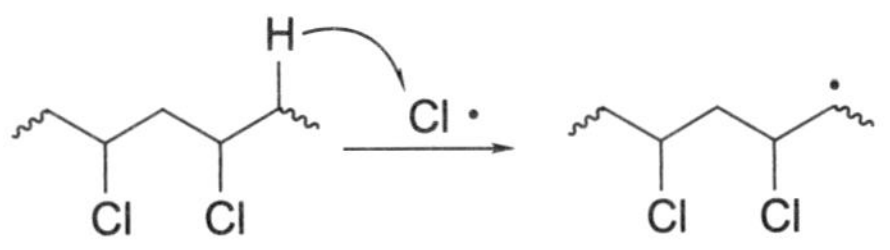

③ 聚氯乙烯链自由基消除氯自由基，形成双键或烯丙基。

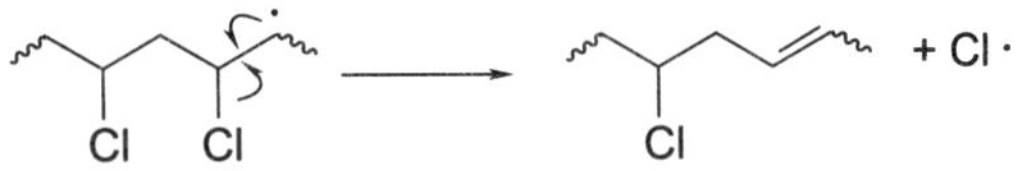

双键的形成会使邻近单元活化，其中的烯丙基氢更易被新生的氯自由基夺取，于是②和③两步反应反复进行，形成"拉链式"连锁脱氯化氢反应。

杂链聚合物主链中碳-杂原子键往往是弱键，容易断裂。例如，聚甲醛热解时，易从羟端基开始，"拉链"式解聚成甲醛（图6-16）。

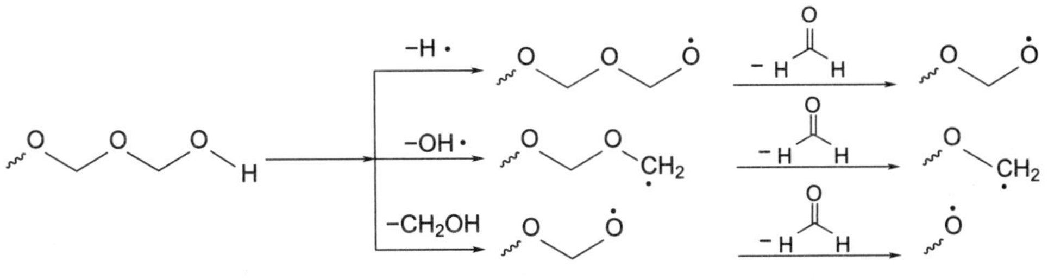

图 6-16 聚甲醛热解机理

涤纶聚酯在干热300℃以上时，经一个六元环中间体状态，裂解为酸和苯甲酸乙烯酯，继续反应并裂解，逐渐释放出乙醛、乙炔等小分子物质。推测反应机理如图6-17所示。

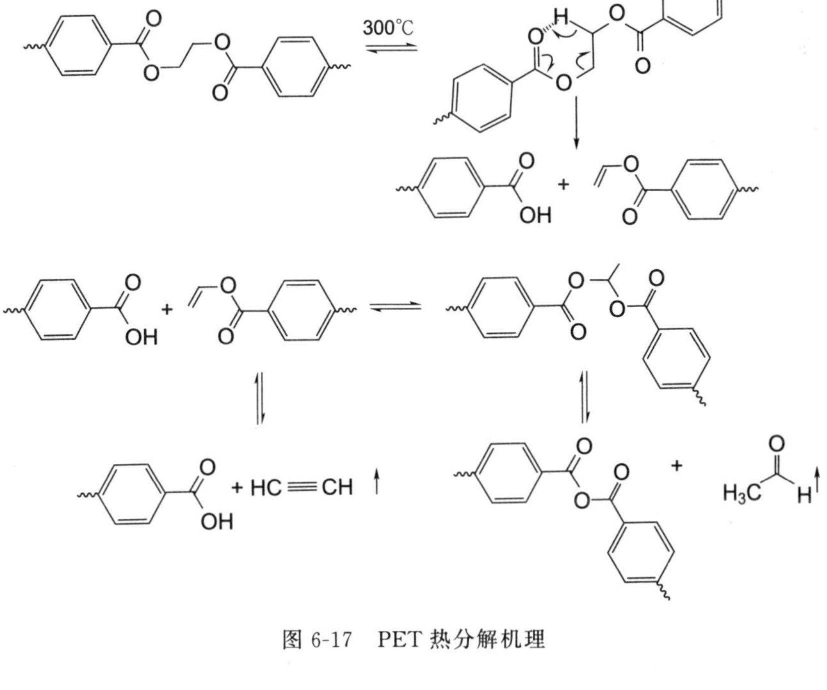

图 6-17 PET 热分解机理

第 6 章 聚合物的化学反应

6.3.2 力化学降解

碳链聚合物中的 C—C 键受到较大机械能时，也能断裂。在聚合物塑炼、熔融挤出或高分子溶液受超声波作用时，极有可能发生降解。例如，天然橡胶分子量高达百万，塑炼过程中受高剪切力作用，分子量会降低至几十万，更便于成型加工。聚合物机械降解时，分子量随时间的延长而降低，但降到某一数值，便不再降低。按力化学降解原理，可制备嵌段共聚物。例如，天然橡胶用甲基丙烯酸甲酯溶胀，然后挤出，由力学作用产生的自由基引发单体聚合和链转移反应，结果形成异戊二烯和甲基丙烯酸甲酯的嵌段共聚物。两种均聚物一起塑炼也可形成嵌段共聚物。

超声波在溶液中能产生周期性的应力和压力，形成"空穴"，其大小相当于几个分子。空穴迅速碰撞，释放出相当大的压力和剪切应力，导致大分子无规断链。

6.3.3 水解和生化降解

缩聚物主链中的碳杂原子键容易发生水解、醇解、酸解、胺解等化学降解反应。涤纶树脂和水在密闭的反应釜中升温至涤纶的熔点以上（如 265℃），就会发生水解反应。利用化学降解的原理，可使缩聚物降解成单体或低聚物，使废聚合物得以回收和循环利用。废涤纶树脂加过量乙二醇可醇解成对苯二甲酸二乙二醇酯或低聚物，聚酰胺经酸或碱催化水解，可得氨基或羧基低聚物，酚醛树脂可用过量苯酚降解成酚或低聚物。聚乳酸易水解，可用作外科缝合线，手术后，无需拆线，在体内生化水解为乳酸，由代谢循环排出体外。许多细菌能产生酶，使缩氨酸和糖类媒介形成水溶性产物。

6.3.4 氧化

高分子材料在长时间使用过程中容易被氧化变质，受热和光照会促进氧化作用。高分子材料对氧化反应的稳定性与其结构相关。碳碳双键、烯丙基和叔碳上的碳氢键容易被氧进攻。碳碳双键和碳氢键被氧化为过氧化物后分解，形成自由基，从而引发连锁降解反应。高分子主链的 C=C 双键可直接与氧加成形成过氧化物，再分解成双氧自由基和醛类。同时，也可以是烯丙基的 C—H 键被氧化形成氢过氧化物，然后分解成自由基，引发降解或交联。

饱和碳链聚合物，如线形聚乙烯热稳定性比较好，耐氧化性相对较强。聚丙烯中的叔氢原子易氧化，不加抗氧剂，几乎无法加工。聚苯乙烯也有叔氢原子，但受苯环共振作用的保护，比聚丙烯稳定，在室温黑暗环境中放置几年，未发现紫外吸收光谱的改变。但在大气中受紫外光照射，经光氧化，产生羰基和羟基，表面很快发黄。继续氧化，聚苯乙烯会降解成苯甲醛、甲基苯基酮。

对 PET 而言，乙二醇部位对氧化作用敏感，会引发如图 6-18 所示的分解反应。

为提高高分子材料的抗氧化能力，首先要设计较稳定的分子链结构，其次是使用涂层等物质隔离空气，再次是添加抗氧剂降低氧化速度。

抗氧剂往往是多种类型配合使用。如链终止剂用来消除自由基，氧化物分解剂可分解过氧化物，过渡金属钝化剂能消除过氧化物的分解催化作用。

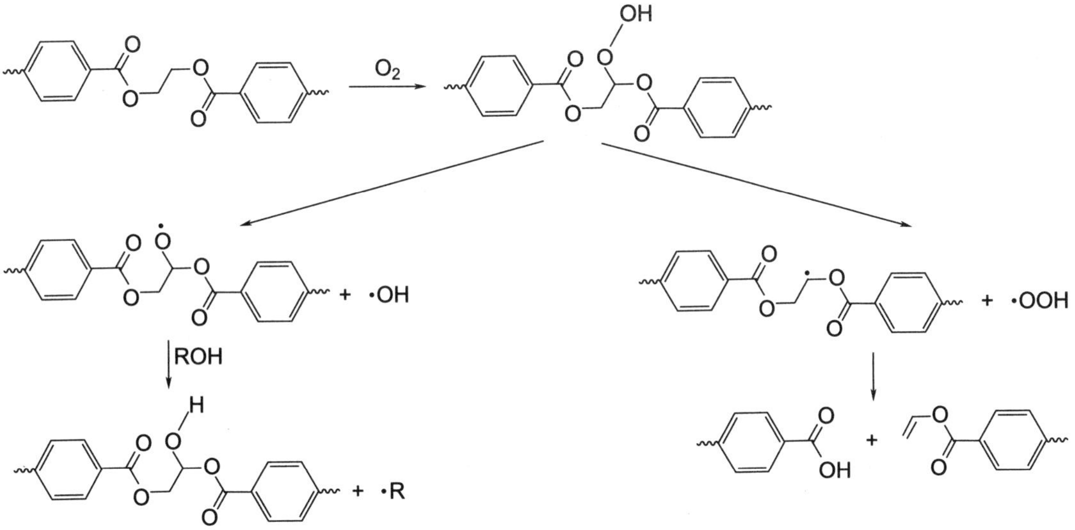

图 6-18 PET 氧化机理

链终止剂型抗氧剂可起到类似阻聚剂或自由基捕捉剂的作用，通过链转移及时将已经产生的初始自由基转变为不活泼的自由基。

终止剂型抗氧剂并不能分解过氧化物使氧化完全停止，因此，通过添加过氧化物分解剂（又称失活剂），与终止剂型抗氧剂合用，可提高抗氧化能力。过氧化物分解剂实质上是有机还原剂，包括硫醇、有机硫化物、三级膦、三级胺等，其作用是使过氧化物还原、分解和失活，而分解剂本身则被氧化。过渡金属钝化剂利用铁、钴、铜、锰、钛等过渡金属对过氧化物分解反应的催化作用，消除或减弱其催化分解作用。上述三类抗氧剂往往复合使用，随待稳定的聚合物而定。常用抗氧剂见图 6-19。

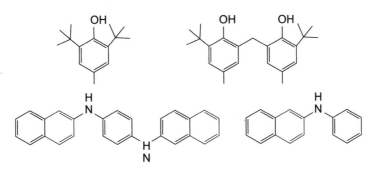

图 6-19 常用抗氧剂

6.3.5 光降解和光氧化降解

聚合物长时间受阳光照射，会发生光降解和光氧化降解反应，使聚合物分子链断裂，导致老化。光的能量与波长有关，波长愈短，能量愈大。日光中短波长远紫外线（120～280nm）可被大气层中臭氧吸收，照射到地面的近紫外部分（300～400nm），能量在 400～300kJ/mol，足以使一些共价键断裂。例如，PET 在光照下会变为琥珀色，变脆，释放气体，其可能的机理如图 6-20 所示。

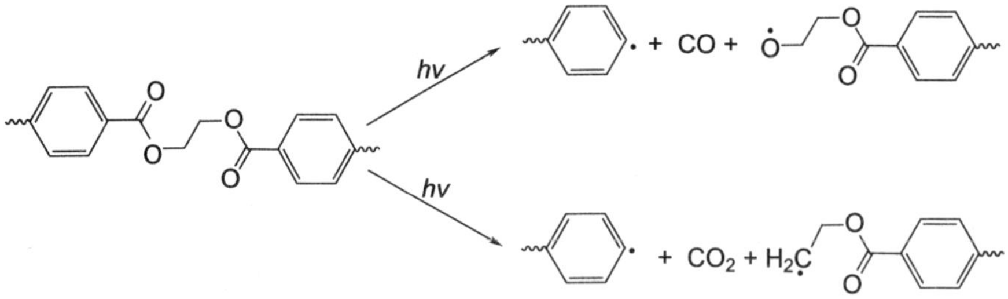

图 6-20 PET 光降解机理

聚合物吸收光能后，一部分分子或基团转变成激发态。处于激发态的原子有可能释放光能（荧光、磷光）或热能，也有部分能量过大，直接导致化学键断裂。

聚（甲基）丙烯酸酯类不吸收紫外线，对光降解很稳定，可用作室外的标牌和外墙涂料。高密度聚乙烯、聚丙烯、天然橡胶、涤纶等容易光氧化降解。常用紫外线吸收剂见图 6-21。

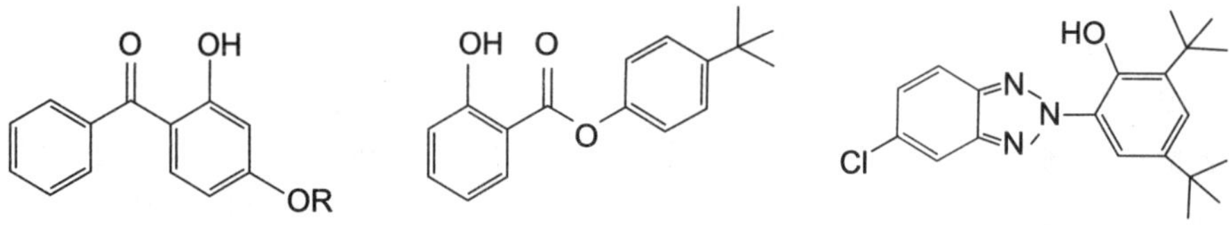

图 6-21 常用紫外线吸收剂

为了防止高分子材料的光降解反应，一般需要使用光稳定剂。光稳定剂按作用机理可分为三类：

① 光屏蔽剂。用于屏蔽或反射紫外线，炭黑（粒度为 15～25nm）、二氧化钛、活性氧化锌和很多颜料都是有效的紫外线屏蔽剂。

② 紫外线吸收剂。能够吸收 290～400nm 的紫外线，放出荧光、磷光或转变成热，常用的紫外线吸收剂有邻羟基二苯甲酮、水杨酸酯类、邻羟基苯并三唑等；

③ 紫外线猝灭剂。与紫外线吸收剂的作用机理相似，都是使激发态的能量以光或热散发出去，然后恢复到基态。但是猝灭剂吸收相邻分子的能量，而吸收剂的光能吸收和转化发生在同一分子内。常用的猝灭剂是二价镍的有机螯合剂或络合物，如双(4-叔辛基苯)亚硫酸镍、硫代烷基酚镍络合物或盐、二硫代氨基甲酸镍盐等。

英语读译资料

(1) Degradation

Here, the term degradation includes any change, decrease, in polymer property because of the impact of environmental factors, namely light, heat, mechanical, and chemicals. Seven polymers represent the majority of the synthetic polymers. These are the various polyethylene, polypropylene, nylons, polyethylene-terephthalate, polystyrene, polyvinyl-chloride, and polycarbonate. Each of these has their own particular mode of degradation. Even so, there are some common generalities for the condensation polymers that contain a noncarbon in the backbone and vinyl polymers that contain only a carbon backbone.

Backbone chain scission degradation can be divided as occurring via depolymerization, random chain breakage, weak-link or preferential site degradation, or some combination of these general routes. In depolymerization, monomer is split off from an activated end group. This is the opposite of the addition polymerization and is often referred to as "unzipping." In general, for vinyl polymers thermal degradation under controlled conditions can result in true depolymerization generally occurring via an unzipping. Polymers such as polymethyl-methacrylate depolymerize to give large amounts of monomer when heated under the appropriate conditions.

Most polymers are susceptible to degradation under natural radiation, sunlight, and high temperatures even in the presence of antioxidants. Thus, low-density polyethylene sheets, Impregnated with carbon black, become brittle after exposure to 1 year's elements in South Florida. High-density polyethylene (HDPE), while more costly, does stand up better to these elements, but again after several seasons, the elements win and the HDPE sheets become brittle and break. Long term degradation is often indicated in clear polymers by a yellowing and a decrease in mechanical properties.

Most hetero-chained polymers, including condensation polymers, are susceptible to aqueous-associated acid or base degradation. This mode of degradation is referred to as hydrolysis. This susceptibility is due to a combination of the chemical reactivity of heteroatom sites and to the materials being at least wetted by the aqueous solution allowing contact between the proton or hydroxide ion to occur.

(2) Cross-linking

Cross-linking reactions are common for both natural and synthetic and vinyl and condensation polymers. These cross-links can act to lock in "memory" preventing free-chain movement. Cross-linking can be chemical or physical. Physical cross-linking occurs in two major modes. First, chain entanglement acts to cause the tangled chains to act as a whole. Second, crystalline formations, large scale or small scale, act to lock in particular structures. This crystalline formation typically increases the strength of a material as well as acting to reduce wholesale chain movement.

Chemical cross-linking often occurs through use of double bonds that are exploited to be the sites of cross-linking. Cross-linking can be effected either through use of such preferential sites as double bonds, or through the use of other especially susceptible sites such as tertiary hydrogens. It can occur without the addition of an external chemical agent or, as in the case of vulcanization, an external agent, a cross-linking agent as sulfur, is added. Cross-linking can be effected through application of heat, mechanically, though exposure to ionizing radiation and nonionizing (such as microwave) radiation, through exposure to active chemical agents, or though any combination of these.

Cross-linking can be positive or negative depending on the extent and the intended result. Chemical cross-linking generally renders the material insoluble. It often increases the strength of the cross-linked material but decreases its flexibility and increases its brittleness.

Most chemical cross-linking is not easily reversible. The progress of formation of a network polymer has been described in a variety of ways. As the extent of cross-linking increases, there is a steady increase in the viscosity of the melt. At some point, there is a rapid increase in viscosity and the mixture becomes elastic and begins to feel like a rubber. At this point, the mixture is said to be "gelled." Beyond this point the polymer is insoluble. The extent of cross-linking can continue beyond this gel point.

(3) Flammability test

Flammability is difficult to measure because the result does not correlate directly the burning behavior in true fire conditions of polymer. Currently, the limiting oxygen index (LOI) of polymer is employed as an indication of flammability of polymer. The LOI is the minimum percentage of oxygen in an oxygen-nitrogen mixture that will initiate and support for three minutes the candle like burning of a polymer sample; that can be expressed by Eq. (6-1). The test can be easily carried out in the laboratory using a small-scale fire.

$$LOI = \frac{V_{O_2}}{V_{O_2} + V_{N_2}} \times 100\% \qquad (6-1)$$

Representative LOI values for some common polymers are given in Table 6-1. Note the large difference in LOI between poly(ethylene oxide) and poly(vinyl alcohol), although they have similar structure. The dehydration of polyvinyl alcohol will cool the polymer during burning. The polymers contained ether linkage exhibit low LOI due to the presence of oxygen atom. Although the polycarbonate contains oxygen, it releases fire extinguishing CO_2 during burning which increases the LOI value. The polymers having aromatic structure especially in the backbone show high LOI, because the aromatic chain is more difficult to break and burn than that of alkyl chain such as poly(phenylene oxide). The inclusion of Si in the polymer increases the LOI because Si is nonflammable as compared to C. The chloride bond of polyvinyl chloride is easy to break under heat and function as an extinguisher to reduce the burning which results in high LOI. The poly(tetrafluoro ethylene) exhibits the highest LOI among the polymers because of the strong C—F bonding and dense structure of the polymer.

Table 6-1 Limiting oxygen indexes (LOI) of some common polymers

Polymers	LOI
poly(oxy methylene)	15
poly(ethylene oxide)	15
poly(methyl methacrylate)	17
polypropylene	17
polyethylene	17
polystyrene	18
poly(1,3-butadiene)	18
poly(vinyl alcohol)	22
polycarbonate	27

Polymers	LOI
poly(phenylene oxide)	28
polysiloxane	30
poly(vinyl chloride)	45
poly(vinylidene chloride)	60
poly(tetrafluoro ethylene)	95

习 题

1. 研究降解的目的有哪些？影响降解的因素有哪些？
2. 研究热降解有哪些方法？简述其要点。
3. 热稳定性与耐热性有何关系和区别？热稳定性与键能有何关系？
4. 哪些基团是热降解、氧化降解、光（氧化）降解的薄弱环节？
5. 热降解有几种类型？简述聚甲基丙烯酸甲酯、聚苯乙烯、聚乙烯、聚氯乙烯热降解的机理特征。
6. 简述聚氯乙烯受热时脱氯化氢的机理，简述热稳定剂的种类和特征。
7. 简单说明聚甲醛、聚硅氧烷、涤纶树脂的热降解特征。
8. 比较聚苯乙烯和对苯二甲酸乙二醇酯的水解稳定性。
9. 比较聚乙烯、聚丁二烯、聚丙烯的氧化降解特征。
10. 比较聚乙烯、聚氯乙烯、聚苯乙烯、涤纶、聚丁二烯的氧化和光（氧化）降解性能。
11. 如何提高 PET 的耐光降解性能？

英语习题

1. What reactions can be used to crosslink the following polymers? Describe the crosslinking reactions by equations.

（1）polyester from ethylene glycol and maleic anhydride;
（2）polyester from ethylene glycol, phthalic acid, and oleic acid;
（3）*cis*-1,4-polyisoprene;
（4）polydimethylsiloxane;
（5）polyethylene;
（6）cellulose;
（7）chlorosulfonated polyethylene。

2. Show by equations the synthesis of each of the following：

（1）cellulose acetate;
（2）cellulose nitrate;
（3）methyl cellulose;
（4）poly(vinyl formal)。

3. Show by equations each of the following:
(1) chlorination of polyethylene;
(2) chlorination of 1,4-polyisoprene;
(3) chlorosulfonation of polyethylene;
(4) synthesis of an anionic ion-exchange polymer from polystyrene;
(5) cyclization of 1,4-polyisoprene by HBr。
4. Show the structure of each of the following:
(1) poly(propene-b-styrene);
(2) poly(styrene-g-methyl methacrylate);
(3) poly(methyl methacrylate-g-styrene);
(4) poly(methyl methacrylate-alt-styrene)。

Describe by equations the methods of synthesizing each of these copolymers.

参 考 文 献

[1] 潘祖仁. 高分子化学（增强版）[M]. 北京：化学工业出版社，2007.

[2] （日）西九保忠臣编. 范星河，谢晓峰译. 高分子化学 [M]. 北京：北京大学出版社，2013.

[3] Geore Odian. Principle of Polymerization [M]. 4th ed. New York：John Wiley&Sons, Inc., 2004.

[4] Charles E Carraher Jr. Carraher's Polymer Chemistry [M]. 9th ed. New York：CRC Press, Taylor & Francis Group, 2013.

[5] Paul C Hiemenz, Tim Lodge. Polymer Chemistry [M]. 2th ed. New York：CRC Press, Taylor & Francis Group, 2007.

[6] 王槐三，寇晓康. 高分子化学教程 [M]. 北京：科学出版社，2002.

[7] Manas Chanda. Introduction to Polymer Science and Chemistry [M]. 2th ed. New York：CRC Press, Taylor & Francis Group, 2006.

[8] Christopher S, Brazel, Stephen I Rosen. Fundamental principles of polymeric materials [M]. 3th ed. New York：John Wiley&Sons, Inc., 2004.

[9] Robert J Young, Peter A Lovell. Introduction to Polymers [M]. 3th ed. New York：CRC Press, Taylor & Francis Group, 2006.

[10] Jean-Luc Wertz, Olivier Bédué, Jean P Mercier. Cellulose Science and Technology [M]. 2th ed. New York：EPFL Press, Taylor & Francis Group, 2010.

[11] Yves Gnanou, Michel Fontanille. Organic and Physical Chemistry of Polymers [M]. New York：John Wiley&Sons, Inc., 2008.

[12] 胡玉洁，何春菊，张瑞军. 天然高分子材料 [M]. 北京：化学工业出版社，2012.

[13] Stephan K. Bio-based plastics：materials and applications [M]. New York：John Wiley&Sons, Inc., 2014.

[14] 高洁，汤烈贵. 纤维素科学 [M]. 北京：科学出版社，1999.

[15] Gandini A, Lacerda T M, Carvalho A J F, et al. Progress of Polymers from Renewable Resources：Furans, Vegetable Oils, and Polysaccharides [J]. Chemical Reviews, 2016, 116 (3)：1637-1669.

[16] Mika L T, Csefalvay E, Nemeth A. Catalytic Conversion of Carbohydrates to Initial Platform Chemicals：Chemistry and Sustainability [J]. Chemical Reviews, 2018, 118 (2)：505-613.

[17] Song J, Chen C, Zhu S, et al. Processing bulk natural wood into a high-performance structural material [J]. Nature, 2018, 554 (7691)：224-228.

[18] Mohanty A K, Vivekanandhan S, Pin J M, et al. Composites from renewable and sustainable resources：Challenges and innovations [J]. Science, 2018, 362 (6414)：536-542.

[19] Shuai L, Amiri M T, Questell-Santiago Y M, et al. Formaldehyde stabilization facilitates lignin monomer production during biomass depolymerization [J]. Science, 2016, 354 (6310)：329-333.

[20] Sun Z, Fridrich B, de Santi A, et al. Bright Side of Lignin Depolymerization：Toward New Platform Chemicals [J]. Chemical Reviews, 2018, 118 (2)：614-678.

[21] Liu X D, Diao H Y, Nishi N. Applied chemistry of natural DNA [J]. Chemical Society Reviews, 2008, 37 (12)：2745-2757.

[22] Becker G, Wurm F R. Functional biodegradable polymers via ring-opening polymerization of monomers without protective groups [J]. Chemical Society Reviews, 2018, 47 (20)：7739-7782.

[23] Bielawski C W, Grubbs R H. Living ring-opening metathesis polymerization [J]. Progress in Polymer Science, 2007, 32 (1)：1-29.

[24] Longo J M, Sanford M J, Coates G W. Ring-Opening Copolymerization of Epoxides and Cyclic Anhydrides with Discrete Metal Complexes：Structure-Property Relationships [J]. Chemical Reviews, 2016, 116 (24)：15167-15197.

[25] Sarazin Y, Carpentier J F. Discrete Cationic Complexes for Ring-Opening Polymerization Catalysis of Cyclic Esters and Epoxides [J]. Chemical Reviews, 2015, 115 (9)：3564-3614.

[26] Shen Y, Fu X, Fu W, et al. Biodegradable stimuli-responsive polypeptide materials prepared by ring opening polymerization [J]. Chemical Society Reviews, 2015, 44 (3)：612-622.

[27] Chen C. Designing catalysts for olefin polymerization and copolymerization：beyond electronic and steric tuning [J].

Nature Reviews Chemistry, 2018, 2 (5): 6-14.

[28] Chen M, Zhong M, Johnson J A. Light-Controlled Radical Polymerization: Mechanisms, Methods, and Applications [J]. Chemical Reviews, 2016, 116 (17): 10167-10211.

[29] Childers M I, Longo J M, Van Zee N J, et al. Stereoselective Epoxide Polymerization and Copolymerization [J]. Chemical Reviews, 2014, 114 (16): 8129-8152.

[30] De Greef T F A, Smulders M M J, Wolffs M, et al. Supramolecular Polymerization [J]. Chemical Reviews, 2009, 109 (11): 5687-5754.

[31] Delferro M, Marks T J. Multinuclear Olefin Polymerization Catalysts [J]. Chemical Reviews, 2011, 111 (3): 2450-2485.

[32] Keddie D J. A guide to the synthesis of block copolymers using reversible-addition fragmentation chain transfer (RAFT) polymerization [J]. Chemical Society Reviews, 2014, 43 (2): 496-505.

[33] Kubisa P. Ionic liquids as solvents for polymerization processes-Progress and challenges [J]. Progress in Polymer Science, 2009, 34 (12): 1333-1347.

[34] Matyjaszewski K. Advanced Materials by Atom Transfer Radical Polymerization [J]. Advanced Materials, 2018, 30 (23).

[35] Nicolas J, Guillaneuf Y, Lefay C, et al. Nitroxide-mediated polymerization [J]. Progress in Polymer Science, 2013, 38 (1): 63-235.

[36] Nishiura M, Hou Z. Novel polymerization catalysts and hydride clusters from rare-earth metal dialkyls [J]. Nature Chemistry, 2010, 2 (4): 257-268.

[37] Ottou W N, Sardon H, Mecerreyes D, et al. Update and challenges in organo-mediated polymerization reactions [J]. Progress in Polymer Science, 2016, 56: 64-115.

[38] Pan X, Fantin M, Yuan F, et al. Externally controlled atom transfer radical polymerization [J]. Chemical Society Reviews, 2018, 47 (14): 5457-5490.

[39] Sarazin Y, Carpentier J F. Discrete Cationic Complexes for Ring-Opening Polymerization Catalysis of Cyclic Esters and Epoxides [J]. Chemical Reviews, 2015, 115 (9): 3564-3614.

[40] Sorrenti A, Leira-Iglesias J, Markvoort A J, et al. Non-equilibrium supramolecular polymerization [J]. Chemical Society Reviews, 2017, 46 (18): 5476-5490.

[41] Stuerzel M, Mihan S, Muelhaupt R. From Multisite Polymerization Catalysis to Sustainable Materials and All-Polyolefin Composites [J]. Chemical Reviews, 2016, 116 (3): 1398-1433.

[42] Teator A J, Lastovickova D N, Bielawski C W. Switchable Polymerization Catalysts [J]. Chemical Reviews, 2016, 116 (4): 1969-1992.

[43] Uemura T, Yanai N, Kitagawa S. Polymerization reactions in porous coordination polymers [J]. Chemical Society Reviews, 2009, 38 (5): 1228-1236.

[44] Yokozawa T, Yokoyama A. Chain-Growth Condensation Polymerization for the Synthesis of Well-Defined Condensation Polymers and pi-Conjugated Polymers [J]. Chemical Reviews, 2009, 109 (11): 5595-5619.

[45] Zetterlund P B, Thickett S C, Perrier S, et al. Controlled/Living Radical Polymerization in Dispersed Systems: An Update [J]. Chemical Reviews, 2015, 115 (18): 9745-9800.